# Minitab Commands Used in This Text ▼▲▼▲▼▲▼▲▼▲▼▲▼▲▼▲▼▲▼▲▼

▶▶▶▶▶▶ **Chapter 6**

```
ZINTerval, confidence level = K, sigma = K,
RN C,...,C, put in C
RNMIs C,...,C, put in C
RMEAn C,...,C, put in C
RMEDian C,...,C, put in C
RSTDev C,...,C, put in C
RMAX C,...,C, put in C
RMIN C,...,C, put in C
RSUM C,...,C, put in C
RSSQ C,...,C, put in C
RCOUnt C,...,C, put in C
```

▶▶▶▶▶▶ **Chapter 7**

```
ZTESt mu = K, sigma = K, data in C;
  ALTernative hypothesis = K.
```

▶▶▶▶▶▶ **Chapter 8**

```
TTESt mu = K, data in C;
  ALTernative hypothesis = K.
TINTerval with conf level = K, data in C
COPY C,...,C into C,...,C;
  USE rows K,...,K;
  USE C = values K,...,K;
  OMIT rows K,...,K;
  OMIT C = values K,...,K.
TWOSample t, K percent confidence, data in C and C;
  ALTernative hypothesis = K;
  POOLed.
```

▶▶▶▶▶▶ **Chapter 9**

```
AOVOneway on data in C,...,C
CODE (K,...,K) to K for C,...,C, put in C,...,C
TABLe data in C,...,C;
  CHISquare [K].
CHISquare on table in C,...,C
```

▶▶▶▶▶▶ **Chapter 10**

```
CORRelation coefficient of C,...,C
REGRess y-value in C on K predictors in C,...,C;
  DW;
  PREDict y-value for x-value set E,...,E.
```

▶▶▶▶▶▶ **Chapter 11**

```
TSPLot [with K periods], data in C
LAG by K, data in C, put in C
UNSTack C into C,...,C;
  SUBScripts in C.
DIFFerence [of lag K] for data in C, put in C
ACF for series in C
PACF for series in C
ARIMA p=K, d=K, q=K [seasonal values sp=K, sd=K, sq=K, s=K];
  [[FOREcast start after period K] forecast length K periods].
```

▶▶▶▶▶▶ **Chapter 12**

No new commands

# Essential Business Statistics: A Minitab Framework

# THE DUXBURY SERIES IN STATISTICS AND DECISION SCIENCES

*Applications, Basics, and Computing of Exploratory Data Analysis,* Velleman and Hoaglin

*Applied Regression Analysis and Other Multivariable Methods,* Second Edition, Kleinbaum and Kupper

*Classical and Modern Regression with Applications,* Myers

*A Course in Business Statistics,* Second Edition, Mendenhall

*Elementary Statistics for Business,* Second Edition, Johnson and Siskin

*Elementary Statistics,* Fifth Edition, Johnson

*Elementary Survey Sampling,* Third Edition, Scheaffer, Mendenhall, and Ott

*Essential Business Statistics: A Minitab Framework,* Bond and Scott

*A First Course in Linear Regression,* Second Edition, Younger

*Fundamental Statistics for Human Services and Social Work,* Krishef

*Fundamental Statistics for the Behavioral Sciences,* Howell

*Fundamentals of Biostatistics,* Second Edition, Rosner

*Fundamentals of Statistics in the Biological, Medical, and Health Sciences,* Runyon

*Introduction to Contemporary Statistical Methods,* Second Edition, Koopmans

*Introduction to Probability and Mathematical Statistics,* Bain and Engelhardt

*Introduction to Probability and Statistics,* Seventh Edition, Mendenhall

*An Introduction to Statistical Methods and Data Analysis,* Third Edition, Ott

*Introductory Statistics for Management and Economics,* Second Edition, Kenkel

*Linear Statistical Models: An Applied Approach,* Bowerman, O'Connell, and Dickey

*Management Science: An Introduction,* Davis, McKeown, and Rakes

*Mathematical Statistics with Applications,* Third Edition, Mendenhall, Scheaffer, and Wackerly

*Minitab Handbook,* Second Edition, Ryan, Joiner, and Ryan

*Minitab Handbook for Business and Economics,* Miller

*Operations Research: Applications and Algorithms,* Winston

*Probability Modeling and Computer Simulation,* Matloff

*Probability and Statistics for Engineers,* Second Edition, Scheaffer and McClave

*Probability and Statistics for Modern Engineering,* Lapin

*Quantitative Management: An Introduction,* Second Edition, Anderson and Lievano

*Quantitative Models for Management,* Second Edition, Davis and McKeown

*Statistical Experiments Using BASIC,* Dowdy

*Statistical Methods for Psychology,* Second Edition, Howell

*Statistical Thinking for Behavioral Scientists,* Hildebrand

*Statistical Thinking for Managers,* Second Edition, Hildebrand and Ott

*Statistics for Management and Economics,* Fifth Edition, Mendenhall, Reinmuth, Beaver, and Duhan

*Statistics: A Tool for the Social Sciences,* Fourth Edition, Ott, Larson, and Mendenhall

*Time Series Analysis,* Cryer

*Time Series Forecasting: Unified Concepts and Computer Implementation,* Second Edition, Bowerman and O'Connell

*Understanding Statistics,* Fourth Edition, Ott and Mendenhall

This book is dedicated to
Robert and Nira Bond
and
Bryan and Lola Scott
our parents

# PREFACE

## TO THE INSTRUCTOR

Our purpose in this book is to develop a substantially new and in some ways radically different approach to the basic one-semester course in business statistics. It is our intention to develop a *completely integrated*, computer-based course in which Minitab is used. This approach is in contrast to that of many available statistics books, in which the use of a particular computer package for a given topic is optional.

In an introductory statistics text for a one-semester course, there is little room for theoretical discussions or proofs. Although these aspects are frequently the joy of the faculty, they are seldom used by business executives in their day-to-day work. Our objective is for the student to obtain a strong intuitive (while theoretically sound) grasp of the material, which will serve them well in their subsequent business careers.

We believe that a good, intuitive grasp of the subjects presented here can be gained in many ways. In this text, we use computer technology for that purpose. However, most of the chapters also contain a section that includes the traditional concepts necessary for performing hand calculations. These sections are *not* essential to the development and flow of the text material and thus may be skipped. Because the hand calculation sections are complete in themselves, some faculty may wish to use some of the sections and omit others.

## TO THE STUDENT

Students have generally been apprehensive about taking statistics courses almost since the first one was developed. Actually, statistics is a very interesting and reasonably simple subject—at least at the introductory level. In this book there are no mathematical manipulations more sophisticated than adding, subtracting, multiplying, dividing, squaring, and taking square roots. Thus, even with the most elementary of today's hand calculators, most introductory statistical material is very simple. Why then all the confusion and uneasiness?

There are probably two reasons. The first is that statistics uses a good number of special symbols (especially Greek letters) to represent certain concepts. These strange-looking marks have tended to make students unsure of themselves. In this book we have attempted to deal with this problem in two ways: First, the number of new symbols used has been held to a minimum; second, the importance of the symbols that are used is stressed.

The second reason for students' fear of statistics is the perceived drudgery. Even a simple statistical problem generally requires a good number of calculations (none of which are complicated). All of these simple calculations need to be correct if the final answer is to be accurate. Further, several homework problems require a lot of "number crunching." This computation also limits students' excitement about the study of statistics. Many times students are so involved in trying to get an answer that they do not know what the answer means once they have found it! Relieving this second problem is one of our major concerns in this book. Therefore—Minitab!

Minitab is a very user-friendly statistical software package that eliminates much of the drudgery in performing long sets of calculations to find a simple answer. The use of the Minitab system allows two very important activities that are impossible in courses using traditional statistics books. First, you can "go exploring" when you are working on a problem; that is, you can try several different approaches to a problem without reentering the data or recalculating various answers. The system allows you to answer the question "What if?" without investing hours of work.

A second advantage of using Minitab is that you can concentrate on the truly important aspects of statistical analysis. In the world of applied statistics—the world in which statistics is used every day to solve actual business problems—the key issues are the quality of the data used in the statistical analysis (input) and the analysis of the results (output). The complicated steps needed to calculate a multiple regression equation is generally not the concern of students in applied courses such as introductory business statistics. To use statistical techniques competently, the quality of the data must first be assured. If the data is poorly developed or biased, your results will not be accurate. Bad data can lead to inaccurate conclusions and very costly mistakes. There is an old saying in the computer field: "Garbage in, garbage out." This simply means that even the most sophisticated computer cannot correct bad data. Thus, our first concern must always be the quality of the input data. Minitab gives you the time to assess this. Since you are not spending all your time crunching numbers and just trying to find an answer, you can spend time thinking about where the data came from.

Similarly, a system like Minitab permits you to concentrate on interpreting output. Finding numeric answers will be quite easy. Thus, with time available, you can determine what the output truly means. The orientation of this book toward input and output will allow you to stop and think about what you have done and the ramifications of the various statistical tests.

# PREFACE

An important point to keep in mind is that *understanding statistics and Minitab is not a process of simply reading this book*. It is the process of *doing* this book. When all is said and done, it is essential to read and work the problems—to sit in front of your CRT or PC and actually input many, if not most, of the Minitab commands shown in the text. If you will spend the time and effort necessary to do the problems and exercises in the book instead of just reading them, you will find your experience of statistics meaningful and exciting.

This textbook is divided into two major parts. The first two chapters are designed to teach the fundamentals of Minitab. However, keep in mind that this is *not* a Minitab course. We will only do enough to get you familiar with the system. Then, starting in Chapter 3 and continuing throughout the rest of the book, new and more powerful Minitab commands will be introduced as they are needed. Since this is a basic statistics course, many of the uses of Minitab will not be covered. For students who need or desire more advanced Minitab commands, there are two excellent books available: *Minitab Handbook*, 2nd ed., by B. Ryan, B. Joiner, and T. Ryan (Boston: Duxbury Press, 1985) and *Minitab Handbook for Business and Economics* by Robert B. Miller (Boston: PWS-Kent, 1988). Additional questions about Minitab can be addressed directly to Minitab, Inc., 3081 Enterprise Dr., State College, PA 16801, telephone (814) 238-3280, telex 881612.

# ACKNOWLEDGMENTS

All texts are the work of many people, not just the authors. We wish to thank Michael Payne, editor of the Duxbury Series in Statistics and Decision Sciences for constant encouragement and assistance on an untraditional project that required both risk and vision. We also wish to thank the reviewers who evaluated the manuscript: Nancy Jo Delaney, Northeastern University; Robert Elrod, Georgia State University; Steven Eriksen, Northeastern University; Nicholas Farnum, California State University at Fullerton; Jerome D. Herniter, Boston University; B. J. Isselhardt, Rochester Institute of Technology; Albert Kagan, University of Northern Iowa; David D. Kreuger, St. Cloud State University; Gerald Miller, Rockhurst College; Kris Moore, Baylor University; R. Natarajan, University of Northern Iowa; Stanley L. Sclove, University of Illinois; Lois M. Shufeldt, Southwest Missouri State; and M. Anthony Wong, Massachusetts Institute of Technology. The computer center at Creighton University has been most cooperative over the years with our mainframe version of Minitab, and their assistance has been greatly appreciated. This book was "classroom tested" on students at Creighton University; we sincerely thank them for their comments and suggestions. Of course, all errors remain our responsibility.

Over the last year and a half Lori H. Tanaka created most of the problem sets and discussion questions in the book. She also developed the solutions for those problems with exacting detail. Although you, as students, might be tired of working problems, Lori's contribution to your learning process should not be underestimated. To Lori a very special thank you!

Finally, we wish to thank our families for their assistance and encouragement over the years of this project.

Kenneth M. Bond
James P. Scott

# CONTENTS

**Introduction: The Go-Fast Running Shoe Company, 1**

## Chapter 1  ELEMENTS OF MINITAB

- **1-1** The Purpose of Minitab, 5
- **1-2** Using the Minitab System, 5
- **1-3** Operations of the Minitab System, 9
  Summary for Chapter 1, 16
  Discussion Questions for Chapter 1, 17
  Problems for Chapter 1, 17

## Chapter 2  BASIC MANIPULATIONS WITH MINITAB ▼▲▼▲▼▲▼▲▼▲▼▲

- **2-1** The LET Command, 24
- **2-2** Columns, 25
- **2-3** Plotting, 29
- **2-4** Index Numbers, 34
  Summary for Chapter 2, 39
  Discussion Questions for Chapter 2, 39
  Problems for Chapter 2, 39

## Chapter 3 DESCRIPTIVE STATISTICS ▼▲▼▲▼▲▼▲▼▲▼▲

3-1 Classification of Measurement Scales, 47
3-2 Measures of Central Tendency, 49
3-3 Visual Presentation of Data, 55
3-4 Numeric Expressions of Dispersion, 61
   Summary for Chapter 3, 66
   Discussion Questions for Chapter 3, 66

   Problems for Chapter 3, 66

## Chapter 4 PROBABILITY AND SAMPLING ▼▲▼▲▼▲▼▲▼▲▼▲▼

4-1 Some General Probability Concepts, 75
4-2 Initial Considerations of Sampling, 92
4-3 Survey Sampling, 95
   Summary for Chapter 4, 96
   Discussion Questions for Chapter 4, 97
   Problems for Chapter 4, 97

## Chapter 5 PROBABILITY DISTRIBUTIONS ▼▲▼▲▼▲▼▲▼

5-1 What Is a Probability Distribution?, 106
5-2 The Bernoulli Distribution, 109
5-3 The Binomial Distribution, 113
5-4 The Poisson Distribution, 120
5-5 The Normal Distribution, 122
5-6 The Central Limit Theorem, 124
   Summary for Chapter 5, 138
   Discussion Questions for Chapter 5, 140
   Problems for Chapter 5, 140

## Chapter 6 STATISTICAL ESTIMATION ▼▲▼▲▼▲▼▲▼▲▼▲

6-1 Point Estimates, 144
6-2 Characteristics of the Normal Curve, 145
6-3 The Standard Normal Curve, 149
6-4 The Confidence Interval, 151
   Summary for Chapter 6, 159
   Appendix for Chapter 6, 160
   Discussion Questions for Chapter 6, 161
   Problems for Chapter 6, 161

## Chapter 7 INTRODUCTION TO HYPOTHESIS TESTING ▼▲▼▲▼▲

7-1 Development of a Hypothesis, 167
7-2 Errors in Testing Hypotheses, 169
7-3 Power of a Test, 170
7-4 Hypothesis Tests, 171
7-5 Formulas and Hand Calculations, 176
   Summary for Chapter 7, 177
   Discussion Questions for Chapter 7, 177
   Problems for Chapter 7, 177

## Chapter 8 STATISTICAL INFERENCE: ONE- AND TWO-SAMPLE CASES ▼▲▼▲▼▲▼▲▼▲▼▲▼▲

8-1 Assumptions of the $t$ Test, 184
8-2 Student's $t$ Test, 185
8-3 Interval Estimates, 187
8-4 Paired $t$ Test, 188
8-5 Two-Sample $t$ Tests, 190
8-6 Formulas and Hand Calculations, 194
   Summary for Chapter 8, 201
   Discussion Questions for Chapter 8, 201
   Problems for Chapter 8, 201

## Chapter 9 STATISTICAL INFERENCE: MULTIPLE-SAMPLE CASES ▼▲▼▲▼▲▼▲▼▲▼▲▼▲

9-1 Analysis of Variance, 210
9-2 Chi-Square Test, 215
9-3 Nonparametric Test Chart, 223
   Summary for Chapter 9, 223
   Discussion Questions for Chapter 9, 223
   Problems for Chapter 9, 223

## Chapter 10 CORRELATION AND REGRESSION ANALYSIS ▼▲▼▲▼

10-1 Correlation Analysis, 237
10-2 Regression Analysis, 239
10-3 An Example of No Correlation, 245
10-4 An Example of Spurious Correlation, 246
10-5 Multiple Regression, 248

**CONTENTS** xiii

**10-6** Formulas and Hand Calculations, 255
Summary for Chapter 10, 257
Discussion Questions for Chapter 10, 257
Problems for Chapter 10, 258

## Chapter 11  FORECASTING

**11-1** Time Series Analysis, 272
**11-2** Classical Decomposition, 273
**11-3** Box–Jenkins Forecasting, 290
Summary for Chapter 11, 295
Discussion Questions for Chapter 11, 296
Problems for Chapter 11, 296

## Chapter 12  PERSPECTIVE

**12-1** Information Ethics, 301
**12-2** Examples of Misuse, 305
**12-3** Long-Range Effects of Statistics, 307
**12-4** Advanced Capabilities of Minitab, 308

## Appendix A, 311

## Appendix B, 313

**Table 1** Normal Curve Areas, 314
**Table 2** Critical Values of Student's $t$ Distribution, 315
**Table 3** Critical Values of the $F$ Distribution, 316
**Table 4** Critical Values of the Chi-Square Distribution, 322
**Table 5** Binomial Probabilities, 323

## Appendix C, 326

## Index, 329

# INTRODUCTION

**T**he Go-Fast Running Shoe Company

Throughout this text we use a set of data from a firm called the Go-Fast Running Shoe Company. A brief description of the company follows. All students should read this background before starting Chapter 1.

In the mid-1800s, the River City Boot Company was founded in Omaha, Nebraska. The firm produced a high-quality working boot for the farmers of the Midwest. The firm grew in size until the 1940s, when employment leveled off at about 100 full-time employees. Since that time, employment has remained steady. Unfortunately, circumstances started to work against the company. The demand for its product gradually started to decline, and the technology used in making the boots remained unchanged, while the cost of leather and other materials rose.

In 1981 the firm was sold to a couple of young college graduates who decided to change the product line completely and rename the company. To capitalize on the increased interest in physical fitness, the firm was renamed the Go-Fast Running Shoe Company and the product line changed to running and jogging shoes.

Historically, the personnel policies of the company had been very good, and the firm had a loyal, solid work force. Pay and benefits were good and the firm had about the normal number of retirements, terminations, and hirings annually. After the switch to the new name and new product line, most employees stayed with the company. The running shoes sold very well, and the same number of employees (100) was able to produce a greater number of shoes with the new technology that was installed.

After a few months, the new owners decided it was time to start gathering more data about their operations. They already knew several things about their employees from employment records, such as their age and educational level. The previous management team did not keep very good records on individual productivity, and the new managers decided to collect productivity data in December of last year to determine which employees were producing the most shoes. In addition, the new managers realized that almost all the employees wore Go-Fast shoes to work, because they had a popular style and were very comfortable. So the new owners decided to test such factors as durability, comfort, and style by allowing employees to wear free samples of new designs. As a general rule, management thought that bigger and taller employees would wear out shoes faster than smaller and shorter employees. However, only the height of each employee was measured since most people were thought to be sensitive about their weight.

For each of the 100 employees, Appendix A of this book contains the measurements of all four variables: age, number of years of school completed, productivity for December, and height. Turn to Appendix A for a minute and scan the columns of numbers. What do you now know about the employees and productivity? If you were president of Go-Fast Running Shoe Company, what would you do with all the numbers in front of you?

Before we can begin to explore the data, we need to develop some basic Minitab skills that will allow us to analyze these values quickly and efficiently. Chapter 1 will review these essential operations.

# ELEMENTS OF MINITAB

**T**he following Minitab commands will be *introduced in this chapter:*

`SET data into column C`
A command to enter data into a single column

`MEAN of values in column C`
A command to find the arithmetic average of a set of values

`SUM values in column C`
A command to find the total value of a group of numbers in a given column

`HISTogram of data in column C`
A command that provides a visual (graphical) summary of a group of values

`READ data into columns C,...,C`
A command that provides a means to enter data into multiple columns at the same time; a companion command to the `SET` command

`ADD E to E, put in E`
A command to add two columns together and store the results as a new column, to add a constant value to a column and store the results in a new column, or to add two constants together and store the results as a constant or column

`SUBTract E from E, put in E`
A command with similar structure to the `ADD` command but for subtraction

`MULTiply E by E, put in E`
A command with similar structure to the `ADD` command but for multiplication

> **DIVIde E by E, put in E**
> A command with similar structure to the **ADD** command but for division
> **SQRT of E, put in E**
> A command to take the square root of a constant or column of values and to store the resulting value(s) as either a constant(s) or column(s)
> **PRINt E,...,E**
> A command to print one or more columns, or one or more constants
> **LET**
> A command used to combine multiple arithmetic steps into a single command and to correct errors in data entry
> **INSErt**
> A command used to insert new data into an existing data set (not on all releases of Minitab)
> **DELEte**
> A command used to remove data from an existing data set (not on all releases of Minitab)
> **END**
> An optional command that signals the end of data entry

The Minitab system offers you maximum opportunity to explore important concepts. In statistical analysis, a great number of calculations are often required to find a single answer. If a second type of information about the same data is needed, then typically a whole new set of calculations must be performed. For a third bit of information, still a third set of calculations is needed. Using the Minitab system, you can enter a data set once and then perform as many calculations on that data set as you like with simple, English-language-oriented commands.

As an illustration of the potential time savings offered by the Minitab system, look at the following simple problem. From Appendix A, we have taken all the ages of the employees in our Go-Fast example (see the Introduction) and drawn a random sample of 20, shown in the following list:

|    |    |    |    |    |
|----|----|----|----|----|
| 32 | 29 | 38 | 36 | 36 |
| 35 | 31 | 26 | 27 | 30 |
| 31 | 32 | 25 | 32 | 28 |
| 31 | 38 | 34 | 31 | 26 |

If you used a standard hand calculator to find the simple average age of this sample of employees, the process would be straightforward. You would simply add up all 20 ages and then divide the total by 20. Although the problem is not complicated, it does require a bit of work. If you counted each keystroke on your calculator, solving this problem would require approximately 125 keystrokes. In Chapter 3, we will introduce the concept of a *standard deviation*; the calculation of this statistic would require approximately 325 additional keystrokes. To find the *trimmed mean*, also explained in Chapter 3, would require a couple hundred more. Visual displays such as histograms, stem-and-leaf displays, and dot plots would all require additional calculations and labor.

By now you should be getting the picture. None of the previous computations are difficult, and each takes only a few minutes to calculate. However, when you need to do several different calculations for several problems, the time and effort quickly accumulate. With Minitab the drudgery of reentering data into a hand calculator is eliminated. After all the original data have been entered, finding the average or standard deviation of the 20 ages, plotting a histogram, or determining any of the other statistics mentioned can be done simply by entering a simple, English-oriented, one-line command.

## 1-1 THE PURPOSE OF MINITAB

Despite its efficiency, the primary purpose of the Minitab system is not to save time. Instead, *it is to allow the exploration, understanding, and development of statistical concepts.* For example, if in our Go-Fast problem about ages of employees we asked you to calculate the standard deviation, it would be simple enough to follow the formula and complete the assignment. However, if when you calculated that number you were not sure about what it really meant, you could plot the data, find the mean, make a histogram, or do whatever else would help you to understand the problem better.

The power of the system is even greater when you want to compare different items. Besides their ages, we also have data on the productivity, the years of schooling, and the height of our employees. What is the relationship between age and productivity? Between schooling and productivity? Normally students do not have the time to explore these concepts because of the hours of work involved. But with Minitab, such comparisons can be made easily and very quickly.

The purpose of statistical analysis is to make a group of observations about a phenomenon more manageable and better understood. The better you understand and appreciate the underlying statistical concepts, the better your ability to present and interpret the meaning of a given group of numbers. Minitab can be an important tool in this process. However, it *cannot* help in manipulating data to make it say what you want. Proper statistical analysis requires high moral and ethical standards. The process of presenting and interpreting data is one of determining the significance of the relationships within a set of data and then explaining this significance to others in a way they can understand.

## 1-2 USING THE MINITAB SYSTEM

The instructions presented below for the use of Minitab are divided into two parts. The first part discusses the computer system; the second explains the use of the Minitab statistical package.

### The Computer System

The instructor for your course will explain the procedures for getting on and off your computer network. Because each one differs, we cannot give you those details. No matter what the system, however, getting on and off is usually not much of a problem. There are two general types of computer environments in use today. The first is the personal computer (PC), and Minitab is available for such machines. If you are using a PC for your Minitab work, then obtaining a printed copy of homework problems is a relatively simple task of turning on the printer and pressing the ⟨Control⟩ and ⟨Print Screen⟩ keys together before executing any particular command. This process will create a printed copy of whatever is on your PC screen. Your instructor will provide you with details.

The second type of computer environment is the mainframe computer with numerous terminals at various locations. At present, the mainframe environment is still the more common on college campuses. In the mainframe environment, you work at a computer terminal that resembles a TV set with a typewriter keyboard. This terminal is called a cathode-ray tube (CRT). If you do not enter your data at a CRT, you will probably have some type of "hard-copy" computer terminal where you also work at a typewriter-like device that prints your results on computer paper as the calculations are performed. Both types of terminals are typically *interactive,* which means that when you type a line of data, the computer will react and prompt you for your next line of data or for further instruction.

Since the interactive format is the most widely used system on college campuses, we will assume its use throughout this text. If your system is not interactive, your instructor will explain the difference. No matter what system you use, once you understand its basic operation, the Minitab commands presented in this text will work as described.

One difficulty is posed by using a CRT to do homework. Normally, CRTs do not provide a printed copy of results, which are only shown on the screen of the terminal. However, at most colleges the computer centers have developed a system by which you can receive hard-copy printouts of results while working on a CRT. Your instructor will provide the necessary guidance. With a little practice, you will probably find the CRT fast and easy to use, and you will be able to do most of your work on it.

You should remember that using Minitab is a two-step process. You must first log on to your computer system and then enter into the statistical program called Minitab. Likewise, when you are finished with your analysis in Minitab, you must first log out of Minitab and then log off your computer system. Once you are logged into the Minitab system, the fun really begins. The basic format of a Minitab command is to indicate what operation you want and then indicate where the desired data is located. Minitab contains about 200 English language commands.

Right now let's look at the sample of employee ages and see how we might get some useful information. Once you see how a few of the commands work, you will quickly understand how the entire system is put together.

## Basic Minitab Commands

Data is entered into columns for Minitab analysis. For our Go-Fast problem we have one set of data, which we want to enter into column 1. If we had two sets of data, we would enter the first data set into column 1 and the second into column 2, and similarly for any additional sets. The total numbers of columns and rows allowed are set by each computer installation. Typically at least 50 columns and several hundred rows per column are available. In this problem we have 20 observations (ages) to enter into the first column.

Either of two commands, **READ** and **SET**, can be used to enter data into a single column. We shall use the **SET** command here. (In Section 1-3 the **READ** command will be explained.) What we must do is enter the 20 numbers into the first column. The command should look like this:

```
SET data into column C1
```

The Minitab system "reads" the keyword **SET** and the place where the numbers are to be entered, namely, column 1. The other words that have been included in the command

## SECTION 1-2 Using the Minitab System

(in lowercase type) are only for your benefit when you first use the Minitab system. They are not needed but will help you understand what is taking place. Thus, this command could also have been written

```
SET C1
```

Look at the two commands closely for a minute. Both tell the computer what to do (**SET** tells the machine that you are about to enter data) and where to place the data (**C1** tells the machine to put the numbers you are about to enter into column 1). Throughout this book, we will capitalize all letters of a command that must be entered. Anything in lowercase letters (as in the first command above) is not essential to the command and is included only so the command makes more sense to you as you learn the system. However, *none* of the commands need to be entered in capital letters. Minitab itself does not distinguish between upper and lowercase letters. We distinguish between the two types here only so that you will know what the essential letters are in Minitab.

Also, when we show an actual command for a specific problem, we will set the command against the left-hand margin showing the prompt given by Minitab. Commands given in a general form will be centered on the page; the Minitab prompt will not be shown.

Returning to our problem of employee ages, we have now informed the computer to enter the data into column 1. The computer will now respond with a prompt (or cursor) for us to begin entering information. When using the **SET** command, several pieces of data are entered on the same line, one data point after another. Data points should be separated by a space, comma, or return. Minitab is a very forgiving computer package because you do not have to be totally consistent in entering data. You can have one space between the first age and the second, and then three spaces between the second age and the third. For example, all of the following data points are correctly entered:

```
DATA>32 29,38 36 36 35, 31
```

(Minitab uses two prompts: The prompt **MTB >** asks for a Minitab command; **DATA>** requires a line of data.)

Despite Minitab's leniency in this regard, we strongly recommend that you get into the habit of entering your data in a consistent manner. Most other computer programs and computer languages will not permit you to be as lax as Minitab in entering data. If you develop this habit early, you will save yourself a lot of time in other computer-oriented courses.

Your complete data for our Go-Fast employee age problem should be entered as follows:

```
MTB >SET data into column C1
DATA>32 29 38 36 36 35 31 26 27
DATA>30 31 32 25 32 28 31 38 34
DATA>31 26
DATA>END
```

Two things need to be noted here. First, you do not enter the **MTB >** and **DATA>** at the start of each line. The computer gives you these prompts. Some computer installations use a single dash and an arrowhead as prompts, and others two dashes and an arrowhead. Your system will give you some type of prompt before you start entering data. The **MTB >** is the most common. Second, note the command **END** has been placed after all the data has been entered. Although this command is not technically necessary, it will make it easier for you to follow the logic of entering your data, ending the data entry, and then starting the various commands to manipulate the data.

As few or as many data points as you wish may be entered on the same line. After you have entered several and are approaching the right-hand border of the terminal, simply press the return, wait for a new cursor, and continue to enter data points. When you come to the end of a line, you may put a comma after the last number if you like before depressing the return key. You may also omit the comma and simply return to the next line and continue to enter data points.

Of course, it is very important to enter data accurately. If you make a mistake when entering a number but have not yet depressed the return for that line of data, simply backspace to your error and retype the correct value. However, if you have already depressed the return key, thus entering those numbers into the computer, Minitab will require some additional steps to correct the problem. One way to correct your error is to reenter the `SET C1` command and start entering your numbers again from the beginning. Other ways to correct your data will be mentioned in Section 1-3 of this chapter. In any event, it is best to start off slowly, examining each line of numbers one last time before you depress the return key; then you will quickly gain confidence and experience.

When you enter data, *do not use commas within a number*. If a value such as 1,234.56 is to be entered as a data point, it must be entered as `1234.56`. If you enter the value as `1,234.56`, you will get two separate entries, `1` and `234.56`, because Minitab will assume that the comma separates one data point from another.

Once all 20 data points have been entered, we are ready to obtain certain types of information about the data. For example, if we want to find the average age of employees in our sample, which is typically called the *mean* age, we must follow the two basic rules: tell the computer what to do and where the data to be analyzed is located. Look at the following command closely and see if it makes sense:

> `MEAN of values in column C1`

We have told the computer what to do and where the data is stored. This command can also be written in a shorthand form by omitting the unessential letters. Since Minitab typically reads the first four letters of any command in its word library, we can write this command as

> `MEAN C1`

When this command is executed, the computer will compute the result and print it out.

If we wanted to add the ages of our 20 employees, the command would be

> `SUM values in column C1`

Notice the format is identical with the previous commands. You first tell the computer what to do (`SUM`) and then the location of the data to be used in the computations (`C1`).

To print a histogram of ages, we type

> `HISTogram of data in column C1`

This, of course, could be abbreviated to

> `HIST C1`

Although you might not know what a histogram is, go ahead and print one. Note that this command takes the same form as the previous commands.

We have mentioned that two sets of commands are needed to enter the statistical package, one to place you on the computer system and one to enter the Minitab routine. When you are finished with all your calculations and want to log off the system, the command **STOP** gets you out of Minitab. Your instructor will tell you how to log off your particular computer system.

In the following sections of this chapter, we will formally describe the commands for adding, subtracting, multiplying, and dividing. Entering more than one column of data at a time and transferring data from one column to another will also be discussed.

For now, go to your computer center, log on one of the terminals, and do the problem given in this chapter. In other words, go explore! At this point you should use the system and become familiar with your computer facilities and the Minitab basics. Enter the data given in this chapter and command the system in **MEAN C1**. Once the system has done that calculation and printed the answer, command it to **HIST C1**, then to **SUM C1**, and finally to **PRINt C1**. You should find the results interesting. Go explore—you have the time and technology to open the immense statistical world for your pleasure. Remember, when you are through playing, type **STOP** to exit the Minitab statistical package before logging off your computer.

We hope you take the time to experiment with your computer system and Minitab before proceeding with the rest of this chapter.

## 1-3 OPERATIONS OF THE MINITAB SYSTEM

The five main topics covered in the rest of this chapter are: (1) the basic framework that Minitab uses to handle data, (2) entering data into the Minitab program, (3) the general rules for manipulating data entered into the system, (4) basic commands for manipulating data, and (5) how to correct data already entered into a column.

### Basic Framework

The easiest way to understand Minitab is to think of a very large worksheet. The worksheet contains 50 or more columns across the top and 200 or more rows down the side. (The actual size of the worksheet will depend upon the computer installation.) The minimum format of 200 rows by 50 columns will be large enough for most problems in this text. When you enter data into Minitab, you must tell the computer which column to use. As you can see, Minitab is a *column*-oriented system. Because of this, a column will normally be a variable of some type, and each observation on that variable will be one of the rows. To illustrate, in our Go-Fast Running Shoe problem, the variable was age. That was our column. Each individual age was entered, one after another, as a row in that column.

When using Minitab, it is not necessary to use columns in consecutive order. You can enter data into column 9, then column 3, and then column 48 if you wish.

With the general rule that columns are variables and that the rows of a column are observations on that variable, let us move on to entering data into Minitab.

## Entering Data ▼▲▼▲▼▲▼▲▼▲▼▲▼▲▼▲▼▲▼▲▼▲▼▲▼▲▼▲▼▲▼▲▼▲▼▲▼

We will discuss two ways to enter data. Either way will work, and the method of entry chosen depends on the problem. We can enter data using either the **SET** command or the **READ** command. In our first illustration, we mentioned the **SET** command. Generally, it is easier to enter a single column of data with the **SET** command and easier to enter several columns of data with the **READ** command.

The basic format for *all* commands is as outlined previously. You must first tell the computer what to do and then which column or columns to use. Recall that the **SET** command took the form

```
SET data in column C1
```

Please notice the form closely. The **SET** command told the computer that you are going to enter data into a single column. You then told the computer where to put the data, namely, **C1**. The **SET** command is most effective when entering a single column of numbers because you can enter several numbers on the same data line.

However, if we want to enter the ages of our 20 employees in column 1, the years of schooling of these employees in column 2, their productivity in column 3, and their height in column 4, then we might want to enter the data differently. We could use the **SET** command and do a **SET C1** and enter age, a **SET C2** and enter schooling, a **SET C3** and enter productivity, and a **SET C4** and enter height. However, a command for entering multiple columns of data is the **READ** command. The **READ** command allows you to enter each of the variables for a single employee on the same line. The input command looks as follows:

```
READ data into columns C1, C2, C3, C4
```

After using this command, you then enter for the first employee the age, schooling, productivity, and height data on the first data line. On the second data line, you enter the same data for the second employee, and so on down to the 20th employee.

For many kinds of problems, this system will be faster than using the **SET** command. Like all of the commands mentioned so far, this command can also be abbreviated. Also, whenever consecutive columns are being used, a dash can be used to indicate the columns. Look closely at the following commands. Each says the same thing, and each is acceptable for entering data:

```
READ data into columns C1, C2, C3, C4
READ data into columns C1-C4
READ C1, C2, C3, C4
READ C1 C2 C3 C4
READ C1-C4
```

Notice the form of the last command. It first tells the computer to enter data (**READ**) and then where to store it (**C1-C4**). If for some reason you want to put the data into columns 1, 5, 10, and 20, then you have to use the command

```
READ data into columns C1, C5, C10, C20
```

You would not use the form

```
READ data into columns C1-C20
```

**SECTION 1-3** Operations of the Minitab System ▶ **11**

*The dash can be used only for consecutive columns.* What you have actually said in this second command is to read 20 columns of data. If you have been following this discussion, it should be obvious that you could also use the `READ` command for a single column of data much like the `SET` command. If you used the command

```
READ data into column C1
```

you could then enter data into that column. However, you could enter only one observation on each line, and you would have to press the carriage return before entering your next data point. Although this system would be slower than using the `SET` command, it would most certainly work.

One final point: if you wanted to enter data in columns 1 through 5 but in a different order, such as columns 2, 4, 1, 5, and 3, then you could *not* use the dash in the `READ` command. Your command line would have to look as follows:

```
READ data into columns C2, C4, C1, C5, C3
```

As with our previous illustrations, any of these examples can be written without the lowercase words. Now that we have entered data into the Minitab worksheet, we shall discuss the general rules for manipulating the data.

Once all your data has been entered and *before* you start your analysis, it is a good idea to check your data one last time to make sure it has all been entered properly. The `PRINt` command will be described shortly, which allows you to check the numbers you actually entered against the original numbers. If you notice errors after all entries have been made, then you will need to correct your data using one of the systems described in the section below called "Correcting Entered Data." For now, let us assume that all your data was entered correctly and you are ready to proceed.

(Note: If you have large blocks of data stored in external files and wish to read them into Minitab for statistical analysis, consult the *Minitab Handbook*. The general command to use for entering data files is

```
READ 'filename' into C,...,C
```

However, since Minitab's interface with your computer system may not match ours exactly, you should check the specifics for using this command group.)

# General Rules▼▲▼▲▼▲▼▲▼▲▼▲▼▲▼▲▼▲▼▲▼▲▼▲▼▲▼▲▼▲▼▲▼▲▼

The following general rules will assist you in using the Minitab system effectively:

**(1)** Minitab reads about 200 different command words. If the first four letters of a command word are spelled correctly, Minitab will proceed to execute the command. In command lines, you may use other words besides command words, but you may *not* use numbers except when specifying which columns are involved in the commands. Either of the following forms of the `SET` command would function properly:

```
SET data into column C1
SET C1
```

However, you could not use the following:

```
SET 20 numbers into column C1
SET data into column 1
```

The first faulty command uses a number (20) in the command line. The **SET** command line may contain a number only when it specifies a column (**C1**). The second faulty command does not specify the column properly. Columns are specified as **C1** or **C10**, not as column 1 or column 10.

**(2)** Each command must be entered on a separate line.

**(3)** You may reuse a column in any subsequent command.

However, whenever a column is reused, any data originally stored in that column will be destroyed before the new data is entered.

## Basic Commands

For all commands in this book, the following symbols will be used. A capital **C** will stand for any column designation. In other words, when we write **READ data in C**, you should interpret the **C** to be whatever column you specify (e.g., **READ data into C1**, **READ data into C10**, etc.). A **K** will be used for constants. Anytime a capital **K** is used in a command, you may substitute any constant you wish. Finally, an **E** will be used when either a column number or a constant is appropriate.

An additional area of concern is the treatment of answers that result from various commands. For example, if you wanted to find the sum of a column, the answer is a *single number*. However, if you had two columns and wanted to add their corresponding row values together, then your answer would also be a column of numbers. Keep in mind the following rule: *When your answer is a single number, the computer will normally print the answer immediately; when your answer is a column of numbers, then you must tell the computer where to store the answers.* If you do not, you will receive an error message, which, depending upon the particular computer installation, will say something like **improper number of arguments**. This message means that you have not told the computer what to do with the results.

As an illustration, assume you have data in column 1 and want to find the average of that column. The proper command would be

**MEAN of values in column C1**

Because this is a single-variable answer, it would be printed immediately after the command was executed.

The general form of the command to average is

**MEAN of values in column C**

It can be abbreviated to

**MEAN C**

Now, assume that you wanted to add the constant 10 to each value in column 1. The proper command would be

**ADD 10 to C1, put in C2**

In general form, this command is

**ADD K to C, put in C2**

Note here that by adding a constant you create a whole new column of numbers. This new column of numbers must be stored somewhere. You cannot simply say **ADD 10 to C1**

### SECTION 1-3 Operations of the Minitab System ▶ 13

because the computer will not know where to store the resulting answers. You must tell it where to put the results.

Note that you could write the **ADD** command as follows:

```
ADD 10 to C1, put in C1
```

This command would take column 1, add 10 to each value, and then reenter the new values in column 1 *after* erasing all the original values. Whenever a column is reused, the old values are destroyed. Sometimes reusing a column is fine, but in some cases you will want both the original data and the new data. As you become more familiar with Minitab and the statistical exercises in this book, it will become easier to determine when to save a column and when to replace it.

The general command to add is written as

```
ADD E to E, put in C
```

Thus, you may add either two columns together or a constant to each row in a column; in either case, you must specify where the answers are to be placed. Each of the following two commands is correct:

```
ADD C1 to C2, put in C3
ADD 1 to C2, put in C3
```

The first command adds columns 1 and 2 together and puts the answers in column 3. The second adds a constant, 1, to column 2 values and stores the new values in column 3.

Now find the error in each of the following:

```
ADD 2,346 to C1, put in C2
ADD column 1 to column 2, put in column 3
```

The first command does not specify the constant properly. The value 2,346 must be entered without the comma. In the second command the columns are not specified correctly. They must be specified as **C1** or **C2**, not as **column 1** or **column 2**.

With this background, all the following commands, although new, should make sense:

```
SUBTract E from E, put in C
MULTiply E by E, put in C
DIVIde E by E, put in C
```

With this group of commands you can add, subtract, multiply, and divide any constant or column by another constant or column. You can use the shorthand form for each of these commands by using just the capitalized letters for each keyword.

Squaring the values in a column, say column 5, would simply involve multiplying that column by itself. For example,

```
MULT C5 by C5, put in C6
```

would square column 5 and put the results in column 6. To find the square root of a column of numbers, the command is

```
SQRT of E, put in C
```

For this command you may enter either a constant or a column number and have the square root of that value calculated and then stored in the specified column.

Two final points should be covered. First, if you calculate a number that has a single answer, such as `MEAN C`, the answer will be printed at the time of calculation. If you wish, you can also store that answer in a column for later use. In that case, the command form is

`MEAN of values in column C, put in C`

As an example, this command might read

`MEAN of values in column C1, put in C2`

This command would both print the average and put that value in the first row of the specified column on your worksheet. The second point concerns printing your results. If you have entered data into several columns, performed various calculations, and stored the results in yet other columns, you may want a printout of various columns when you are through. The command is

`PRINt C,...,C`

As you can see, one or more columns may be printed with this command. All of the following commands would work if there was proper data in each column:

```
PRIN C1, C6, C9
PRIN C6, C1, C9
PRIN C1-C5
PRIN C3
```

The first command would print columns 1, 6, and 9 in that order and skip all other columns. The second would print the same three columns but in a different order. The third would print the first five columns. The last example is slightly different. In this case, only one column of numbers is specified. When this command is entered, the results are printed as a *row* of numbers instead of a column. The reason is simply to save space. The first group of numbers in your column will be printed as row 1, the next group as row 2, and so on. Whenever two or more columns are requested, the results will be printed in the column format. (Note that if you wish to print a single column of numbers as a column—that is, vertically—use the command `PRIN C1 C2`, where `C2` is an empty column.)

## Correcting Entered Data ▼▲▼▲▼▲▼▲▼▲▼▲▼▲▼▲▼▲▼▲▼▲▼▲▼

We have cautioned you to be sure your data is correct before you press the return and enter those values. We also suggested that after entering all data you execute a `PRINt` command and again make sure the data entered into your columns is correct. Occasionally you will enter data incorrectly and need to correct it. There are four ways to correct inaccurate data.

If you have made many errors in entering your data, starting over might be easiest. To do this, reenter the `READ` or `SET` command, making sure you specify the correct column(s) you wish to use. If you have entered several columns of data and only one column is inaccurate, then use the `SET` command for the bad column and reenter only that column. Remember that whenever you reuse a column, the old data is erased before the new data is entered.

If only one or two numbers in your data are inaccurate, the `LET` command is a simple way to correct the errors. The `LET` command is very powerful and is discussed in more detail in Chapter 2. For the present, we can use the `LET` command to correct a wrong entry

## SECTION 1·3 Operations of the Minitab System ▶15

as follows: Assume that the fifth number of column 7, which should be 345.6, is wrong. Then on a new line (after the cursor), you would enter the following command:

LET C7(5)=345.6

What you have specified is that the number 345.6 should be placed in column 7 (**C7**) and row 5 (**(5)**). The general form of the command is

LET C(K)=K

Thus the column (**C**), the row (**K**), and the new number (**K**) are specified. Notice that the row is placed in parentheses. If you only had a few errors in your data set, this command would be an easy way to correct them.

A third method of correction is used when a row of data has been left out of the input. You might have left the numbers out of a single column or several columns. Here the **INSErt** command should be used. Assume that between rows 4 and 5 of column 7 you omitted the value of 186.3. To use the **INSErt** command entails three operations. You first tell the computer where to enter the data, then what value(s) to enter, and then to **END** the process. Referring to the example just developed, examine the following illustration:

```
MTB >INSErt between rows 4 and 5 of column C7
DATA>186.3
DATA>END
```

After the cursor, we first told the computer where to insert the value needed; on the second line we told it what value to enter; finally we terminated the process.

As a second illustration, assume that between rows 7 and 8 of columns 4, 5, 6, and 7 we omitted three rows of data. Our problem is then to insert three new sets of numbers. Examine the form of the command closely.

```
MTB >INSErt between rows 7 and 8 of columns C4-C7
DATA>K K K K
DATA>K K K K
DATA>K K K K
DATA>END
```

The command first tells the computer which rows and then which columns are involved. After that command, we enter the numbers desired. Of course, any rows of data after the insert line are moved down. In this example, our old row 10 of data would become row 13 since three new lines would be added before row 10. The general form of the command is as follows:

INSErt between rows K and K of columns C-C

This command can be abbreviated to

INSE K K C-C

It is important to note that the columns do not have to be consecutive. You can specify different columns by separating them with either a space or a comma. If you have forgotten data at the start of your worksheet, then insert between rows 0 and 1.

The final way to edit data is with the **DELEte** command. This command allows several options. The basic form is as follows:

DELEte row(s) K, K, K from column(s) C, C, C

With this command a single row may be deleted, multiple rows from a single column may be deleted, or multiple rows from multiple columns may be deleted. Assume we wish to eliminate the fourth row from column 7. The command would be

>DELEte row 4 from column C7

which could be reduced to

>DELE 4 C7

If we wished to eliminate several nonconsecutive rows such as rows 4, 7, and 9 from column 7, the command would be

>DELEte rows 4, 7, and 9 from column C7

Consecutive rows are done in a slightly different manner from consecutive columns. We use a colon (:) between the first and last row. For example, if we wanted to eliminate rows 1 through 9 of column 7, the column would be

>DELEte rows 1:9 from column C7

The final set of **DELEte** illustrations involve multiple columns. If the same row is to be eliminated from several nonconsecutive columns, such as row 4 from columns 3, 7, and 11, our command would be

>DELEte row 4 from columns C3, C7, C11

For consecutive columns, the dash is used as in past examples. For row 4 of columns 5 through 13, we would enter

>DELEte row 4 from columns C5-C13

Finally, multiple rows from multiple columns can also be eliminated. Rows 4, 7, and 11 from columns 2, 5, and 15 may be deleted as follows:

>DELEte rows 4, 7, 11, from columns C2, C5, C15

To delete consecutive rows 4 through 11 from columns 2 through 15, the command would be

>DELEte rows 4:11 from columns C2-C15

These examples illustrate only part of the power of various Minitab commands. There are also other ways to edit data, but for our purposes these commands are sufficient. For those wanting or needing more information for specific applications, we recommend the *Minitab Handbook*.

# SUMMARY

This chapter described the basic commands in Minitab. Data can be entered by either of two commands, **READ** or **SET**; a variety of calculations, such as **ADD**, **MULT**, **SUBT**, **DIVI**, and **SQRT**, can be performed; and data can be shown by using the **PRIN** command.

All commands follow the basic pattern of telling the computer (1) which function to perform, (2) where the data is located (or what the constant value is) on which to perform the activity, and (3) in some cases, where to store the results of the calculations.

# DISCUSSION QUESTIONS

1. What advantage does Minitab have over "number crunching"?
2. What is the difference between a `SET` and a `READ` command?
3. Enumerate and explain each way of correcting data.

# PROBLEMS

**1-1** In 1983, the 10 highest-paid executives in the United States received well over $3 million each. Their incomes are given in the following list in thousands of dollars:

| | | | | |
|---|---|---|---|---|
| 13,229 | 4,221 | 6,921 | 3,915 | 7,292 |
| 3,718 | 3,783 | 4,349 | 4,301 | 6,083 |

a) What must be omitted when entering data in the form given in this problem?
b) Enter the data on the executives' incomes using the `SET` command.
c) `PRINt` the data to check your accuracy.
d) What is the mean (average) income for these executives?

**1-2** Since the mid-1970s, research and development (R&D) spending has been rising. Shown are the R&D expenditures for five East Coast and five West Coast companies in thousands of dollars:

| East Coast | West Coast |
|---|---|
| 2,175 | 2,126 |
| 707 | 1,764 |
| 691 | 879 |
| 834 | 2,053 |
| 710 | 781 |

a) Using the `SET` command, enter the data on the R&D expenditures for the East Coast in column 1 and the West Coast in column 2. Then list the data to check your accuracy.
b) What is the total expenditure for the East Coast? What is the mean?
c) What is the total for the West Coast? What is the mean?
d) Assuming you `SET` your data in `C1` and `C2`, use the `READ` command and enter the same values in `C3` and `C4`. List all four columns and check your data for accuracy.

**1-3** The producer of WQBX-TV is interested in the number of TV viewing hours of teenagers in grades 9 through 12. This information will help in determining the types of programs to schedule. A sample of the number of viewing hours per day for teenagers follows:

| | | | | | | |
|---|---|---|---|---|---|---|
| 6 | 5 | 2 | 6 | 2 | 3 | 6 |
| 5 | 6 | 6 | 3 | 6 | 6 | 6 |
| 2 | 5 | 2 | 4 | 6 | 6 | 3 |
| 2 | 4 | 6 | 6 | 6 | 2 | 6 |
| 3 | 3 | 6 | 5 | 4 | 3 | 5 |

a) Enter this data using the `SET` command. If you choose to, enter the data using the `READ` command. (Note that the `READ` command in this case involves more work.)

**b)** Create a histogram for this data using the **HIST** command (even though you may not yet understand what a histogram means). It should look like the following:

```
MIDDLE OF      NUMBER OF
INTERVAL      OBSERVATIONS
    2              6       ******
    3              6       ******
    4              3       ***
    5              5       *****
    6             15       ***************
```

**c)** Based on this data, how many hours a day on the average do teenagers watch TV?

**1-4** The market share held by a local whiskey distributor for the years 1976 and 1984 is shown below.

| Liquor Category | 1976  | 1984  |
|---|---|---|
| Bourbon  | 13.8% | 18.5% |
| Canadian | 12.9  | 13.1  |
| Scotch   | 11.4  | 13.6  |
| Blends   | 7.9   | 15.2  |

**a)** Enter this data into **C1** and **C2** using the **READ** command.
**b)** Subtract 1976 figures from 1984 figures to determine the change in market share for each type of whiskey. Be careful about which column is subtracted from which.
**c)** What is the total change in market share?
**d)** If the distributor succeeds in raising his market share in 1985 by 3% over the 1984 figures in each category of liquor, what will the new market share figures be?
**e)** Use a **PRINt** command to check your results.

**1-5** The ratio of the price of gold to silver often hovers around 50 to 1. Listed is a sample of gold prices in dollars.

```
350    290    389    410    310    364
325    420    288    370    411    400
```

**a)** Enter the data using the **SET** command.
**b)** Divide each price by 50 to obtain the approximate price of silver for each corresponding gold price.
**c)** Assume an error was made and that the price of 325 was actually 375. Use the **LET** command to make the correction.
**d)** With the correction made, determine the mean dollar price of gold and silver. (Remember, you will have to go back and change the silver price.)
**e)** What would have to be done differently if the **READ** command was used to enter the data?

**1-6** A recent male graduate in international marketing was offered four overseas jobs. For each country, he determined the annual living expenses needed for himself, his wife, and his two children. His figures are shown on the next page.

## PROBLEMS

| City | Housing | Transportation | Goods and Services |
|---|---|---|---|
| Hong Kong | $52,360 | $15,878 | $12,713 |
| Geneva | 22,667 | 5,674 | 17,397 |
| Paris | 15,199 | 4,786 | 13,098 |
| Mexico City | 8,175 | 2,893 | 10,139 |

**a)** Enter the three different costs for each city using the **READ** command.
**b)** What are three different shorthand ways to write the **READ** command in this problem?
**c)** Assume that you forgot to include the living expenses for the city of London. Use the **INSErt** command to enter the following data between the information on Geneva and Paris:

| London | 15,269 | 6,254 | 13,846 |

**d)** Explain the difference between a **LET** and an **INSErt** command.

**1-7** Shown in the following table are average income data on a sample of elderly individuals, ages 65 to 75.

| Age | Monthly Social Security Income | Other Income |
|---|---|---|
| 65 | $500 | $250 |
| 66 | 525 | 320 |
| 67 | 460 | 400 |
| 68 | 420 | 150 |
| 69 | 500 | 165 |
| 70 | 550 | 90 |
| 71 | 475 | 22 |
| 72 | 430 | 5 |
| 73 | 510 | 30 |
| 74 | 515 | 50 |
| 75 | 435 | 10 |

**a)** What is the average amount of Social Security income received by the elderly aged 65 to 75?
**b)** What is their average monthly income from other sources?
**c)** Determine the average total income for each age group in the sample.
**d)** Divide each Social Security income figure by its respective total income figure to determine the percentage of total income obtained from Social Security.
**e)** What can we infer from these percentages?

**1-8** Sam, a building contractor, has just submitted bids for five urban homes. The information on these homes follows:

| Home | Costs | Profit |
|---|---|---|
| 1 | $76,000 | $23,000 |
| 2 | 54,500 | 10,000 |
| 3 | 39,800 | 16,000 |
| 4 | 62,300 | 18,200 |
| 5 | 73,000 | 25,000 |

**a)** Determine the average profit for the five homes.
**b)** If Sam finishes two weeks ahead of schedule, profit per home will increase by $1,500. Compute the new profit figures for each home.
**c)** If Sam finishes four weeks behind schedule, profit per home will decrease by $2,600. Compute the new profit figures.
**d)** What is the new mean profit on all five homes if Sam finishes two weeks earlier? Four weeks later?

**1-9** Refer to the Go-Fast Running Shoe data in Appendix A.
  **a)** Use the `READ` command to enter the data on age, schooling, productivity, and height of the first 20 employees. List the data to check your accuracy.
  **b)** What is the average age of these 20 employees?
  **c)** What is the average number of school years completed?
  **d)** For this group of employees, what is the average number of units produced per person?
  **e)** What is the average height of these employees?
  **f)** Use the `HISTogram` command to display the data on the number of school years completed. Your histogram should look as follows:

```
MIDDLE OF        NUMBER OF
INTERVAL        OBSERVATIONS
   11               6          ******
   12               8          ********
   13               4          ****
   14               2          **
```

**1-10** The Cartwright Farm is known for its fine thoroughbred racehorses. At a recent auction in Kentucky, five of its thoroughbreds sold for record amounts. The figures, shown below, are in thousands of dollars.

| Horse | Selling Price |
| --- | --- |
| Angel's Halo | 1,200 |
| North Wind | 975 |
| Band o' Gold | 1,350 |
| Rainbow's End | 1,050 |
| Sixth Sense | 875 |

  **a)** What was the average amount received for the five horses?
  **b)** The Cartwright Farm managed to sell yet another thoroughbred after the auction. Use the `INSErt` command to add Creative Genius to the end of the list. Creative Genius sold for $1,300,000. List the data to check your accuracy.
  **c)** What is the new average amount received by the Cartwright Farm?
  **d)** Because North Wind's buyer could not fulfill her obligations, North Wind was returned to the Cartwright Farm. Use the `DELEte` command to remove this thoroughbred from the list.
  **e)** Without North Wind, what was the average amount received?

**1-11** Japanese automobile manufacturers have dominated the small car market in the United States for the past five years. Assume that the number of small passenger cars sold in the years 1981 through 1985 by U.S. and Japanese manufacturers are represented by the figures given (in millions of units).

## PROBLEMS

| Year | United States | Japan |
|------|---------------|-------|
| 1981 | 1.20 | 1.50 |
| 1982 | 1.15 | 1.60 |
| 1983 | 1.30 | 1.40 |
| 1984 | 1.25 | 1.55 |
| 1985 | 1.20 | 1.60 |

a) Use the MEAN command to find the average number of small cars sold annually by U.S. manufacturers.
b) What is the annual average sold by Japanese manufacturers?
c) What is the total number of small cars sold by U.S. and Japanese manufacturers from 1981 through 1985?
d) Use the SUBTract command to determine how many more units were sold by Japanese manufacturers in each year. Use the PRINt command to check your answers.
e) From 1981 through 1985, how many more cars were sold by Japanese manufacturers?

**1-12** Assume that the five states that pay their public school teachers the highest salaries are Alaska, the District of Columbia, Michigan, Hawaii, and Washington. Data on salaries and the respective number of pupils per teacher for these states follow:

| State | Salary | Pupils per Teacher |
|-------|--------|--------------------|
| Alaska | $29,000 | 26 |
| District of Columbia | 27,000 | 34 |
| Michigan | 24,300 | 20 |
| Hawaii | 23,700 | 26 |
| Washington | 23,500 | 30 |

a) Use the READ command to enter the data into C1 and C2.
b) For these states, what is the average salary for a teacher?
c) What is the average number of pupils per teacher?
d) For each state, determine the amount of money a teacher receives per child. (DIVIde C1 by C2 and place the results in C3.)

**1-13** Bill, a sales clerk, receives a bonus for every sale over $5. A 2% bonus is applied to the amount exceeding $5 on the sales receipt. Bill's sales (in dollars) for yesterday are as follows:

| | | | |
|---|---|---|---|
| 5.25 | 7.81 | 5.50 | 17.50 |
| 10.60 | 66.00 | 8.95 | 32.50 |
| 7.25 | 5.45 | 6.06 | 6.60 |
| 31.00 | 29.00 | 12.31 | 9.97 |

a) Using the SET command, enter the data into C1.
b) What was Bill's average sale?
c) What was his sales total?
d) Subtract $5 from each receipt and place the data in C2.

**CHAPTER 1** Elements of Minitab

    e) Use the `MULTiply` command and multiply C2 by 2% to determine Bill's bonus. Enter the data into C3.
    f) What was Bill's total bonus?

**1-14** Oil companies have recently discovered that it is cheaper to purchase oil rather than develop their own oil fields. The cost per barrel of oil for four companies follows:

| Company | Purchased Oil | Company's Oil |
|---------|---------------|---------------|
| 1 | $6.83 | $ 8.10 |
| 2 | 6.72 | 8.32 |
| 3 | 5.35 | 11.93 |
| 4 | 5.82 | 13.60 |

    a) `READ` the data into C1 and C2.
    b) What is the mean purchase price for the four companies?
    c) What is the mean cost per barrel for their own oil?
    d) Subtract each purchase price from the respective company's oil price to determine the cost advantage of purchasing the oil. Enter the data into C3. Use the `PRINt` command to check your answers.

# BASIC MANIPULATIONS WITH MINITAB

*The following Minitab commands will be introduced in this chapter:*

`LET`
A command used for arithmetic functions and with other Minitab commands

`STACK E on top of E, put in C`
A command to join two columns, a column and a constant, or two constants and to place the result in a new column

`SORT data in C, put in C`
A command to arrange the values in a given column in numerical order and to store the ordered values in a new column

`PLOT values in C vs C`
A command to plot variables

`TSPLot of data in C`
A command to create a time series plot, in which each value in a column is plotted sequentially

`NAME C 'YOURNAME'`
A command to name a column

`ROUNd E, put in E`
A command to round off the values in a column or the value of a constant and to store the new value(s) in a new column or constant

# CHAPTER 2 Basic Manipulations with Minitab

In Chapter 1, we introduced the first elements of entering data in Minitab. We also covered the format for issuing most commands. In this chapter, we will see how to manipulate data within a column and between columns. We will also cover a few additional commands for naming columns and rearranging data within a column.

## 2-1 THE LET COMMAND

In the last chapter, the commands to add, subtract, multiply, and divide were mentioned. The reason those commands were not covered in greater detail is that they are seldom used. In most cases, a much more powerful command, **LET**, will be used to perform all of the arithmetic operations. In our experience, a student who has become familiar with the **LET** command uses it to perform many operations. In Chapter 1, the **LET** command was introduced to assist in correcting data. In this chapter, much greater use will be made of the command's power.

Our discussion of the **LET** command will be broken into two parts. The first will deal with using the **LET** command for simple arithmetic functions, the second with using other Minitab commands within the **LET** command.

### Algebraic Functions

To illustrate the algebraic flexibility of the **LET** command, assume that we have data in each of the first three columns on our worksheet, **C1**, **C2**, and **C3**. Using the standard algebraic expressions and rules, we can specify a new column as any function of the old columns or a constant. For example, examine each of the following:

```
LET C4 = C2 + C3
LET C5 = C3 - C1
LET C6 = C4 - 10
```

If you look closely at each of these examples, you can see a common format. First, you specify the new column in which the results of the manipulations are to be placed—in these cases, **C4**, **C5**, and **C6**, respectively. Our first example requests that the values in columns 2 and 3 be added together and placed in column 4. Our second illustration subtracts the corresponding values in column 1 from the values in column 3 and places the results in column 5. Finally, in the third illustration, 10 is subtracted from each value in column 4 and the results placed in column 6.

Now examine the following set of illustrations:

```
LET C4 = C1 * C2
LET C5 = C3 / C4
LET C6 = (C3 + C4) / C2
LET C7 = (C4 ** 2) * C3 / (C2 - C1)
```

These commands might look complicated, but they really are quite simple. In the first illustration we are using the sign for multiplication, namely **\***. Thus, this command indicates that we should multiply each value in column 1 by the corresponding value in column 2 and then place the resulting values in column 4. Our second illustration involves division, **/**, and has the results of dividing each value in column 3 by the corresponding value in column 4 and placing the results in column 5. Our third illustration uses parentheses for the first time. These parentheses are used as in algebra. The operations inside the parentheses are

### SECTION 2-2 Columns

performed first, then the operations outside. In this case, we are adding the corresponding values in columns 3 and 4 together and dividing the result by the value in column 2. The answers are then placed in column 6. In our final illustration we are using the exponentiation sign, `**`. In this case the values in column 4 are to be squared (`**2`), and those squared values are to be multiplied by the corresponding values in column 3. The totals determined by that calculation are divided by the result of subtracting column 2 from column 1. Our final answer is placed in column 7.

We should note here that the spacing in the **LET** command line is not essential. We have added it only for ease of comprehension. Except for a space after the word **LET**, the rest of the command line need not contain spaces. For example, our last command line could be expressed in either of the following forms:

```
LET C7=(C4**2)*C3/(C2-C1)
LET C7 = (C4**2) *C3/ (C2-C1)
```

Once you have spent a little time using it, you will find the **LET** command very helpful for performing simple operations.

## Column Operations and Functions

Numerous column operations within the Minitab system can also be used within the **LET** command. We will illustrate most of these operations in later chapters as they are needed, but for now we will provide one example. In Chapter 1, the command **MEAN** was described for finding the arithmetic average. If we wanted to find the mean value of column 1 and then add that value to each value in column 2, our command would be

```
LET C3 = MEAN(C1) + C2
```

When using the **LET** command in this form, it is important to keep several rules in mind:

**(1)** Parentheses are necessary with functions and column operations. In our example, you must use **MEAN(C1)**, *not* **MEAN C1**.

**(2)** No extra text may be used on a **LET** command line.

**(3)** Functions may be expressed by the first four characters in the function's name or written out in full.

We will be covering the use of functions in the **LET** command more extensively as such commands are needed. For now, it is important that you realize that such a set of options is available. The *Minitab Handbook* covers all aspects of the **LET** command and should be consulted if you need any further information.

## 2-2 COLUMNS

### Storing Output Values from Calculations

The Minitab system uses a worksheet concept for storing data. Consider the following task: We have 10 values in column 1 and want to find the mean value of those 10 values and store the resulting calculation. Notice that the result of the calculation is a single value, not a column. There will be several cases where the result of a command is just a single value rather than a column of numbers.

The proper way to handle such a calculation is to use the system of constants. In the example above, the command would be

```
MTB >MEAN C1, put in K1
```

Here we used **K1**, but we could just as easily have used **K5** or **K10** or **K20**.

If we now wanted to subtract the calculated mean (**K1**) from each value in column 1, we would use the **LET** command or the **SUBTract** command. Since **C1** is a column of values from which we are going to subtract a specific amount, the results of such a calculation will also be a column of numbers. The correct way to perform this subtraction would be with either of the following commands:

```
MTB >LET C3 = C1 - K1
MTB >SUBTract K1 from C1, put in C3
```

Either command would subtract **K1** from each value in **C1** and place the results in **C3**.

Whenever an answer is a single value, it is possible to use a column command

```
MTB >MEAN C1, put in C2
```

Here column 2 will simply have a single value in row 1. Minitab will *not* give you any kind of error message because the command can be executed. However, if you later used this column 2 in a calculation such as

```
MTB >SUBTract C1 from C2, put in C3
```

you would get an error message informing you that the columns are of unequal length. Only the first row in column 1 and the first (and only) row in column 2 would be subtracted. No other values would be involved.

Thus, if you perform a calculation and the resulting answer has a single numeric value, develop the habit of storing such a value as a constant (**K**). It is possible to store such values in columns, but such a practice can lead to more problems than it is worth. Not many calculations result in single-value answers, and you will quickly learn which commands do. Also, get in the habit of using the **LET** command instead of the arithmetic commands whenever possible. Equivalent commands of both types are shown in Table 2-1.

**TABLE 2-1** Equivalent LET and Arithmetic Commands

| LET | Arithmetic |
| --- | --- |
| LET C3 = C1 + C2 | ADD C1 to C2, put in C3 |
| LET C3 = C1 - C2 | SUBTract C2 from C1, put in C3 |
| LET C3 = C1 * C2 | MULTiply C1 by C2, put in C3 |
| LET C3 = C1 / C2 | DIVIde C1 by C2, put in C3 |

With the **LET** command, we can also calculate and store other values in a single command. All of the following commands are valid uses of the **LET** command:

```
MTB >LET K1 = MEAN(C1)
MTB >LET C2 = C1 - K1 / 2
MTB >LET K2 = 5
MTB >LET K4 = K1 - K2
MTB >LET K3 = MEAN(C2) / K2
```

**SECTION 2-2** Columns      ▶ **27**

## Joining Columns Together ▼▲▼▲▼▲▼▲▼▲▼▲▼▲▼▲▼▲▼▲▼▲▼▲▼▲

Occasionally when using various statistical techniques (i.e., indexing), it is convenient to place all the values in one column at the end of all the values in another. Notice, we are not *adding* the columns together but *joining* them. Thus, if both of the starting columns had 8 values each, the new column would have 16 values.

The command to join one column to another is the **STACk** command, and it has the form

            STACk E on top of E, put in C

This command says that you can stack any column or constant on top of any other column or constant and place the results in a new column. As an illustration, assume we have values in column 4 and in column 7 and we wish to join them together and place the results in column 11. Our command would be

```
MTB >STACk C4 on top of C7, put in C11
```

Be careful to specify the order intended. You would get a different result with the command

```
MTB >STACk C7 on top of C4, put in C11
```

It is possible to join three or more columns in a single command:

```
MTB >STACk C1 to C2 to C3, put in C10
```

## Entering Repetitive Data ▼▲▼▲▼▲▼▲▼▲▼▲▼▲▼▲▼▲▼▲▼▲▼▲▼

At times you may want to enter the same value into a large number of rows of a single column. For example, assume you wanted to enter the value of 8 in the first 25 rows of column 5 and the value of 14 in the next 20 rows of column 5. Our final column would then have 45 values. It is certainly possible to simply enter the values manually. However, it would be much faster to use a special form of the **SET** command. For repeated entries, you place the number to be repeated in parentheses immediately after the value for the number of times the item is to be repeated. Examine the following illustration:

```
MTB >SET values in C5
DATA>25(8), 20(14)
```

As desired, this places the value of 8 in the first 25 rows of column 5 and then the value of 14 in the next 20 rows of column 5.

Another special form of data entry involves sequential data. In this case you enter a colon between the first and last numbers of the sequence. Examine each of the following illustrations:

                1:4
                8:25
                30:6
                5:18, 65:48

The first illustration enters the values 1 through 4. In actual form, the command set would be

```
MTB >SET C1
DATA>1:4
```

The second example enters the values 8 through 25; the third enters the values 30 through 6 in descending order; the final one enters the values 5 through 18 in ascending order and then the values 65 through 48 in descending order.

## Other Commands for Columns

There are many other Minitab commands for manipulating data in columns, but for the present we will mention only two more. Others will be introduced as they are needed.

Occasionally it is convenient to put all the values in a column in order from the smallest value to the largest. This can be done with the command **SORT**:

SORT data in C, put in C

This command could be handy with problems such as the ages of the employees at the Go-Fast Running Shoe Company. If we wanted to know the age of the youngest or oldest employee, we could use the **SORT** command on the column in which the age data was stored. After the **SORT** command we would use a **PRINt** command and simply inspect the column of numbers printed. The first number printed would be the smallest value and the last number the largest. (The **DESCribe** command, discussed in Chapter 3, also prints the minimum and maximum values.)

The use of the **SORT** command brings up an important point about using Minitab or any other type of technology in performing statistical analysis. Frequently the most efficient way to solve a problem is to do part of the problem with technology and part of the problem by hand. Minitab will not answer every statistical question. For example, if in our Go-Fast example you wanted to know how many of the employees were of a specific age, the easiest way to find out would be to enter the data, **SORT** the values, **PRINt** the column, and then manually count how many times the particular value of interest is found. Since all such values would be listed together, the operation would be quick and simple. It is technically possible to perform a number of operations and have Minitab tell you the frequency of a given number, but a balanced combination of computer assistance and human effort will best solve this problem. It is always important to consider when Minitab or any other computer system will save you work and when it will be faster and easier to do the work manually.

The final command discussed in this section is the **NAME** command. As the number of columns increases in a problem, it is frequently difficult to remember which variable is in which column. The **NAME** command allows you to assign up to eight characters to a column. However, there are two restrictions: First, you cannot begin or end a name with a space; second, you cannot use a single quote (apostrophe) within a name.

For example, in our Go-Fast Running Shoe Company example, if ages were in column 1, years of schooling in column 2, productivity in column 3, and height in column 4, the following commands could be used:

```
MTB >NAME C1 = 'AGE'
MTB >NAME C2 = 'SCHOOL'
MTB >NAME C3 = 'PROD.'
MTB >NAME C4 = 'HEIGHT'
```

More than one name can also be placed on the same line:

```
MTB >NAME C1 = 'AGE', C2 = 'SCHOOL', C3 = 'PROD.', C4 = 'HEIGHT'
```

### SECTION 2-3 Plotting

The general form for this command is always the same: The column is specified and then the name is enclosed in single quotes. Once a column has been named, that name will be used on all output involving that column. You may specify a named column by either its number or its name. However, the name must always be enclosed in single quotes. For example, assuming that we had named all four columns as shown above, we could use either of the following commands:

```
PRINT C1
PRINT 'AGE'
```

However, the command **PRINT HEIGHT** would not work because the column name is not enclosed in single quotation marks.

Column names may be specified in either uppercase or lowercase letters, or any combination. The computer will not distinguish between cases. However, when printing column operations, the name will be printed in the same sequence of uppercase and lowercase letters that you used to enter the name. Also, you may name columns before they have values. For example, if you knew you were going to put ages into column 1, then you could start your Minitab analysis by naming column 1 **'AGE'** before entering data. However, you *cannot* use a column name before you have assigned that name to a column. Thus, you could issue the commands

```
MTB >NAME C1 = 'AGE'
MTB >SET 'AGE'
```

but *not* the commands

```
MTB >SET 'AGE'
MTB >NAME C1 = 'AGE'
```

Finally, column names may be used in **LET** commands. If column 1 was named **'AGE'** and column 2 named **'SCHOOL'**, then both of the following commands would perform the same function:

```
LET C5 = C1 * C2 / MEAN(C1)
LET C5 = 'AGE' * 'SCHOOL' / MEAN('AGE')
```

Notice that all the standard rules for the **LET** command remain the same.

We have found that most students name their columns so that their printouts look sharp and are easy to interpret. However, except when a problem is particularly complicated, most students use column numbers in their command lines.

## 2-3 PLOTTING

Minitab will plot data for you in several different ways, but we will mention only two here. You have probably plotted values on a standard Cartesian coordinate system. In Figure 2-1, points have been plotted in each of the four quadrants.

Minitab will also plot points but in a slightly different way. Examine Figure 2-2 for a moment, which plots the same points shown in Figure 2-1. You will notice that the two axes

**FIGURE 2-1** Traditional Cartesian Coordinate Graph

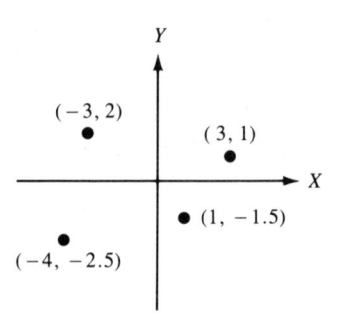

```
                        Y
                        ▲
              (−3, 2)   │
                 ●      │    (3, 1)
                        │       ●
         ───────────────┼──────────────▶ X
                        │  ● (1, −1.5)
                 ●      │
           (−4, −2.5)   │
                        │
```

**FIGURE 2-2** Data Points in Figure 2-1 Plotted by Minitab

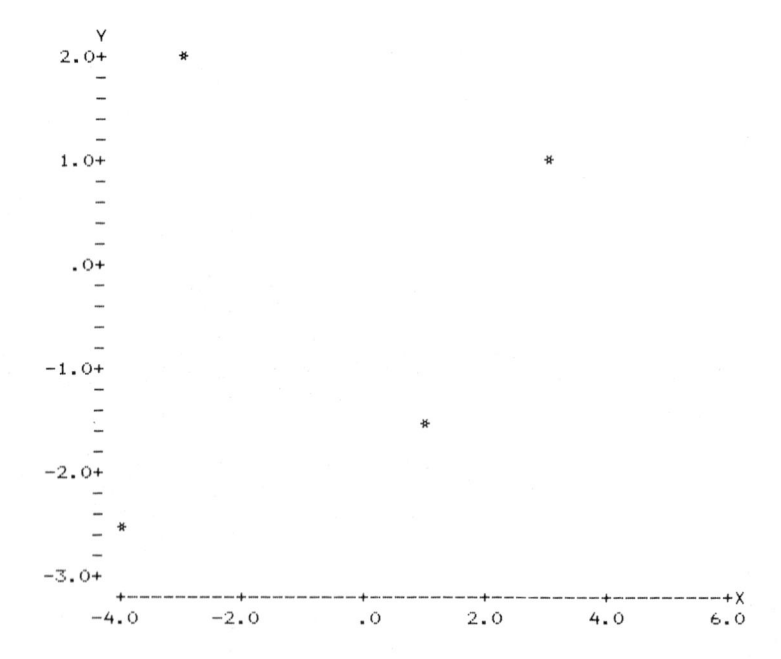

```
      Y
   2.0+        *
      -
      -
      -
      -
   1.0+                                    *
      -
      -
      -
      -
    .0+
      -
      -
      -
      -
  -1.0+
      -
      -
      :                       *
      -
  -2.0+
      -
      -  *
      -
  -3.0+
      +---------+---------+---------+---------+---------+X
     -4.0      -2.0       .0       2.0       4.0       6.0
```

do not cross in the middle but run from a positive value through 0 to a negative value in two continuous lines. To find the (0, 0) point (the middle of a traditional Cartesian coordinate system), you would simply sight up from the X-axis at 0 and across from the Y-axis at 0. This point would be equivalent to the origin in traditional graphs. Minitab will automatically set the values on each axis according to the data. If all values are nonnegative, then the axes of the plot will be the boundaries of the positive quadrant, or upper right-hand quadrant, of the Cartesian graph.

# SECTION 2-3 Plotting

## The PLOT Command ▼▲▼▲▼▲▼▲▼▲▼▲▼▲▼▲▼▲▼▲▼▲▼▲▼▲▼▲▼▲▼

To do plotting in Minitab you will use the **PLOT** command. This easy command will allow you to get a quick look at clustering in your data. The form of the command is

```
PLOT values in C vs C
```

In using this command you must be sure to specify the columns in the proper order. The first column specified will be plotted along the *Y*-axis, and the second column along the *X*-axis. When plotting it is often helpful to name your columns. Then when your plots are printed out, the columns will be named and you will be less likely to make errors in interpretation.

To illustrate, we have entered 16 numbers into both columns 1 and 2. Then we enter the following two **PLOT** commands:

```
MTB >PLOT values in C1 vs C2
MTB >PLOT values in C2 vs C1
```

Figure 2-3 is the result of the first command, and Figure 2-4 the result of the second. Both of these plots are correct, but notice the difference in their appearance. Be sure that you specify the order of your columns in the way you intend.

**FIGURE 2-3** Plot of 16 Points with Column 1 Specified First

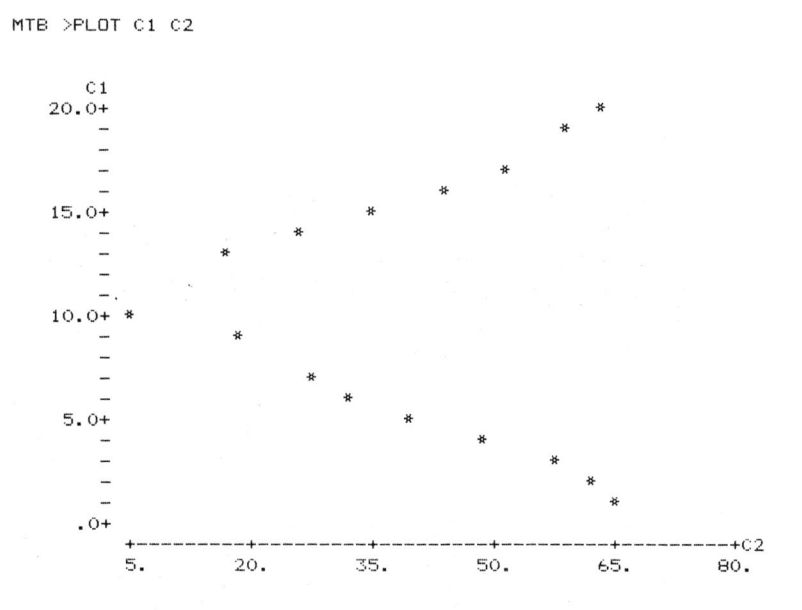

```
MTB >PLOT C1 C2

      C1
  20.0+                                                      *
      -                                                  *
      -
      -                                            *
      -                                      *
  15.0+                           *
      -                    *
      -              *
      -
      -.
  10.0+  *
      -           *
      -                 *
      -                    *
   5.0+                         *
      -                              *
      -                                   *
      -                                        *
    .0+
      +---------+---------+---------+---------+---------+C2
      5.       20.       35.       50.       65.       80.
```

In our next illustration of the **PLOT** command, assume that we want to read the ages of the first 20 employees of Go-Fast into column 1 and their number of years of schooling into column 2. In addition, we want to name column 1 `'AGE'` and column 2 `'SCHOOL'`. The command group would look as follows:

**FIGURE 2-4** Plot of Points in Figure 2-3 with Column 2 Specified First

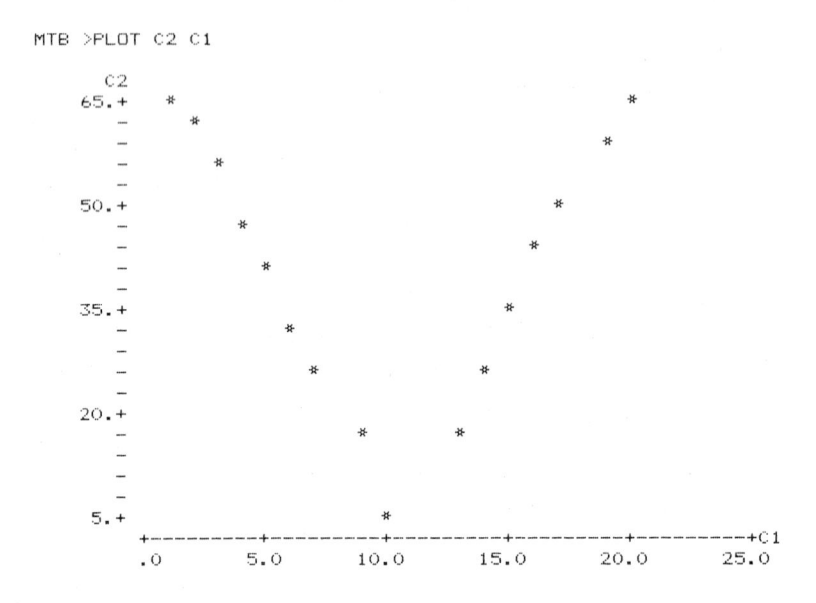

```
MTB >PLOT C2 C1
     C2
   65.+    *                                      *
       -        *
       -                                              *
       -     *
       -
   50.+
       -        *                           *
       -                                       *
       -          *
       -
   35.+
       -             *                  *
       -
       -                   *       *
       -
   20.+
       -                *       *
       -
       -
       -
    5.+
       -                      *
        +---------+---------+---------+---------+---------+C1
       .0       5.0       10.0      15.0      20.0      25.0
```

```
MTB >READ C1 C2
DATA>35, 11
DATA>33, 13
  :
DATA>32, 12
DATA>NAME C1 'AGE', C2 'SCHOOL'
MTB >PLOT C2 C1
```

When this group of commands is executed, it produces the graph shown in Figure 2-5. You will notice that most of the points are indicated by a **\***, as in the previous plots. However, there are also four cases where the number **2** appears. In this type of plotting, Minitab will use a number to represent a place where a common set of values appears more than once. In this case, for example, there are two employees who are both 33 and have 13 years of schooling. If there had been five employees with these same traits, then a **5** would have appeared at that intersection. In this way, Minitab shows you data point frequency.

# Time Series Plots ▼▲▼▲▼▲▼▲▼▲▼▲▼▲▼▲▼▲▼▲▼▲▼▲▼▲▼▲▼▲▼▲▼▲▼

On occasion you will want to plot a series of points against time. For example, you might want to plot the last 12 months' production against the numbers 1 through 12 to see whether there is a trend over time. Minitab will automatically plot any single column of values against consecutive numbers using the command

> TSPLot of data in C

The command **TSPLot** stands for "time series plot." The *X*-axis will always represent the series of consecutive numbers. If the column of numbers specified contains 30 observations, then the **TSPLot** command will plot 30 values on the *X*-axis; if you have 50 observations,

## SECTION 2-3 Plotting

**FIGURE 2-5** Plot of Data on Age and Schooling for 20 Go-Fast Employees

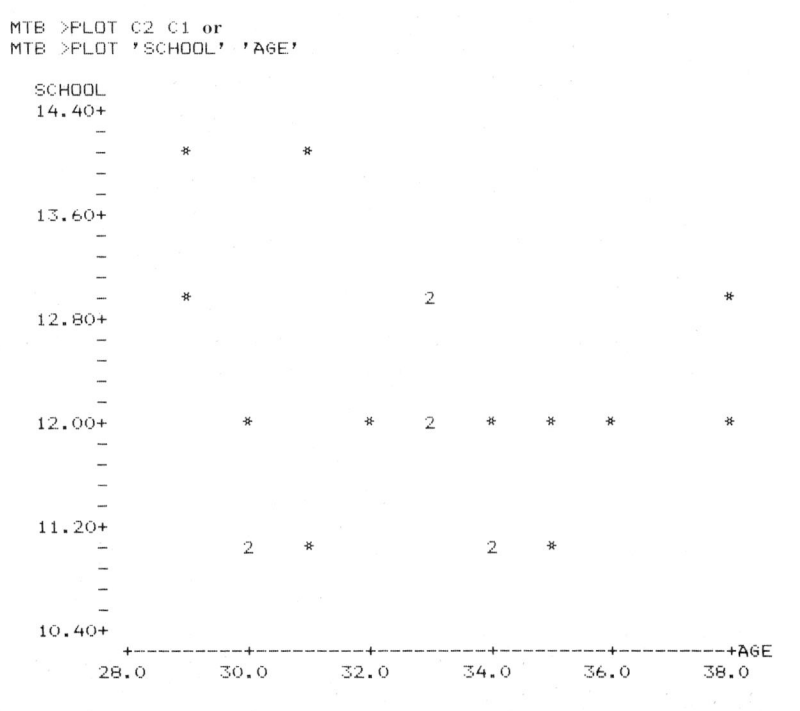

```
MTB >PLOT C2 C1 or
MTB >PLOT 'SCHOOL' 'AGE'

  SCHOOL
   14.40+
        -
        -         *           *
        -
        -
   13.60+
        -
        -
        -
        -         *                    2                         *
   12.80+
        -
        -
        -
        -
   12.00+         *           *        2    *    *    *          *
        -
        -
        -
        -
   11.20+
        -              2      *                  2    *
        -
        -
        -
   10.40+
        +---------+---------+---------+---------+---------+AGE
       28.0      30.0      32.0      34.0      36.0      38.0
```

it will plot 50 values. In many business applications data is obtained weekly, monthly, or yearly. Issuing a **TSPLot** command is a quick way to see how such data changes over time.

As an illustration, assume that you have 24 months of productivity data on a single Go-Fast employee and you are interested in seeing whether there is a pattern in this employee's productivity level over time. Assume that the productivity level has been entered into column 1, named **'AMOUNT'**. The following form of the **TSPLot** command should be used:

`MTB >TSPLot of data in C1`

Notice that this command differs from the previous **PLOT** commands in that only one column is specified. Since all the values in the column are going to be plotted sequentially, Minitab automatically adds a second column, of equal length to the first, that is plotted along the X-axis. The result of our command is shown in Figure 2-6.

Note that this Minitab command does not plot the data points with the usual **\***. Instead, each point is represented by a number. Thus, the location of a particular data point is shown by the location of its number. For example the 3rd value (row 3 in column 1) is represented by the number 3 and indicates a production amount of about 235, the 4th value is indicated by the number 4 and indicates a production rate of about 220, and the 10th value is represented by a 0 and indicates a production rate of 240. Then the sequence of numbers is repeated. Thus, the 14th value is indicated by a 4 and indicates a production level near 210. Notice that because the first two data points represent equal values, the **1** and **2** are close together and almost look like the value 12. However they represent two separate data points.

**FIGURE 2-6** A Graph Produced by `TSPLot` for 24 Months of Productivity Data on a Single Go-Fast Employee

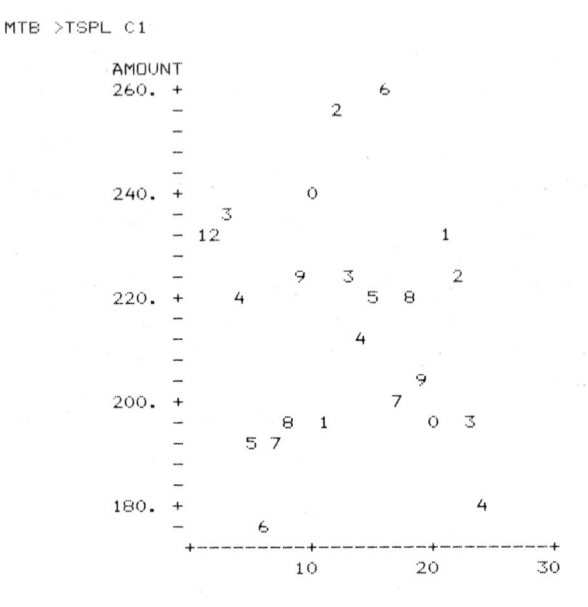

```
MTB >TSPL C1

AMOUNT
260. +                              6
     -                         2
     -
     -
     -
240. +                    0
     -         3
     -    12                             1
     -
     -                 9   3             2
220. +         4            5   8
     -                   4
     -
     -                              9
200. +                           7
     -                 8   1        0   3
     -              5  7
     -
     -
180. +                                       4
     -            6
     +---------+---------+---------+
               10        20        30
```

## 2-4 INDEX NUMBERS

One example of the power of Minitab is found in the computation of index numbers, or business indicators. Index numbers are used to compare price information from different years. In business the ability to compare prices in different years is frequently essential. This is because in many instances one set of prices is based on, say, 1984 data, and another set of prices on 1980 data. In such a case, you cannot directly compare the prices. Index numbers provide a means of converting several different prices to a common base so that they can be compared more accurately.

### Simple Index Numbers

As an illustration, consider the following example. In Table 2-2 we have the price of rubber in each of five years based upon an index. For now, don't worry about how the index was calculated; we will discuss that in the next section.

**TABLE 2-2** Price Index for Rubber (base year = 1984)

| Year | Index |
|------|-------|
| 1981 | 89    |
| 1982 | 93    |
| 1983 | 96    |
| 1984 | 100   |
| 1985 | 101   |

## SECTION 2-4 Index Numbers

Notice that the base year for the data is 1984. This means that rubber materials that would have cost $100 in 1984 would have cost $96 in 1983 and $93 in 1982. If the industry projects a price index of 105 (base 1984) for 1986, then we would expect that the same amount of materials that cost $100 in 1984 would cost $105 in 1986. Such data will help a company to prepare budget data and to control costs more effectively.

Another use of index numbers is to project a product's manufacturing cost at some future time. For example, the Go-Fast Running Shoe Company might use a large amount of rubber, nylon, and leather in its running shoes. If industry index numbers for price were available for each of those commodities (or if the data was stable enough and you could calculate the index numbers yourself), any projections about the price indices for rubber, nylon, or leather could be incorporated into the budget projections for Go-Fast.

Now that you have seen some of the usefulness of index numbers, we can use Minitab to calculate them. Minitab does not have a specific command to compute index numbers, but we can use the **LET** command to calculate their values. Consider the hypothetical industry data shown in Table 2-3.

**TABLE 2-3** Average Selling Price of Athletic Shoes and Quantities (in thousands) Sold, 1981–1985

| Year | Running Shoes Price | Running Shoes Quantity | Training Flats Price | Training Flats Quantity | Racing Shoes Price | Racing Shoes Quantity |
|---|---|---|---|---|---|---|
| 1981 | $30 | 1,000 | $40 | 800 | $50 | 200 |
| 1982 | 35 | 2,000 | 50 | 2,000 | 55 | 300 |
| 1983 | 38 | 2,200 | 51 | 2,500 | 60 | 300 |
| 1984 | 40 | 2,400 | 51 | 2,700 | 60 | 350 |
| 1985 | 41 | 2,600 | 52 | 3,000 | 61 | 350 |

To compute a price index for running shoes, we would first select a *base year* (let's use 1983) and then divide each year's price by the base year price and multiply the result by 100. That is, we would represent each year's price as a percentage of the base year's price. As you can see, index numbers are really nothing more than percentages of a chosen year's value, which is always represented by 100 (actually 100%).

Using Minitab and the data in Table 2-3, we would enter the years into **C1** and the prices for each type of shoe into **C2**, **C3**, and **C4**. Our commands would look as follows:

```
MTB >READ C1, C2, C3, C4
DATA>1981, 30, 40, 50
DATA>1982, 35, 50, 55
DATA>1983, 38, 51, 60
DATA>1984, 40, 51, 60
DATA>1985, 41, 52, 61
DATA>END
```

With the data entered, the next step would be to calculate the base year index and then make every other price in that column a percentage of that base price. For the prices of running shoes (column 2), with the base year 1983, the calculation would be as follows:

```
MTB >LET C6 = (C2 / 38) * 100
```

Column 6 would then contain the index numbers for running shoes with 1983 as the base year. The convention when using index numbers is to round off decimals and only use whole numbers. We can accomplish this with a simple Minitab command known as the rounding command. Examine the following step:

```
MTB >ROUNd C6, put in C6
```

Before we printed and discussed the results, we could easily calculate the price indices for training flats and racing shoes. Notice that instead of dividing by 38 (the 1983 price for running shoes), we would divide by 51 for training flats and 60 for racing shoes. The commands would be:

```
MTB >LET C7 = (C3 / 51) * 100
MTB >LET C8 = (C4 / 60) * 100
```

The values now found in C7 and C8 would then be rounded off as those in C6 were.

We could have taken a bit of a short cut and done the calculations and rounding all at once. The following two commands would have been used:

```
MTB >LET C7 = ROUNd ((C3 / 51) * 100)
MTB >LET C8 = ROUNd ((C4 / 60) * 100)
```

If you did not know that you could do both steps together, don't worry about it. As you gain more experience with Minitab, such insights will come automatically.

You can now print the results of these calculations with the following command:

```
MTB >PRINt C1, C6, C7, C8
```

The results are shown in Table 2-4.

**TABLE 2-4**  Simple Index Numbers for Sales of Running, Training, and Racing Shoes (base year = 1983)

| C1 | C6 | C7 | C8 |
|---|---|---|---|
| 1981 | 79 | 78 | 83 |
| 1982 | 92 | 98 | 92 |
| 1983 | 100 | 100 | 100 |
| 1984 | 105 | 100 | 100 |
| 1985 | 108 | 102 | 102 |

We have now computed the index numbers (base 1983) for each of the three types of shoes that Go-Fast sells. If the federal government had projected that the general price index (base year 1983) would be 112 in 1986, you would have known that the costs of raw materials would probably rise. On the other hand, if the index in 1986 had been projected to be 101, prices for raw materials would have probably been stable or fallen slightly.

## Aggregate Index Numbers

The price index just calculated is known as the *simple index number*. Another common indexing procedure involves aggregating (adding) the prices across classes of shoes and indexing those aggregates (or sums). This technique produces an index for athletic shoes in general rather than for just one product (i.e., training flats).

Before we discuss the procedure for the aggregate method, a comment about its weakness is appropriate. If the quantity of each product sold is about the same, then this type of index is fairly useful. However, if the quantity sold of one product in the aggregate

## SECTION 2-4 Index Numbers

is very small and the quantity of another is very large, then bias will be introduced by overweighting the small seller and underweighting the big seller.

With this caution in mind, we will now calculate two different aggregate indices. The first is obtained by averaging *individual* product index values to obtain an industry average price index. Assuming that our data is still stored as described above, we would proceed as follows:

```
MTB >LET C12 = ROUNd (C6 + C7 + C8) / 3
MTB >PRINt C1, C12
```

Table 2-5 shows the results of these commands. This procedure is known as a *simple average of index numbers*. For Go-Fast this index is a rough industry average of their product line.

**TABLE 2-5** Simple Averages of Index Numbers for Aggregate Shoe Sales (base year = 1983)

| C1 | C12 |
|---|---|
| 1981 | 80 |
| 1982 | 94 |
| 1983 | 100 |
| 1984 | 102 |
| 1985 | 104 |

The second calculation is a *simple aggregate price index*. We would start by selecting a base year. For our example, we will continue to use 1983. Next, for each year, we would sum the price values for each product. Since we have three products, and if our original data is still stored in columns 1 through 4, the command would be

```
MTB >LET C10 = (C2 + C3 + C4)
MTB >PRINt C1 C10
```

Table 2-6 shows the results of this calculation. Notice the value for 1983.

**TABLE 2-6** Gross Aggregate Prices for Aggregate Shoe Sales

| C1 | C10 |
|---|---|
| 1981 | 120 |
| 1982 | 140 |
| 1983 | 149 |
| 1984 | 151 |
| 1985 | 154 |

Since 1983 is our base year, we now take the aggregate value for 1983 (which is 149), divide all values by 149, and multiply the results by 100 to obtain our simple aggregate price index. The commands are

```
MTB >LET C11 = (C10 / 149) * 100
MTB >ROUNd C11, put in C11
MTB >PRINt C1, C11
```

Table 2-7 shows our simple aggregate price index. This index indicates the overall price of the Go-Fast product line for each of the years represented.

**TABLE 2-7** Simple Aggregate Price Index for Aggregate Shoe Sales (base year = 1983)

|  C1  | C11 |
|------|-----|
| 1981 |  81 |
| 1982 |  94 |
| 1983 | 100 |
| 1984 | 101 |
| 1985 | 103 |

## Laspeyres-Type Index Numbers

As mentioned, the shortcoming of the previous methods of computing industry index numbers (across classes) is that it does not weight differences in sales. We can correct this by weighting according to the units sold in a standard year. For instance, we can use our base year as the standard year and produce a *Laspeyres-type index number*. In this case, we weight each column by the number of units sold. Then we use the weighted value representing 1983 for calculating our index values. The calculations would be as follows:

```
MTB >LET C13 = (C2*2200 + C3*2500 + C4*300)
MTB >PRINt C1, C13
```

We would then obtain the table shown in Table 2-8.

**TABLE 2-8** Weighted Values of Price Times Quantity for Aggregate Shoe Sales

|  C1  |  C13   |
|------|--------|
| 1981 | 181000 |
| 1982 | 218500 |
| 1983 | 229100 |
| 1984 | 233500 |
| 1985 | 238500 |

Notice the value for 1983 in Table 2-8. This value, 229,100, is what we use to calculate our price index. With this value now calculated, we can finish computing the index by using the following commands to obtain the results in Table 2-9.

```
MTB >LET C14 = (C13 / 229100) * 100
MTB >ROUNd C14, put in C14
MTB >PRINt C1, C14
```

**TABLE 2-9** Laspeyres-Type Index Numbers for Aggregate Shoe Sales (base year = 1983)

|  C1  | C14 |
|------|-----|
| 1981 |  79 |
| 1982 |  95 |
| 1983 | 100 |
| 1984 | 102 |
| 1985 | 104 |

## Other Index Numbers

You should now have a sense of how index numbers can be used. Since we intended only to introduce the basic concept, we will end our discussion here. However, there are many other types of index number. Some of these, such as the *Paasche-type* index, allow some year other than the base year to be used in the weighting process. For example, if it seemed appropriate, one might use 1984 quantities and 1983 prices.

We have now demonstrated the ability of Minitab to compute index numbers. If you want to appreciate the amount of time saved by using the **LET** command, recalculate by hand any of the index numbers for a different base year.

# SUMMARY

In this chapter, we have concluded our discussion of how to use Minitab. The various methods for entering data, manipulating values within and between columns, and naming columns will provide you with all the means necessary to begin our study of statistics. New commands will be introduced as they are needed throughout the remainder of the text.

# DISCUSSION QUESTIONS

1. Explain the problems that might arise when performing calculations involving columns if there is only one observation in one of the columns.
2. What rules must be remembered when executing **LET** commands?
3. What is the limitation of the simple average index number? How does the Laspeyres-type index number try to overcome this limitation?

# PROBLEMS

**2-1** Use the following data and, where possible, simple **LET** commands to answer the questions below.

| C1 | C2 | C3 | C4 |
|----|----|----|----|
| 6  | 33 | 9  | 12 |
| 25 | 26 | 11 | 3  |
| 9  | 20 | 9  | 6  |
| 19 | 14 | 2  | 16 |
| 21 | 15 | 33 | 25 |
| 3  | 6  | 31 | 28 |
| 5  | 17 | 29 | 35 |

a) Use the **READ** command to enter the data into **C1** through **C4** exactly as shown. List the data to check your accuracy.

b) Add column 2 to column 3 and place the results in column 5. (Remember to use the **LET** command.)

c) Subtract column 1 from column 3 and place the results in column 6.

d) Multiply column 5 by column 6 and put the results in column 7.

e) Divide column 7 by column 1 and place the results in column 8.

f) Square all values in column 4 and divide the new values by column 7 minus column 2; place the results in column 9.

g) **PRINt** all columns involved and visually check your work.
h) In a **LET** command, where are spaces required?

**2-2** Over the years, U.S. and European garment workers have asked for higher wages and more fringe benefits. As a result, the ability of the garment industry in those countries to produce inexpensive, quality clothes has declined substantially. Consequently, the industry has had to turn to other countries to produce its garments. Sample data for workers in the garment industry in different countries appears in the following table:

| Country or Region | Average Hourly Wage | Average Age | Average Years of Schooling |
|---|---|---|---|
| United States | $3.50 | 47 | 12 |
| Europe | 3.05 | 49 | 11 |
| Hong Kong | 0.75 | 30 | 5 |
| Taiwan | 0.60 | 27 | 6 |
| Sri Lanka | 0.05 | 19 | 2 |

a) Use the **READ** command to enter your data.
b) Use the **NAME** command to title your columns `'WAGE'`, `'AGE'`, and `'SCHOOL'`.
c) Use the **LET** command to multiply each wage rate by 40 to obtain the total weekly wage. Place the data in column 4.
d) For these five areas, what is the mean wage? The mean age? The mean number of years of schooling?
e) Use the **LET** command to store the mean wage as a constant. Then multiply the constant (**K1**) by 40 hours and place in **K2**.

**2-3** A champion dog breeder keeps careful records of his potential prize-winning show dogs. Toby (a Doberman), Bruno (a chow chow), and Clarence (a beagle) come from good lineages. Shown are their weights in pounds.

| Age (months) | Toby | Bruno | Clarence |
|---|---|---|---|
| 2 | 5 | 4 | 3 |
| 4 | 14 | 11 | 7 |
| 6 | 35 | 24 | 15 |
| 8 | 59 | 37 | 23 |
| 10 | 84 | 51 | 29 |
| 12 | 94 | 58 | 33 |

a) Use the **READ** command to enter the data.
b) Use the **NAME** command to title the columns.
c) Using the column names, enter the **PRINt** command to check your accuracy.
d) Plot Toby's weight using the **TSPLot** command.
e) Plot the weights for Bruno and Clarence using the **TSPLot** command.

**2-4** Because electricity costs have skyrocketed in recent years, utility companies have sought cheaper sources of energy. Some companies have turned to sources with surplus power, such as Quebec's hydroelectric generating system. The data that follows compares the cost per kilowatt-hour in 1984 and in 1985 (when surplus power was purchased) in four U.S. cities.

# PROBLEMS

| City | 1984 | 1985 |
|---|---|---|
| Minneapolis | 9.8¢ | 2.8¢ |
| Boston | 15.8 | 12.2 |
| New York City | 16.4 | 8.5 |
| Washington, D.C. | 10.5 | 6.7 |

**a)** For each city, use the **LET** command to determine the change in cost per kilowatt-hour between 1984 and 1985.
**b)** Use the **LET** command to store 100 as a constant (**K1**).
**c)** Multiply the 1984 data by **K1** using the **LET** command and store the data in **C4**. Do the same for the 1985 data and store in **C5**.
**d)** Average **C4** and average **C5** to determine the electricity bill for a person in each city using 100 kilowatt-hours in 1984 and 1985.

**2-5** Computer Check-Up, Inc., a company that specializes in repairing personal computers, hired Marketing, Inc., to gather information on its current customers. The purpose of this study was to develop a profile of Computer Check-Up's average customer. Here is a sample of the data that was collected.

| Customer | Age | Income |
|---|---|---|
| 1 | 42 | $28,900 |
| 2 | 54 | 31,200 |
| 3 | 36 | 21,000 |
| 4 | 32 | 37,000 |
| 5 | 31 | 26,500 |
| 6 | 27 | 27,200 |
| 7 | 36 | 32,000 |
| 8 | 37 | 45,000 |

**a)** Enter the data for age into **C1** and for income into **C2**. Also, use the **NAME** command on the data. List the data to check your accuracy.
**b)** What is the average income of a Computer Check-Up customer?
**c)** Using the **LET** command, divide income by 52 to determine each customer's weekly income.
**d)** Plot **'AGE'** on the Y-axis and **'INCOME'** on the X-axis. Is there a relationship between age and income?

**2-6** Dr. Eleanor Kage, a teacher of organizational behavior at a West Coast college, has these records in her grade book.

| 9:00 Class || 12:00 Class ||
| Age | Test Score | Age | Test Score |
|---|---|---|---|
| 21 | 40 | 18 | 46 |
| 25 | 34 | 25 | 43 |
| 19 | 36 | 24 | 45 |
| 18 | 37 | 21 | 36 |

*(continued on next page)*

## CHAPTER 2 Basic Manipulations with Minitab

|  | 9:00 Class |  | 12:00 Class |
|---|---|---|---|
| Age | Test Score | Age | Test Score |
| 25 | 39 | 18 | 36 |
| 22 | 49 | 18 | 44 |
| 21 | 40 | 19 | 49 |
| 19 | 43 | 23 | 48 |
| 19 | 38 | 25 | 27 |
| 24 | 39 | 23 | 34 |
| 21 | 45 | 19 | 38 |
| 22 | 41 | 22 | 45 |
| 22 | 32 |  |  |
| 26 | 30 |  |  |

a) Enter all four columns of data and use the **NAME** command to title the columns. Be sure to distinguish between the ages and test scores of the two classes.

b) Using the **LET** command, multiply each test score by 2 for a 100-point score (the original test had 50 points). Place the data for the two classes in **C5** and **C6**, respectively.

c) From **C5** and **C6**, find the mean score for each class. Place the mean in **C7** for the 9:00 class and in **C8** for the 12:00 class.

d) Without referring to Minitab, how long will columns 7 and 8 be?

e) Use the **STACK** command to combine **C1** with **C3** and **C5** with **C6**; place the data in **C9** and **C10**, respectively.

f) Use the **SORT** command on **C9** and **C10**. Print the data to check your accuracy.

**2-7** Hundreds of orders are received each day for a fast-selling lightweight phone. Order blanks use a code to specify color: 1 for red, 2 for black, 3 for white, and 4 for beige. Here is a sample of the color orders:

```
3  4  3  4  3  1  3  1
4  4  2  4  4  2  3  4
3  2  4  4  3  2  4  2
3  4  2  3  4  3  4  2
2  1  2  2  4  3  3  1
1  1  1  2  4  4  3  4
```

a) Manually count the number of orders for each color.

b) Enter the data, using the **SET** command, into **C1**.

c) Use the **SORT** command on **C1** and put the results in **C2**.

d) Count the number of orders after printing **C2**.

e) Enter the data into **C3** using the **SET** command for repeated entry. For example,

MTB >SET C3
MTB >4(1)

f) Use the **HISTogram** command on this data and then visually inspect the relative popularity of the different phone colors.

**2-8** Todd Thomas's hand-thrown vases have gained much respect and popularity in the art world. At a recent showing in New York City, the following items were sold. Todd classifies his vases by their height.

# PROBLEMS

| Height (inches) | Price |
|---|---|
| 8 | $105 |
| 8 | 220 |
| 8 | 160 |
| 10 | 240 |
| 10 | 116 |
| 10 | 107 |
| 12 | 87 |
| 12 | 95 |
| 19 | 106 |
| 19 | 215 |
| 19 | 235 |
| 19 | 245 |
| 19 | 550 |

a) Use the **SET** command for repeated entry to enter the data on height in **C1**.
b) Use the **SET** command to enter the price data in **C2**.
c) Give **C1** the name **'HEIGHT'** and **C2** the name **'PRICE'**.
d) Plot the data. Enter height on the X-axis and price on the Y-axis.
e) Use the **SORT** command on the price column. Print the data to check your accuracy.

**2-9** At the 10th reunion of Harvard Business School's class of 1960, the historian collected the following information on 15 alumni:

| Alumnus | Age | Salary |
|---|---|---|
| 1 | 35 | $18,000 |
| 2 | 35 | 36,000 |
| 3 | 32 | 21,000 |
| 4 | 32 | 25,500 |
| 5 | 37 | 26,000 |
| 6 | 36 | 76,400 |
| 7 | 32 | 31,000 |
| 8 | 32 | 30,000 |
| 9 | 33 | 29,700 |
| 10 | 33 | 23,000 |
| 11 | 35 | 32,000 |
| 12 | 32 | 36,000 |
| 13 | 32 | 27,700 |
| 14 | 32 | 21,400 |
| 15 | 39 | 35,000 |

a) Enter the data on age and salary. Use the **READ** and **NAME** command.
b) What is the mean salary for this sample?
c) Plot the data with age on the X-axis and salary on the Y-axis. Use the names of the columns when plotting the data.
d) Using the **LET** command, divide the salaries by 12 to determine the monthly salary for each alumnus. Print the data to check your answers.

## CHAPTER 2 Basic Manipulations with Minitab

**2-10** The American public is rapidly becoming more health conscious. As a result, smart eating and light drinking are in fashion. Following are data for yearly per person consumption of certain foods:

|  | Coffee (cups) | | Meat (lbs) | |
| --- | --- | --- | --- | --- |
| Year | Caffeinated | Decaffeinated | Beef | Chicken |
| 1981 | 807 | 91 | 78.5 | 48.7 |
| 1983 | 650 | 115 | 76.0 | 51.2 |
| 1985 | 566 | 160 | 61.0 | 54.8 |

a) Use the READ command to enter the data.
b) Do a TSPLot on each category.
c) What do you notice?

**2-11** Doctor March recently graduated from dental school. In order to establish her practice, she chose to lease her equipment instead of purchasing it. Her annual expenditures are as follows:

| Equipment | Cost |
| --- | --- |
| X-ray equipment | $10,200 |
| Dental chairs | 14,725 |
| Laboratory equipment | 9,800 |

a) Use the NAME command to title the column 'COSTS'.
b) Find the total expenditure for Doctor March. Store the total as a constant using the LET command.
c) Multiply Doctor March's total expenditure by 10 to determine the amount she would spend in 10 years if she continued to lease. (Use the LET command.)

**2-12** Refer to the Go-Fast Running Shoe Company data in Appendix A.
a) Use the READ command to enter the data for age, schooling, productivity, and height, respectively, for the first 10 employees into C1 through C4.
b) Use the NAME command to title the respective columns 'AGE', 'SCHOOL', 'PRODUCT', and 'HEIGHT'.
c) Divide the 'HEIGHT' column by 12 to translate the measurement from inches to feet. (Use the LET command.)
d) Plot 'PRODUCT' on the Y-axis and 'SCHOOL' on the X-axis.
e) Use the SORT command on C1 through C4, placing the results in C6 through C9, respectively. Print the data to check your accuracy.

**2-13** The Golf Swing Company prides itself on processing its orders within four hours of receipt. Five employees were observed to check on their productivity. The following data on orders packed per hour was collected:

|  | Employee | | | | |
| --- | --- | --- | --- | --- | --- |
| Week | 1 | 2 | 3 | 4 | 5 |
| 1 | 15 | 22 | 10 | 26 | 21 |
| 2 | 19 | 22 | 24 | 18 | 21 |
| 3 | 21 | 20 | 15 | 23 | 15 |

# PROBLEMS

a) If productivity were to be cubed, how many orders per hour would the fourth employee pack? (Use the **LET** command for exponents.)
b) Use the **TSPLot** command on each employee's output over the three weeks.
c) Examine the time series plots. Which employee has improved the most?

**2-14** Go back to Problem 1-6 concerning overseas living expenses.
a) Enter the data using the **READ** command. (Do not include the London data.) Use the **NAME** command to title your columns.
b) Use the **PRINt** command to check your data. Be sure to refer to each column by its name.
c) Using the **LET** command, what are the total living expenses for each city?
d) Assume that the family will be stationed in a foreign country for seven years. Use the **LET** command to multiply the totals by 7. List the data to check your accuracy.

**2-15** Personal care companies face stiff competition within their industry. They must advertise extensively to be tops in their field. Recently this group of companies has been performing quite well in the stock market. As a result, they are being closely watched by investors. Following are data on this industry:

| Company | Stock Price | Net Income (millions) |
|---|---|---|
| 1 | $33.75 | $112.1 |
| 2 | 58.50 | 145.9 |
| 3 | 33.50 | 119.5 |
| 4 | 47.50 | 28.3 |

a) Enter the data using the **READ** command.
b) Title your columns.
c) Use the **PLOT** command, placing stock price on the X-axis and net income on the Y-axis.
d) Use the **PLOT** command again, this time with stock price on the Y-axis and net income on the X-axis.

**2-16** Use the data in Table 2-3 on page 35.
a) Calculate a simple index number for each type of shoe using 1981 as the base year.
b) Calculate a simple average of index numbers for all types of shoes using 1982 as the base year. Recalculate the simple average using 1985 as the base year. Is either set of calculations a good index number set? Why or why not?
c) Calculate a simple aggregative price index using 1982 as the base year. Does this index provide a good measure of price changes for Go-Fast over the years?
d) Calculate a Laspeyres-type index number set using 1981 as the base year. Recalculate the Laspeyres-type index set using 1985 as the base year. Does either set of calculations provide good index numbers? Why or why not?

# 3

# DESCRIPTIVE STATISTICS

*T*he following Minitab commands will be introduced in this chapter:

`DESCribe data in C,...,C`
A command that produces 11 descriptive statistics for a column or group of columns (see Table 3-1) (The extended calculations are not contained in all releases of Minitab; earlier releases provide only the number of observations, the mean, and the standard deviation. Many of the descriptors issued by this command can be calculated individually as single column commands.)

`N data in C, [put in K]`
A command that gives the number of nonmissing observations in a column; the resulting value may be stored as a constant

`NMISs data in C, [put in K]`
A command that gives the number of missing observations in a column; the resulting value may be stored as a constant

`MEAN data in C, [put in K]`
A command that gives the mean value of a column; the resulting value may be stored as a constant

`MEDIan data in C, [put in K]`
A command that gives the median value of a column; the resulting value may be stored as a constant

**SECTION 3-1** Classification of Measurement Scales

> `STDEv data in C, [put in K]`
> A command that gives the standard deviation; the resulting value may be stored as a constant
> `MAX data in C, [put in K]`
> A command that gives the largest value in a column; the resulting value may be stored as a constant
> `MIN data in C, [put in K]`
> A command that gives the smallest value in a column; the resulting value may be stored as a constant
> `SUM data in C, [put in K]`
> A command that determines the total of all values in a column; the resulting value may be stored as a constant
> `SSQ data in C, [put in K]`
> A command that squares each value in a column and then sums all squared values; the resulting value may be stored as a constant
> `COUNt data in C, [put in K]`
> A command that determines the total number of entries in a column (the sum of missing and nonmissing observations); the resulting value may be stored as a constant
> `STEM-and-leaf of data in C,...,C`
> A command that calculates and creates a stem-and-leaf diagram for data in specified columns
> `DOTPlot of data in C,...,C`
> A command that calculates the same basic output as the `HISTogram` command but displays the results horizontally instead of vertically
> `BOXPlot of data in C,...,C`
> A command that creates a display of the distribution, indicating the median, the middle half of the data, and likely outliers.

In the first two chapters, we concentrated on learning the basics of Minitab. However, since this is not a computer course but a statistics course, it is time to turn our attention to statistics. Minitab is a tool for assisting us in statistical analysis. Like any good computer tool, Minitab will do pretty much whatever we tell it. The critical thing to remember is that it is always management that decides whether we have used our techniques properly. Selecting the right statistical test is therefore essential. There are many types of statistical tests with many different assumptions behind them. Much of the remaining text will be concerned with understanding these differences and selecting the proper test.

In statistical analysis, we are constantly trying to understand the relationships within a group of numbers. There are two general measures of relationship. The first is a measure of the average value; the second is a measure of how spread out the numbers within a group are from the average. However, before we discuss these concepts, it is necessary to discuss the different types of numbers.

## 3-1 CLASSIFICATION OF MEASUREMENT SCALES

For our purposes, numbers are really measurement scales that are used to classify data or relationships among different types of data. We typically discuss numbers representing either nominal, ordinal, or interval and ratio scales. Each of these scales have different characteristics and must be treated differently in statistical analysis.

## Nominal Data

*Nominal* data (sometimes called categorical) is the kind of data for which numbers do not really fit. Examples are male and female, the 50 states, or the countries of the world. However, in many of these cases, we enter some type of numerical code, such as 0 for male and 1 for female. Computers do not care what codes you use; only *you* must understand what they mean.

Thus, for the 50 states, we could list them alphabetically and assign the value of 1 to the first state (Alabama) and 50 to the last (Wyoming). What do these numbers really mean? Is the state with the assigned value of 20 (Maryland) in some way twice the state with an assigned value of 10 (Georgia)? Obviously, this type of number has a special kind of relationship with the numbered object. However, the computer does not know if the 50 numbers for the states represent nominal data or not. If you wanted the computer to average those 50 values, it most certainly would. But what would the obtained average of 25.5 mean? Nothing!

Nominal numbers are used solely to classify, and care must be taken to treat these values acordingly.

## Ordinal Data

The second kind of data is *ordinal*. For example, in a university, there are four academic ranks: instructor, assistant professor, associate professor, and full professor. If you assign the value of 1 to the instructor, 2 to the assistant professor, 3 to the associate professor, and 4 to the full professor, you have a set of ordinal classifications because each academic rank is higher than the previous one. The problem with this type of data is that the distance between ranks is not necessarily the same. Is the relationship between instructor and assistant professor (a numeric difference of 1) the same as the relationship between associate professor and full professor (also a difference of 1)? As you can see, ordinal numbers convey a different set of meanings than nominal data, but a one-unit difference cannot be considered equal in all cases.

Ordinal data is typically encountered in attitudinal surveys. Any survey where there are, say, five possible answers to a question ranging from 1 = "strongly agree" to 5 = "strongly disagree" will produce an ordinal group of numbers. A great number of things can be done with this type of research design. However, care must be taken that the kind of data you are working with will allow the calculation of the statistical measure you wish to use.

## Interval and Ratio Data

The final types of data are *interval* and *ratio* (also called *metric*). These are the most common kind and are probably what you first think of when someone starts talking about measures. Speed, weight, distance, height, money, and volume are all examples of ratio data. Ratio data are those that have true zero values. As a result, certain mathematical relationships apply. Thus, 8 miles per hour is twice as fast as 4 miles per hour, and 80 miles per hour is twice as fast as 40. Interval data lacks a true zero. Thus, interval scales have arbitrary zero points. A good example is temperature (as usually measured). The temperature of 0 °C is an arbitrary value. As a result, 20 °C is not twice as warm as 10 °C.

# SECTION 3-2 Measures of Central Tendency

It is important to realize that these different types of data must be treated differently. The values must be described in different ways, and different statistical tests must be used. Remember that a computer does not know if a group of numbers is nominal, ordinal, or ratio or metric. If you command Minitab to find the mean of a set of values, it will do so. But it is your responsibility, as the user of statistical analysis, to ensure that someone has not performed inappropriate statistical tests. For the greater part of this text, we will be using metric data. This is certainly the most common kind of data. When other types of numbers are used, we will be careful to explain the differences.

## 3-2 MEASURES OF CENTRAL TENDENCY

Central tendency is a fairly simple concept. It is a single value or very small group of values that we can use to represent a larger group of numbers. In our Go-Fast Running Shoe Company example, we have the productivity for each employee for the month of December (Appendix A). What can we use as a good measure of the monthly production?

### Mean, Median, and Mode

Three measures of central tendency have been used over the years: the mean, the median, and the mode. The most common of these measures is the mean. This measure is technically referred to as the *arithmetic mean* and is found by summing all observations and then dividing the total by the number of observations. You probably already know this value as the "average." The *median* is the value that appears in the middle of the data group. If there is an odd number of items in a data set, the median value is the middle value; if there is an even number, the two middle values are averaged to find the median. Finally, the *mode* is the value that occurs most frequently. If one value is found the most, then the data has a single modal value. If two values have an equal frequency, the data is bimodal, and so on. If no values are repeated in a data set, there is no modal value.

The mean, median, and mode may or may not have about the same value for a given group of numbers. If a distribution of numbers is fairly *symmetric* (i.e., the distribution resembles a smooth hill with most values near the center and about the same numbers of values on either end), then the mean, median, and mode will be about the same. A distribution of 501 student grades is shown in a histogram in Figure 3-1. Here you see this even hill shape, or what is usually called a bell-shaped curve. In this distribution, the mean is 50.138, the median is 50.000, and the mode is also 50.000.

A different type of distribution, formally known as a *Poisson* (Figure 3-2), will be discussed later. In this figure, the peak of the hill is not in the center but is skewed towards the lower values. The mean of these 2,000 observations is 125.94, the median is 100.00, and the mode is 100. In this case, there is a noticeable difference between the mean and the other two measures.

The distribution in Figure 3-3 is *bimodal;* that is, there are two humps (or hills) in the distribution. This is the kind of distribution you might obtain from measuring the weights of men and women; in general, women will cluster at a lower weight than men. In the figure shown, the mean is 50.313 and the median is 51.000, but the figure is bimodal with mode values at 35 and 65.

**FIGURE 3-1**  A Normal Distribution

```
MTB >HIST C1

Histogram of C1    N = 501
Each * represents 2 obs.

    Midpoint    Count
        34.       1     *
        36.       3     **
        38.       5     ***
        40.       7     ****
        42.      20     **********
        44.      36     ******************
        46.      42     *********************
        48.      68     **********************************
        50.      80     ****************************************
        52.      76     **************************************
        54.      65     *********************************
        56.      47     ************************
        58.      25     *************
        60.      18     *********
        62.       6     ***
        64.       1     *
        66.       1     *
```

**FIGURE 3-2**  A Poisson Distribution

```
MTB >HIST C4

Histogram of C1    N = 2000
Each * represents 15 obs.

    Midpoint    Count
         0.     160    ***********
        50.     414    ******************************
       100.     515    ***********************************
       150.     414    ******************************
       200.     264    ******************
       250.     142    **********
       300.      65    *****
       350.      19    **
       400.       7    *
       450.       1    *
```

As can be seen from these illustrations, different measures of central tendency do not always have the same value. The problem with the most frequently used measure, the mean, is that it is drastically affected by any extreme values at one end of a given distribution. When the distribution is skewed or bimodal, the mean is a less effective measure. For example, in a business where the owner received a very large salary and unskilled employees received very low salaries, the mean could be $10,000, the median $7,500, and the mode $4,000. If most unskilled workers received $4,000 per year, then that would be the modal salary. If the salary in the middle of the entire group of salaries was $7,500, then that salary would be the median. Finally, if the president received a salary of several hundred thousand dollars, his or her salary could pull the mean salary up to $10,000.

Which measure of central tendency is most useful? That depends on what you are trying to measure. If I was part of management and wanted to show how well my employees were paid, I would use the mean. If I was part of labor and wanted to show how badly employees were paid, I would use the mode. As a user of statistical analysis, you must be

**SECTION 3-2** Measures of Central Tendency  ▶ **51**

**FIGURE 3-3**  A Bimodal Distribution

```
MTB >HIST C3

Histogram of C1    N = 1000
Each * represents 5 obs.

    Midpoint    Count
      20.          2     *
      25.         24     *****
      30.        131     ***************************
      35.        181     *************************************
      40.        126     **************************
      45.         30     ******
      50.          7     **
      55.         31     *******
      60.        100     ********************
      65.        193     ***************************************
      70.        134     ***************************
      75.         37     ********
      80.          4     *
      85.          1     *
```

aware that different interest groups will manipulate data in exactly this way. It is your responsibility to evaluate any statistical analysis and determine what particular cause the users of the analysis may be trying to promote.

# The DESCribe Command ▼▲▼▲▼▲▼▲▼▲▼▲▼▲▼▲▼▲▼▲▼▲▼▲▼▲▼▲▼

Minitab has a single command that will calculate several descriptive measures of a data set, including the mean and median. That command is the **DESCribe** command. The general form of the command is

    DESCribe data in C,...,C

With this command you can get information on a single column, several nonconsecutive columns, or several consecutive columns. The following three examples illustrate each of these possibilities, respectively:

```
MTB >DESCribe data in C3
MTB >DESCribe data in C3, C5, C9
MTB >DESCribe data in C3-C9
```

Table 3-1 lists the data that the **DESCribe** command will print out.

This output contains several helpful measures. If you were working with a data set containing several columns of numbers, some of the columns might not be of equal length or might have missing values. The first number printed, **N**, tells you how many values were used in the calculations that follow. The number, which is the *sample size*, will be very important in later discussions. For now, you can use the general rule that larger samples give more accurate results than smaller sample sizes. In our Go-Fast example, if we took a sample of five employees and found their average production for December, would that figure provide as good an estimate of employee productivity as the figure for a sample of 50? Probably not. For this reason, Minitab first tells you how many observations are actually involved in the calculations. **NMISS** refers to the number of missing values in a column (if any). The **MEAN** and **MEDIAN** are the traditional measures we have been discussing.

**52** ◀  **CHAPTER 3** Descriptive Statistics

**TABLE 3-1**  Output of the DESCribe Command

| Name | Description |
|---|---|
| N | Number of nonmissing values in a column |
| NMISS | Number of missing values in a column (omitted if there are no missing values) |
| MEAN | Arithmetic mean |
| MEDIAN | Middle value |
| TRMEAN | 5% trimmed mean |
| STDEV | Standard deviation |
| SEMEAN | Standard error of the mean |
| MAX | Maximum value |
| MIN | Minimum value |
| Q3 | Third (upper) quartile |
| Q1 | First (lower) quartile |

*Note:* These statistics are printed for each column specified.

TRMEAN is a very handy measure that will be discussed later. TRMEAN, which stands for trimmed mean, eliminates the top and bottom 5% of the observations and calculates the mean value for the remaining 90%. This has the effect of eliminating the values at both extremes and frequently gives a more consistent measure of central tendency. STDEV and SEMEAN, the standard deviation and the standard error of the mean, will also be discussed later. MAX and MIN list the largest and smallest values; subtracting one from the other gives you a measure of the *range* of the data. We will return to the importance of those measures very soon. Finally, the value for Q3 represents the upper quartile, or top 25%, of the values and Q1 represents the lower quartile, or the bottom 25%, of the values. The difference between these two quartiles (called the *interquartile range*) provides a measure of the range of the middle 50% of the data, as opposed to the range between the MAX and MIN values, which provides the range of the entire data set.

The production data for all 100 Go-Fast employees has been entered into column 3 and a DESCribe command has been executed. The results of the computations are shown in Figure 3-4.

As you can see, there is some interesting information contained in Figure 3-4. Our mean production is 205.50 pairs of shoes per employee in December. Several other descriptive measures are also important. If we subtract the smallest value (120) from the largest (265), we see how spread out the values are for shoe production. This spread (which will be called *variation* for the rest of this book) is extremely important. In our illustration, the difference between our lowest producer and our highest producer is more than double the lowest producer's value. Management would probably want to investigate such a large vari-

**FIGURE 3-4**  Results of DESCribe Command on Go-Fast Productivity Data

```
MTB >DESC C3

            N      MEAN    MEDIAN    TRMEAN     STDEV    SEMEAN
C1        100    205.50    205.00    206.56     28.32      2.83

          MIN       MAX        Q1        Q3
C1     120.00    265.00    190.00    228.75
```

**SECTION 3-2** Measures of Central Tendency  ▶**53**

ance in production. However, if the difference had been only 4, then the variance would have been quite small. A small variance would also be valuable information for management, because it indicates uniform productivity among the employees.

There will be occasions when you do not want all the statistics calculated by the `DESCribe` command. Numerous individual statistics can also be calculated. A list of commands to obtain single statistics are listed below.

```
N     data in C,[put in K]
NMISs data in C,[put in K]
MEAN  data in C,[put in K]
MEDIan data in C,[put in K]
STDEv data in C,[put in K]
MAX   data in C,[put in K]
MIN   data in C,[put in K]
SUM   data in C,[put in K]
SSQ   data in C,[put in K]
COUNt data in C,[put in K]
```

Each of these commands has two forms: with and without storage of the statistic. For example, you can enter the `MEAN` command as any of the following:

```
MTB >MEAN data in C
MTB >MEAN data in C, put in C
MTB >MEAN data in C, put in K
```

We recommend that you store single statistics as constants rather than as columns.

There are three commands in the list of commands to obtain single statistics that are not included in the `DESCribe` command. We will need these three statistics in later chapters, where they will be discussed fully. Those commands are `SUM`, which simply adds all the values in a column; `SSQ`, which squares each value and then adds them all together; and `COUNT`, which gives the total number of entries in a column (the sum of `N` and `NMISS`).

## Hand Computations and Formulas▼▲▼▲▼▲▼▲▼▲▼▲▼▲▼▲▼▲▼▲▼▲▼

Many statisticians believe that without knowing how to compute a statistical test, a student cannot fully comprehend its usefulness. In this section, and in similar sections throughout the book, we will show the actual computations performed by Minitab.

▶ ▷ ▶ ▷ ▶ ▷ **The Mode**  The mode is defined as the value that occurs most frequently. As an example, a college faculty might have the following distribution of hair color:

| Color | Frequency |
|---|---|
| Blond | 15 |
| Brown | 40 |
| Black | 7 |
| White | 12 |
| None | 2 |

In this case, the modal hair color is brown, because it is the color occurring most frequently.

### The Median

The median is the middle value in a distribution that is rank-ordered. The starting lineup of the Tank Town Tigers might be rated by the football coach as follows:

| Rank | Name | Rating |
|------|------|--------|
| 1 | Ed Smith | Superior |
| 2 | John Jones | Excellent |
| 3 | Fred Flintstone | Excellent |
| 4 | Jim Johnson | Excellent |
| 5 | George Bush | Excellent |
| 6 | Larry Holmes | Excellent |
| 7 | Jerry Brown | Good |
| 8 | Dan Dumbbell | Good |
| 9 | Hank Williams | Good |
| 10 | Tiny Tim | Fair |
| 11 | Tom Thumb | Fair |

We could now locate the average (median) player, Larry Holmes. Five players are ranked above him and five ranked below; thus, Larry represents the median or middle value. Likewise, the coach would have to say that "on the average," Tank Town has an excellent starting lineup.

### The Arithmetic Mean

The mean is the traditional average that we learned in grade school and is what is usually meant when someone says they have calculated an average. However, in the world of statistics, you must be careful that the average you are talking about is the same average that someone else is using.

To calculate the arithmetic mean, we sum a group of numbers and then divide by the number of numbers in the group. If we observe algebraic tradition and call each number in the group $x_i$ and the number of numbers $n$, then the formula for calculating the mean is

$$\text{mean} = \bar{x} = \frac{\sum_{i=1}^{n} x_i}{n}$$

where $\sum_{i=1}^{n} x_i$ means the sum of each individual value of $x$. For example, to average the following five numbers:

| $x_1$ | $x_2$ | $x_3$ | $x_4$ | $x_5$ |
|-------|-------|-------|-------|-------|
| 7 | 3 | 4 | 6 | 10 |

We would find the sum of the numbers:

$$\sum_{i=1}^{5} x_i = 7 + 3 + 4 + 6 + 10 = 30$$

and then divide by the number of numbers (5):

$$\bar{x} = \frac{30}{5} = 6$$

## 3-3 VISUAL PRESENTATION OF DATA

The concept of variation can be demonstrated both mathematically and visually. Two widely used mathematical measures of variation are the variance and the standard deviation. The standard deviation is simply the square root of the variance. However, before we discuss the mathematical value of these terms, it will be helpful to look at visual representations of our data. Minitab makes it easy to look at data with several different commands: namely, `HISTogram`, `STEM-and-leaf`, `DOTPlot`, and `BOXPlot`.

### The Histogram

The most commonly used of these commands is `HISTogram`. A *histogram* is a visual representation of data showing the frequency of values in a certain range. Because the histogram is used primarily to summarize or describe a large amount of data, it needs to be both complete and compact. In general, the number of classes (or groups) used in each histogram ranges from 5 for small data sets to 15 for large data sets. Although there are exceptions to this rule of thumb, a histogram with fewer than 5 classes lacks detail, while one with more than 15 is hard to visualize. Thus Minitab will generally follow this rule of thumb (and a few others), although we can override its assumptions when necessary. The general form of the `HISTogram` command is

```
HISTogram of data in C,...,C
```

Assuming our productivity data is in column 3, the following command would be entered:

```
MTB >HISTogram of data in C3
```

The results are shown in Figure 3-5.

**FIGURE 3-5**  Histogram of Go-Fast Productivity Data

```
MTB >HIST C3

Histogram of PRODUCT   N = 100

Midpoint    Count
    120        1   *
    140        4   ****
    160        4   ****
    180       12   ************
    200       31   *******************************
    220       23   ***********************
    240       21   *********************
    260        4   ****
```

A histogram provides several pieces of information. First, it breaks data into groups or intervals. The width of each interval and the number of intervals are important. The first column in Figure 3-5 indicates that each interval is 20 units wide, and there are eight groups of data. To determine the actual range of each class, we must go an equal distance on each side of the midpoint value. For example, our fourth class has a midpoint of 180. Since all classes are 20 units wide, the fourth class extends from 170 to 190. Thus, 12 employees had productivity between 170 and 190 units during the month of December.

An important point must be noted here. *We cannot have classes that overlap.* Thus, we cannot have one class from 170 to 190 and the next class from 190 to 210. What do you do with a value of exactly 190? The class widths are really 170 to 189.9999 and 190 to 209.9999. However, because such numbering is cumbersome, Minitab breaks the classes into even units with the following rule: If a value falls exactly on the boundary between two classes, it is always placed in the higher class. Thus any value of exactly 190 would be placed in the 190–210 range.

The `HISTogram` command shows not only the number of observations within a class but also a visual representation of that number. However, note that we do not know the actual values within a class, only the numbers of values within each class. For example, we know that 12 employees produced between 170 and 190 pairs of shoes in the month of December, but that is all. How many actually produced 175 or 180 pairs? From the histogram, we cannot tell.

Figure 3-4 shows a great deal of diversity in the data. Our range is from 110 to 270, since our first midpoint is 120 and our last midpoint is 260. Therefore our data seems to have a great deal of variance, or spread.

## Optional Arguments for the `HISTogram` Command

In using the `HISTogram` command (and many other commands), Minitab uses what are called "default" values. In other words, Minitab automatically decides what ranges and what midpoints to use. However, you can specify the range width and starting midpoint yourself. This is important when you want to compare data from two groups and Minitab prints out its default histograms with different scales. You cannot compare results when the scales are different. The technical term for a parameter that is specified in a command is *argument*. Thus, you first specify the command and then provide the necessary arguments. For example, in our productivity data, suppose we want a histogram with a range (or increment) of 10 and a starting midpoint of 110. Our command group would be

```
MTB >HISTogram of data in C3;
MTB > INCRement = 10;
MTB > STARt = 110.
```

Several things are important to note here. First, a semicolon is used at the end of the command and each argument except the last, which is followed by a period. The period tells Minitab that the last argument has been entered and that calculations should begin. This format of ending a command line with a semicolon and specifying additional arguments with additional semicolons is common in Minitab. Be sure to remember that the last argument must be followed by a period.

If you accidentally put a semicolon at the end of the last argument, you can simply put a period on the next line. For example, in the illustration above you could have entered

```
MTB >HISTogram C3;
MTB > INCRement 10;
MTB > STARt 110;
MTB > .
```

Also, each argument must go on a separate line, but it can go anywhere on that line. We have indented each argument for clarity and suggest that you do the same.

## SECTION 3-3 Visual Presentation of Data

If you specify a starting point that is higher than some of your data points, those data points will be eliminated. In our case, if we had specified a starting point of 180 and an increment of 10, then all values below 175 would have been eliminated from the histogram.

Figures 3-6, 3-7, and 3-8 are three histograms of the same data set: Minitab's default histogram (Figure 3-6), a histogram built with an increment of 10 and a start of 120 (Figure 3-7), and a histogram with a start of 120 and an increment of 40 (Figure 3-8). Note that all three histograms are for the same data and therefore are made from exactly the same information. However, they certainly look different.

**FIGURE 3-6** Histogram of Go-Fast Productivity Data Using Minitab Default Arguments (same as Figure 3-5)

```
MTB >HIST C3

Histogram of PRODUCT   N = 100

Midpoint    Count
   120.       1     *
   140.       4     ****
   160.       4     ****
   180.      12     ************
   200.      31     *******************************
   220.      23     ***********************
   240.      21     *********************
   260.       4     ****
```

**FIGURE 3-7** Histogram of Go-Fast Productivity Data Using INCRement = 10, STARt = 120

```
MTB >HIST C3;
SUBC> STAR 120;
SUBC> INCR 10.

Histogram of PRODUCT   N = 100

Midpoint    Count
   120.0      1     *
   130.0      1     *
   140.0      2     **
   150.0      2     **
   160.0      2     **
   170.0      2     **
   180.0      7     *******
   190.0     11     ***********
   200.0     19     *******************
   210.0     10     **********
   220.0     11     ***********
   230.0     18     ******************
   240.0      9     *********
   250.0      1     *
   260.0      3     ***
   270.0      1     *
```

Whenever you compare statistical data between two or more groups, it is essential that the data be measured on a common scale. Violating this rule is one of the most frequent mistakes when using histograms, plots, and other types of graphs. Sometimes such a violation is intentional in order to deceive. Although most companies and individuals do not intentionally change the basis of comparison, less ethical people frequently attempt this type of trick to make their product or service look better than the competition's.

**58** ◀ **CHAPTER 3** Descriptive Statistics

**FIGURE 3-8**  HISTogram of Go-Fast Productivity Data Using INCRement = 40, STARt = 120

```
MTB >HIST C3;
SUBC> STAR 120;
SUBC> INCR 40.

Histogram of PRODUCT   N = 100

   Midpoint    Count
      120.0       2    **
      160.0      11    ***********
      200.0      47    ***********************************************
      240.0      37    *************************************
      280.0       3    ***
```

Of course, not all data for different variables must be measured on the same scale. In our Go-Fast example, we obviously would not measure ages on the same axis as productivity. In this case, because the numbers are not comparable, we do not use the same scale for our graphs. However, if the numbers are equivalent and the purpose is to compare two or more similar groups, then you must use comparable scales for your charts.

## Stem-and-Leaf Displays ▼▲▼▲▼▲▼▲▼▲▼▲▼▲▼▲▼▲▼▲▼▲▼▲▼▲▼▲▼▲▼▲▼▲▼

The stem-and-leaf plot differs from a histogram because the actual values in the data are used for the plots instead of just frequencies. The form of the Minitab command is

```
STEM-and-leaf of data in C,...,C
```

Assuming our production data is still in column 3, the following **STEM-and-leaf** command would produce the plot in Figure 3-9.

```
MTB >STEM-and-leaf of data in C3
```

To interpret this figure, first consider the name, *stem-and-leaf*. All of our values are broken into two parts. The stem part consists of all digits but the units digit (i.e., *15*0, *16*5, *17*8, etc.), and the leaf is the units digit number (i.e., 15*0*, 16*5*, 17*8*, etc.). In Figure 3-9, the second column is the stem and the third column, which spreads out on the page, is the leaf. For example, in the sixth row of the figure, the value in the second column is 17. Therefore 17 is combined with *each* value in the leaf section of that same line. The leaf values are 0, 5, 5, and 5. That means that there were four employees with productivity values of 170, 175, 175, and 175. On the fifth line we see that three employees in the 160–170 range had productivity values of 160, 160, and 165.

The stem-and-leaf plot gives information that we did not know from the histogram: All of our values are measured in units of 5. People produced 175 or 180 or 185 units, but never 183. An alert manager might want to verify the accuracy of such a reporting system. These figures may be quite correct; for instance, the production process might create five units at a time. The point is that we now know this fact and we did not know it before.

Now to return to Figure 3-9. The first column is a measure indicating how many values lie on that row *and* on rows further below the mean. For example, in the fifth row of the figure, the value in the first column is 9; this means that there are nine values on that line and on other lines that are further below the mean. Since that line represents the values for

## SECTION 3-3 Visual Presentation of Data

**FIGURE 3-9** Stem-and-Leaf Display of Go-Fast Productivity Data

```
MTB >STEM C3

Stem-and-leaf of PRODUCT     N  = 100
Leaf Unit = 1.0

    1     12 0
    2     13 0
    5     14 005
    6     15 0
    9     16 005
   13     17 0555
   21     18 00005555
   38     19 00000005555555555
  (14)    20 00000000055555
   48     21 00000555
   40     22 000000005555555
   25     23 0000000000055555
    9     24 00005
    4     25 5
    3     26 005
```

160, this means the number of values in the 160s, 150s, and lower. At the other end of the figure the interpretation is the same. On the third row up from the bottom of the figure, the line for production of 240 units, we also have the figure of 9 in the first column. Again, this indicates that there are nine measurement points on that line and further away, namely, in the 250s and 260s. Also, notice the value of 14 in parentheses. That indicates that the median value is in that group. The frequency values in the first column lead up to that value from *both* directions.

The very top of our figure contains the units of measurement. Decimal points are not used in the **STEM-and-leaf** command. Thus our value for productivity of 120, or an age of 12, or years of service of 1.2 would all be shown the same way. The **Leaf Unit** at the top of the figure indicates the placement of the decimal point. The next line is an example. In our case, the value **12** should be read as 120. However, if the decimal unit had been 0.1, then the value of **12** would have represented 12. If the decimal unit had been 10, then 12 would have represented 1,200.

Some data will have many leaf values for a single stem. In these cases, the stem values will extend over two, three, or more lines. The leaf portion will be in numeric order from the smallest value to the largest. For example, if we had 150 observations between values of 60 and 65, then the stem of the first line would be 6 and the leaf might consist of 0's and 1's; the next line would have the same stem (6) and possibly the 2's and 3's of the leaf, and so on through the final leaf of 8's and 9's.

With the **STEM-and-leaf** display, we can quickly see the frequency of various values. In Figure 3-9 we can see that the value of 230 occurs 11 times and is the most frequent value. Four values—120, 130, 150, and 255—occur only once. However, you should compare this figure with the histograms. Notice that all have the same general shape. If you smooth out the minor variations, all these graphs resemble a hill. They start small on each end and become taller in the middle where the values are most frequent. Some histograms are taller and have less spread on each end. Others are very flat and have a small number of values spread out a long way on each end. Throughout this text, such variance in data will be a central concern.

## Dot Plots

The histogram in Figure 3-5 has a vertical orientation. This same data could be printed along the horizontal axis so that the data points go up instead of out as in the histogram. The `DOTPlot` command prints the data horizontally and takes the form

>DOTPlot of data in C,...,C

The same set of optional arguments used with the `HISTogram` command can be used with the `DOTPlot` command. Thus you can specify an increment and starting point other than those specified by the default settings.

For our productivity data stored in column 3, the `DOTPlot` command would be

MTB >DOTPlot data in C3

The results of this command are shown in Figure 3-10.

**FIGURE 3-10**  Dot Plot of Go-Fast Productivity Data

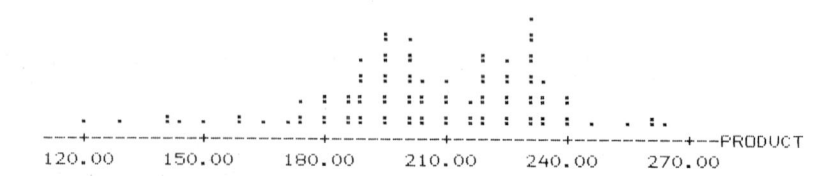

```
MTB >DOTP C3

                                      :  .        :
                           . : :        :  . :
                           : : :.   .   : : :.
                     . : :: : :: : .: : :: :
        .   .     :. . :.  .: : :: : :: : :: : .  . :.
        ---+---------+---------+---------+---------+---------+--PRODUCT
         120.00    150.00    180.00    210.00    240.00    270.00
```

## Box Plots

Like the histogram, stem-and-leaf, and dot plot, the box plot is a popular tool for analyzing data. The box plot is a plot of a portion of the sample distribution and is produced by the command

BOXPlot of data in C,...,C

Thus our production data in column 3 could be printed with the command

MTB >BOXPlot of data in C3

The results of such a command are shown in Figure 3-11.

**FIGURE 3-11**  Box Plot of Go-Fast Productivity Data

```
MTB >BOXP C3
                                      ---------------
         *  *   ------------------I    +    I-----------
                                      ---------------
        ----+---------+---------+---------+---------+---------+--PRODUCT
          120       150       180       210       240       270
```

**SECTION 3-4** Numerical Expressions of Dispersion ▶ **61**

The box plot includes a + to show the median of the distribution, a box that encloses the middle half of the data, and "whiskers" that include the expected range of the distribution. *Possible outliers* (values far enough from the mean that the statistical rules of the **BOXPlot** command call them out for special notice) are marked by a **\***. *Probable outliers* (values extremely far away from the mean according to statistical rules) are marked by **O**. Such outliers should always be checked to insure that an error was not made in collecting the data or entering the data into the computer.

## Comparing Two Independent Sets of Data ▼▲▼▲▼▲▼▲▼▲▼▲▼▲▼▲▼

In the discussion of the **HISTogram** command, we stated that the data sets must have common scales if they are to be compared. In Minitab there is a subcommand for the **HISTogram**, **STEM-and-leaf**, and **DOTPlot** commands which will ensure that the basis of comparison is the same.

When the **HISTogram** command was first introduced, the command to make histograms from two or more columns was shown to be

```
MTB >HISTogram C1, C2, C4
```

This command will calculate and print three separate histograms. Each will be placed on the most convenient scale for that particular group of data points. To ensure that the data groups are placed on the same scale, the following subcommand should be used:

```
MTB >HISTogram C1, C2, C4;
SUBC> SAME.
```

A similar set of commands would be used to make comparable dot plots:

```
MTB >DOTPlot C1, C2, C4;
SUBC> SAME.
```

The command set for the **STEM-and-leaf** command would be similar. The printouts created by the **HISTogram** and **STEM-and-leaf** commands with the subcommand **SAME** will look exactly like the printouts already shown except that the scales of measurement will be identical for all the created graphs. However, the printout created by the **DOTPlot** command with the subcommand **SAME** will place all the plots in the same display rather than in separate displays. As an illustration of this difference, 100 observations have been read into columns 10, 14, and 22. The output from the **DOTPlot** command without the subcommand **SAME** is shown in Figure 3-12, and the output with the subcommand **SAME** is shown in Figure 3-13.

## 3-4 NUMERIC EXPRESSIONS OF DISPERSION
▼▲▼▲

In our last section, we tried to get a feel for the grouping of values. In this section, we will attempt to express such variation in the data numerically. Consider the following simple problem. Assume you have two small sets of data as follows:

data set 1:   4, 4, 4, 4, 4, 6, 6, 6, 6, 6
data set 2:   1, 1, 1, 1, 1, 9, 9, 9, 9, 9

**FIGURE 3-12**  Dot Plots Created without SAME Subcommand

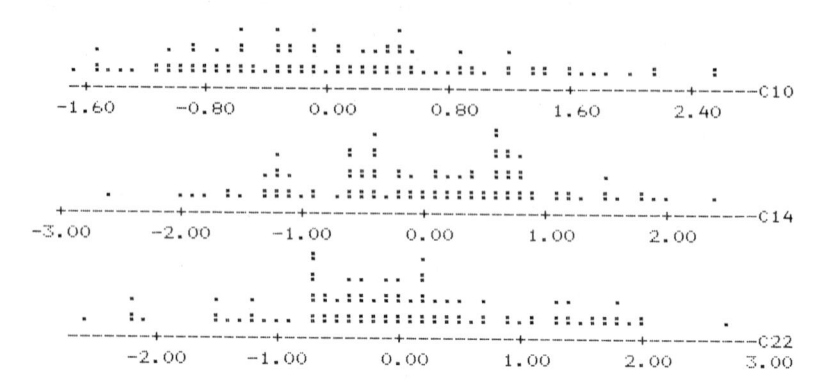

```
MTB >DOTP C10 C14 C22
                         .    .
             .   :   .   :    ::  :  :     ..::.
     . :... ::::::::.::::.:::::::::...::. : :: :... . :           :
    -+---------+---------+---------+---------+---------+-----C10
   -1.60     -0.80      0.00      0.80      1.60      2.40

                                       .         :
                       :  :           ::.
                      .:.         ::: :.  :..: :::
       .         ... :. :::.: .:::.:::::::::::: ::. :. :...    .
    +---------+---------+---------+---------+---------+-------C14
  -3.00      -2.00     -1.00      0.00      1.00      2.00

                                       :
                                       :  ..  ..  :
              .        .  .           ::.::.::.:...  .      ..     .
        .    :.     :..:... :::::::::::::::.: :.: ::.:::.:         .
    ---------+---------+---------+---------+---------+---------+---C22
          -2.00     -1.00      0.00      1.00      2.00      3.00
```

**FIGURE 3-13**  Dot Plots Created with SAME Subcommand

```
MTB >DOTP C10 C14 C22;
SUBC> SAME.
                                    .     .
                                 .  :  .    .:
                             ::  .::  :::  :.:::   .   . .:
                      . ::..::::::.:::.:::::.:.::.: : :.: . :         :
    ---------+---------+---------+---------+---------+---------+---C10
                                       .         :
                       :  :           ::.
                      .:.         ::: :.  :..: :::
       .         ... :. :::.: .:::.:::::::::::: ::. :. :...    .
    ---------+---------+---------+---------+---------+---------+---C14
                                       :
                                       :  ..  ..  :
              .        .  .           ::.::.::.:...  .      ..     .
        .    :.     :..:... :::::::::::::::.: :.: ::.:::.:         .
    ---------+---------+---------+---------+---------+---------+---C22
          -2.00     -1.00      0.00      1.00      2.00      3.00
```

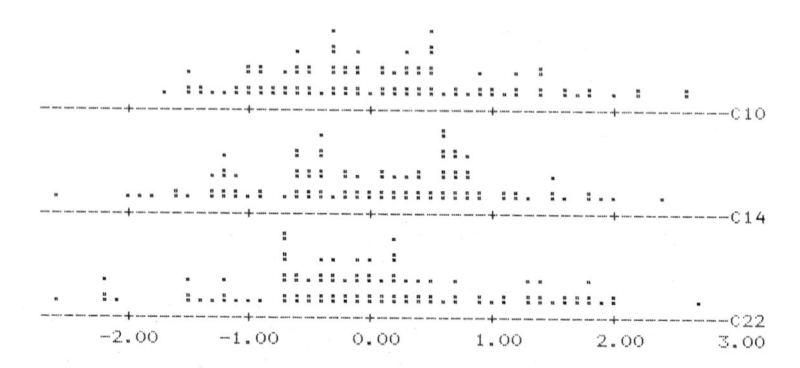

The mean of each of these data sets is 5.0. However, the first data set has values very close to the mean, while the second data set has values further away. If the mean were the only measure we had, then it would seem that the data sets were very similar. However, you might intuitively feel that the mean provides a better estimate of the values in the first data set than in the second, since the mean value of 5 is closer to the actual values in the first data set.

# Calculating the Variance

One way to reveal which mean is most representative of a data set is to calculate the variance. For example, in data set 1, we could calculate the difference between each individual observation and the mean. These individual differences in a sense represent the "distance" between the mean and each observation. After subtracting each value from the mean, these differences could then be added together. However, if we did that, a problem would arise.

### SECTION 3-4 Numerical Expressions of Dispersion

We would be adding five values of −1 and five values of +1, and the total would be 0. If we did the same thing for data set 2, exactly the same thing would happen. Our measures of the variance would still be 0, because we would add together five values of −4 and five values of +4. Thus, this approach does not differentiate the variance between the two groups.

However, let us continue this same line of logic. If after we subtracted each value from the mean we squared the differences, the minus signs would drop out. If we then added all the values together, the totals would be quite different. For data set 1, these calculations would yield a value of 10; for data set 2, we would get a total of 160. We now seem to be on the right track, but there is still a problem. If one data set had 10 values in it and the other had 500 values, the second data set would almost certainly have a higher variation simply because of the greater number of values.

## The Standard Deviation

Our measure of total variation is not yet exactly what we want. A better measure is the *average* variation per observation, since that eliminates the problem of larger samples necessarily having larger variations. It seems logical that a sample of 500 observations in which all the observations are very close to the mean should have a very small variation, and that a sample of 10 observations in which each is far away from the mean should have a larger variation. Thus, by dividing the total of the squared differences by the number of observations minus 1, we would obtain an unbiased measure of the amount of variation per observation. We divide by the number of observations minus 1 ($n − 1$) to correct for a tendency of the variance of a subset (or sample) to be smaller than that of its respective set (or population). This provides an "unbiased estimate" of the population variance.

But what does the variance measure? What is the unit of measure? If our data sets are in dollars, then our variance measure is squared dollars. That does not make a lot of sense to most people. However, if we take the square root of the variance, the new value will be in the same units as the original data. Thus, if data sets 1 and 2 are values in dollars, then the square root of the variance, which is 1.05 for data set 1 and 4.22 for data set 2, is also in dollars. The square root of the variance (which is the sum of the squared differences between the individual values and the mean) is called the *standard deviation*.

The standard deviation is commonly used to measure the variability of data. To summarize for data group 1, the mean is 5 and the standard deviation is 1.05. If this data was the individual sales in dollars at a store, we could say that the average sale was $5.00 and the average distance of each sale from the mean (standard deviation) was $1.05. In sample data set 2 for a second store, the average sale was also $5.00, but the average variability was $4.22. When our data is converted back to common units and expressed as averages, we can compare variability measures and know with certainty that a higher standard deviation means greater variance.

The usefulness of this concept should now become apparent. In our example of the data sets for two stores, would you be more comfortable in predicting the amount of the next sales purchase in store 1 or in store 2? The amount of variability in the data thus affects our ability to use the information. Information with very low variance is more consistent, and we can place more trust in such information. If the fundamental question in statistical analysis is to ask "What can we say about the phenomenon represented by this group of values?", then it is obvious that as the variance becomes smaller and smaller, we can say more and have greater certainty in what we say.

## The Trimmed Mean

Our `DESCribe` command prints a 5% trimmed mean. In our productivity data, and from the histogram in Figure 3-5, you can see that there are five people at the low end of the productivity scale and a few people at the high end. If we eliminate these extremes of production, we get another measure of production that can be used to estimate the average. The `TRMEAN` variable produced by the `DESCribe` command does exactly that. It eliminates the extremes of the data set so that very high values or very low values do not overly influence the calculation of the mean. Many people feel that the 5% trimmed mean is a better estimate of central tendency than the untrimmed mean. The productivity data reveals that there are 1 person producing 120 pairs, 1 person producing 130 pairs, and 2 people producing 140 pairs. If just those 4 were eliminated from the calculation of the mean, as well as a similar group at the high end of the distribution, we would have a slightly different and possibly truer measure of the average. The `TRMEAN` calculation simply eliminates the bottom 5% and top 5% of the observations and then computes the mean from the remaining values.

By running Minitab on the data for December productivity by employees at Go-Fast, we obtain a regular mean of 205.50 with a standard deviation of 28.32 pairs of shoes per employee. Based on the `TRMEAN` calculation, the mean is 206.56 and the standard deviation 21.40. The significant drop in standard deviation from 28.32 to 21.40 pairs per employee should not be surprising. Since the `TRMEAN` calculations drop off observations furthest from the mean, the new standard deviation for the central 90% of the observations should be lower.

## Using Minitab to Obtain the Standard Deviation

You already know that the `DESCribe` command calculates the standard deviation. However, you can also use the `STDEv` command directly in the form

```
STDEv data in C, [put in K]
```

Both calculations assume a *sample* situation and thus always divide the total variance by $n - 1$ observations. This is the most frequently used measure of the spread of observations in any group of data.

You should now complete the same set of calculations on data sets 1 and 2, which were discussed earlier in this section. After you have completed those calculations by hand, enter the data in Minitab and execute both a `DESCribe` command and a `STDEv` command. The results of all three calculations should be identical.

## Hand Calculations of Measures of Dispersion

There are several common measures of dispersion typically done by hand calculation.

## SECTION 3-4 Numerical Expressions of Dispersion

▶▶▶▶▶ **The Range** The range is simply the difference between the largest and the smallest observation. In our calculation of the mean at the end of Section 3-2, we used five numbers ranging in value from 3 to 10. Thus, the range for these values is

$$\text{range} = 10 - 3 = 7$$

Often we state the range in terms of the largest and smallest observations. In our example of the football players, the performance of the starting lineup ranged from "fair" to "superior." Thus, the range can be used for both ordinal and metric data.

▶▶▶▶▶ **The Variance** The variance and the standard deviation are clearly the most important measures of dispersion. The variance is computed by finding the differences between each observation and the mean, squaring those differences, and then dividing the sum of those squares by the size of the population minus 1. The formula is

$$s^2 = \frac{\sum_{i=1}^{n} (\bar{x} - x_i)^2}{n - 1}$$

Thus for the same group of five numbers discussed at the end of Section 3-2,

| $i$ | $x_i$ | $\bar{x}$ | $\bar{x} - x_i$ | $(\bar{x} - x_i)^2$ |
|---|---|---|---|---|
| 1 | 7 | 6 | −1 | 1 |
| 2 | 3 | 6 | 3 | 9 |
| 3 | 4 | 6 | 2 | 4 |
| 4 | 6 | 6 | 0 | 0 |
| 5 | 10 | 6 | −4 | 16 |

So the sum of our squared deviations from the mean is

$$\sum_{i=1}^{5} (\bar{x} - x_i)^2 = 1 + 9 + 4 + 0 + 16$$
$$= 30$$

and the mean squared deviation (or variance) is

$$s^2 = \frac{30}{5 - 1} = 7.5 \text{ square units}$$

▶▶▶▶▶ **The Standard Deviation** The standard deviation is computed by taking the square root of the variance. Although the formula looks a bit imposing, it is really quite simple when taken one step at a time. You first compute the variance exactly as described above and then simply take the square root of that value. Remember that you always perform the operations inside a square root sign (radical) before performing the square root calculation itself. Our formula is thus

$$s = \sqrt{\frac{\sum_{i=1}^{n}(\bar{x} - x_i)^2}{n - 1}}$$

Since we already know that the variance is 7.5, we can find the standard deviation by simply taking the square root of this value:

$$s = \sqrt{7.5} = 2.739$$

# SUMMARY

This chapter has discussed the clustering (central tendency) and the dispersion (variance) of data. Various methods for displaying data have also been demonstrated. However, all measures described in this chapter are meaningless if the proper underlying assumptions are not met. Many inaccurate and improper statistical conclusions have been passed off on unsuspecting people by skipping over such details as how the sample was obtained.

# DISCUSSION QUESTIONS

1. What is the difference between nominal, ordinal, and metric data?
2. What is meant by central tendency?
3. Why and when does the mean become a less effective measure of central tendency?
4. What is the difference between a histogram and a stem-and-leaf display?
5. What is necessary when comparing the histograms for two groups of data?
6. What is the trimmed mean?
7. What are the variance and the standard deviation?

# PROBLEMS

**3-1** Sky's the Limit, Inc., installed phones in 10 Boeing 747 airplanes. Because this is an innovative venture, Sky's the Limit is extremely interested in the initial data on the use of these phones. Following is a sample of the number of minutes used per call during the first five months of service:

| 5  | 6  | 9  | 7  | 3 | 3  | 11 |
|----|----|----|----|---|----|----|
| 14 | 7  | 7  | 5  | 3 | 10 | 6  |
| 6  | 5  | 5  | 15 | 8 | 6  | 9  |
| 6  | 5  | 6  | 6  | 7 | 13 | 4  |
| 7  | 8  | 6  | 11 | 3 | 6  | 6  |
| 12 | 5  | 12 | 6  | 9 | 6  | 7  |
| 1  | 29 | 6  | 12 | 6 | 7  | 6  |

a) Use the **SET** command to enter the data into **C1**.
b) Title the column **'MINUTES'** using the **NAME** command.

## PROBLEMS

c) Use the DESCribe command on 'MINUTES'.
d) Sky's the Limit's chief executive officer (CEO) expects phone use to average five minutes per customer. Will he be pleased with the initial data?
e) Is the mean a good representation of the data? (Use the standard deviation.)
f) Will the CEO still be pleased with the average?

**3-2** Refer to Problem 1-1 on executive salaries.
a) Enter the data, using the SET command, into C1.
b) Use the DESCribe command on the data.
c) What is the trimmed mean?
d) Is there much difference between the mean and the trimmed mean in this example? Explain.

**3-3** A new car plant has been in operation for two years. Managers expect output per worker to increase as the workers become accustomed to the plant. Data on output for a sample of employees for the two years follows:

| Employee | Year 1 | Year 2 |
|---|---|---|
| 1 | 90 | 112 |
| 2 | 95 | 97 |
| 3 | 87 | 102 |
| 4 | 54 | 100 |
| 5 | 105 | 103 |
| 6 | 89 | 121 |
| 7 | 149 | 125 |
| 8 | 95 | 131 |
| 9 | 86 | 114 |
| 10 | 89 | 120 |
| 11 | 97 | 119 |
| 12 | 97 | 129 |
| 13 | 86 | 99 |
| 14 | 57 | 100 |
| 15 | 95 | 107 |
| 16 | 97 | 102 |
| 17 | 55 | 124 |

a) Use the READ command and enter data into C1 and C2.
b) Title C1 and C2 'OUTPUT 1' and 'OUTPUT 2', respectively.
c) Use the DESCribe command on each column.
d) Which data set is more spread out or has more variance? Explain.
e) Has worker output increased in year 2?

**3-4** In 1984, a company repurchased 6 million shares of its common stock. The price per share ranged from $28.87 to $56.62. A sample of the buy-back stock prices follows:

| | | | | | |
|---|---|---|---|---|---|
| $53.45 | $30.07 | $49.92 | $30.01 | $35.54 | $56.10 |
| 32.47 | 55.51 | 52.62 | 53.10 | 45.92 | 49.40 |
| 55.60 | 47.62 | 54.40 | 55.00 | 51.21 | 29.10 |
| 29.95 | 55.02 | 29.80 | 48.82 | 54.23 | 56.62 |

a) Enter the data using the **SET** command into **C1**.
b) Calculate the following individual statistics (do not use the **DESCribe** command):

        MEAN C1
        STDEv C1

c) Do you feel confident saying that the mean is a good representation of the buy-back prices? (Use the concept of standard deviation.)
d) On the average, how much was spent to repurchase the 6 million shares?

**3-5** The Sea King Corporation noticed a sharp decline in the number of lobsters caught per outing. To supplement this information, Sea King recorded the weight of the total catch for each outing. The sample data (in pounds) follows:

| 256 | 352 | 427 | 435 | 462 | 326 |
|-----|-----|-----|-----|-----|-----|
| 357 | 423 | 324 | 464 | 433 | 469 |
| 325 | 436 | 259 | 427 | 465 | 320 |
| 259 | 329 | 257 | 250 | 358 | 463 |
| 465 | 466 | 431 | 251 | 252 | 429 |
| 433 | 253 | 321 | 467 | 422 | 461 |

a) Enter the data into **C1** using the **SET** command.
b) Create a histogram of **C1**.
c) Use the **STEM-and-leaf** command to create a stem-and-leaf display.
d) What does each major part of the stem-and-leaf display mean?
e) Create a dot plot of **C1**.
f) Compare the shape of the stem-and-leaf display, the dot plot, and the histogram.

**3-6** Lewis Cadillac of Pennsylvania operates two showrooms. As an incentive to his employees, the president, M. Lewis, offered a $10,000 bonus to the showroom that attracted the most customers during a four-week period. Showroom 1 advertised its discounts on TV, while showroom 2 advertised its discounts on the radio. The number of customers attracted per day by each showroom is displayed in the following:

| Showroom 1 |    |    |    | Showroom 2 |    |    |    |
|----|----|----|----|----|----|----|----|
| 7  | 12 | 15 | 7  | 21 | 20 | 12 | 20 |
| 9  | 5  | 9  | 9  | 3  | 5  | 20 | 5  |
| 4  | 10 | 14 | 7  | 5  | 3  | 21 | 10 |
| 10 | 9  | 6  | 7  | 6  | 18 | 3  | 5  |
| 6  | 10 | 9  | 12 | 20 | 21 | 3  | 5  |
| 6  | 10 | 12 | 9  | 18 | 6  | 6  | 20 |
| 7  | 7  | 9  | 10 | 21 | 6  | 5  | 11 |

a) Enter the data into **C1** and **C2** using either the **READ** or the **SET** command.
b) Use the **DESCribe** command on each set of data.
c) What is the range for each set of data?
d) Create both histograms and dot plots for each set of data. Be sure to use the **SAME** subcommand.
e) What type of distribution (normal, bimodal, etc.) does the data for showroom 1 resemble? For showroom 2?

# PROBLEMS

**f)** Compare the two means. Which seems to be a more effective medium of advertising, TV or radio (assuming equal advertising budgets)?

**g)** For each set of data, use the **SUM** command to total the number of customers attracted during the four-week period.

**h)** Which showroom received the bonus?

**3-7** Women tend to patronize Shopper's Stop convenience stores in the day rather than at night. Consequently, Shopper's Stop is in favor of a congressional bill that would begin daylight saving time in March instead of April. This extension of daylight time could increase the convenience stores' profits considerably. Following is a sample of the actual annual profits for 11 Shopper's Stop stores and their expected annual profits if the bill passes:

| Present Profits | Expected Profits |
|---|---|
| $84,000 | $98,000 |
| 30,000 | 34,000 |
| 27,000 | 31,000 |
| 20,000 | 23,000 |
| 25,000 | 29,000 |
| 24,000 | 28,000 |
| 19,000 | 21,000 |
| 24,000 | 27,000 |
| 21,000 | 25,000 |
| 27,000 | 28,000 |
| 32,000 | 36,000 |

**a)** Use the **READ** command to enter the data into **C1** and **C2**.
**b)** Use the **DESCribe** command on **C1** and **C2**.
**c)** Is there much difference between the mean and the trimmed mean? Explain.
**d)** Determine the amount of increased profit each store anticipates. (Let **C3** = **C2** − **C1**.)
**e)** What is the range of the increased profit in **C3**?
**f)** What could cause the range to be this large?
**g)** Is this large range an indication that something is wrong?

**3-8** From April through June, Picture Perfect, Inc., a yearbook publishing company, has three crews working round the clock to meet publication deadlines. For each crew, a sample of worker output per hour follows measured by the number of pages produced:

| Crew 1 | | | Crew 2 | | | Crew 3 | | |
|---|---|---|---|---|---|---|---|---|
| 85 | 96 | 106 | 119 | 119 | 139 | 120 | 101 | 127 |
| 101 | 126 | 125 | 135 | 139 | 125 | 101 | 105 | 100 |
| 114 | 92 | 84 | 135 | 135 | 116 | 105 | 110 | 105 |
| 114 | 112 | 126 | 105 | 117 | 115 | 125 | 125 | 120 |
| 83 | 126 | 130 | 119 | 119 | 119 | 100 | 115 | 105 |
| 115 | 115 | 116 | 125 | 132 | 132 | 101 | 115 | 95 |
| 126 | 85 | 110 | 125 | 115 | 125 | 96 | 96 | 101 |
| 80 | 85 | 85 | 115 | 125 | 119 | 120 | 105 | 101 |
| 80 | 112 | 115 | 125 | 135 | 125 | 106 | 96 | 101 |
| 92 | 92 | 92 | 132 | 132 | 132 | 95 | 112 | 111 |

a) Enter the data into C1, C2, and C3.
b) Use the DESCribe command on C1, C2, and C3.
c) Which crew produces the most as indicated by the mean? By the maximum value?
d) Which crew has the greatest range?
e) Create both a histogram and a dot plot for each set of data without using the SAME subcommand.
f) Create both a histogram and a dot plot for each set of data using the SAME subcommand.
g) What type of distribution does each set of data represent?
h) What is the mode for each crew?

**3-9** In the early 1980s, Flowers by Joan, Inc., expanded its operations to include rose cultivation. Each variety has stems long enough to be accepted for sale by florist shops. For her records, Joan took the following samples of the length in inches of the stems for each variety:

| Velvet Pink | | Lavender Lace | | American Beauty Red | |
|---|---|---|---|---|---|
| 20 | 20 | 8 | 8 | 12 | 14 |
| 15 | 11 | 7 | 10 | 14 | 16 |
| 16 | 20 | 10 | 8 | 11 | 12 |
| 20 | 20 | 14 | 7 | 16 | 15 |
| 20 | 14 | 8 | 6 | 12 | 12 |
| 11 | 20 | 8 | 9 | 11 | 12 |
| 8 | 7 | 10 | 8 | 10 | 11 |
| 16 | 10 | 8 | 8 | 8 | 12 |
| 18 | 20 | 15 | 8 | 11 | 10 |
| 18 | 20 | 16 | 15 | 12 | 12 |

a) Enter the data into C1, C2, and C3.
b) Use individual commands to compute the mean and standard deviation for each set of data. Do not use the DESCribe command.
c) For which variety of rose would you feel most comfortable guaranteeing that an order had a stem length equal to its respective mean? (Use the concept of standard deviation.)
d) What is the mode for each variety?

**3-10** Color Blow-Up's policy is "20-minute color enlargements or your money back." Because of this policy, Color Blow-Up needs to know during which hours of business it will draw the most customers, because the busier the time of day, the more workers it needs. Here is a sample of the data obtained on nine days.

| Day | 8:00 | 9:00 | 10:00 | 11:00 | 12:00 | 1:00 | 2:00 | 3:00 | 4:00 |
|---|---|---|---|---|---|---|---|---|---|
| 1 | 2 | 5 | 9 | 23 | 2 | 1 | 3 | 11 | 26 |
| 2 | 5 | 6 | 8 | 14 | 7 | 3 | 5 | 17 | 16 |
| 3 | 9 | 2 | 9 | 19 | 15 | 17 | 9 | 12 | 12 |
| 4 | 2 | 2 | 14 | 18 | 19 | 22 | 6 | 11 | 15 |
| 5 | 4 | 4 | 6 | 27 | 16 | 22 | 4 | 12 | 19 |
| 6 | 5 | 2 | 9 | 12 | 24 | 16 | 11 | 10 | 24 |
| 7 | 4 | 3 | 16 | 26 | 13 | 9 | 6 | 19 | 37 |
| 8 | 2 | 3 | 22 | 20 | 12 | 21 | 7 | 12 | 12 |
| 9 | 3 | 2 | 9 | 24 | 21 | 22 | 9 | 9 | 18 |

## PROBLEMS

a) Enter the data into C1 through C9 using the READ command.
b) What is the average number of customers for each hour? Use the individual command MEAN.
c) What is the standard deviation for each hour? Use the individual command STDEv.
d) For which time period do you feel most confident saying that the mean is representative of the respective data for that hour? Explain.

**3-11** Many people find home businesses appealing. Mrs. Peel began her home pie-baking business because it eliminated commuting, allowed for a flexible schedule, and provided time for child care. Business was slow at first, but it has picked up considerably since then. The number of pies sold per month in 1982, 1983, and 1984 follow:

| Month | 1982 | 1983 | 1984 |
|---|---|---|---|
| January | 22 | 20 | 87 |
| February | 14 | 22 | 95 |
| March | 12 | 19 | 125 |
| April | 16 | 29 | 170 |
| May | 20 | 35 | 160 |
| June | 15 | 39 | 248 |
| July | 25 | 46 | 301 |
| August | 21 | 69 | 270 |
| September | 22 | 85 | 265 |
| October | 30 | 72 | 216 |
| November | 19 | 88 | 312 |
| December | 22 | 93 | 214 |

a) Enter the data into C1, C2, and C3.
b) Use the DESCribe command on C1, C2, and C3.
c) When referring to the DESCribe command output, how do you determine what the middle 50% of the data is?
d) What is the middle 50% of the data for C1, C2, and C3?
e) Use the SORT command and place the data in C4, C5, and C6, respectively. Print the data and manually count down to the middle of each data set to check the 50% calculation.

**3-12** Charting the number of customers per day for each stylist is part of Classic Cutter's system for performance appraisal. The following sample data is on two of Classic Cutter's stylists:

| Lynn | | | Daren | | |
|---|---|---|---|---|---|
| 9 | 12 | 12 | 9 | 10 | 11 |
| 12 | 12 | 8 | 10 | 14 | 11 |
| 12 | 10 | 12 | 14 | 8 | 9 |
| 13 | 12 | 12 | 17 | 9 | 14 |
| 12 | 13 | 15 | 17 | 9 | 15 |
| 15 | 8 | 10 | 10 | 9 | 9 |
| 10 | 10 | 13 | 8 | 8 | 9 |
| 7 | 15 | 13 | 8 | 15 | 10 |
| 9 | 12 | 15 | 9 | 21 | 11 |
| 9 | 15 | 17 | 11 | 9 | 10 |
| | | 18 | | | 15 |

a) Enter the data into C1 and C2.
b) What is the average number of customers worked on per day by each stylist?
c) Create a stem-and-leaf display for each stylist.
d) What type of distribution does each display represent?
e) Which mean is a better representative of its data set? Explain. (Refer to the distribution type.)

**3-13** First National Bank of Chicago recently installed an automatic transfer system with 16 terminals located throughout the city. Customer acceptance has been overwhelming with the 10th Street terminal receiving the most use. A sample of the number of transactions per day at the 10th Street terminal follows:

| 120 | 116 | 215 | 215 | 165 | 215 | 116 |
| 165 | 100 | 365 | 314 | 116 | 120 | 365 |
| 215 | 455 | 116 | 202 | 100 | 116 | 120 |
| 310 | 114 | 120 | 455 | 110 | 310 | 116 |
| 120 | 160 | 165 | 200 | 120 | 365 | 165 |

a) Use the SET command to enter the data into C1.
b) Use the DESCribe command on this sample.
c) If the mean is 102 and the standard deviation is 160 for the other 15 terminals combined, does the 10th Street terminal seem to receive substantially more use than the rest?
d) Is there much difference between the mean and the trimmed mean? Why or why not?

**3-14** Jean's, a fine restaurant, is implementing a new inventory system. In order to save on storage costs, this system requires that an item be ordered a day before it is needed. As a result, accurate estimates of the items needed for the next day's service is important. Here is a sample of the number of bottles of red wine ordered for use the next day:

| Monday | Tuesday | Wednesday | Thursday | Friday | Saturday | Sunday |
|--------|---------|-----------|----------|--------|----------|--------|
| 6      | 2       | 15        | 3        | 32     | 40       | 31     |
| 9      | 9       | 14        | 29       | 36     | 38       | 29     |
| 5      | 3       | 16        | 6        | 29     | 36       | 28     |
| 4      | 6       | 14        | 10       | 19     | 38       | 30     |
| 5      | 2       | 12        | 15       | 32     | 37       | 30     |
| 3      | 23      | 15        | 12       | 33     | 40       | 31     |

a) Enter the data into C1 through C7 using the READ command.
b) Use individual commands to determine the mean and standard deviation of C1 through C7.
c) If the amount ordered equals the mean, how confident are you that the respective mean is a good representative of each day's order? (Use the standard deviation.)
d) Compare the means. The most wine is consumed on which days?

**3-15** In 1984, five auto industry executives were sent to a driving school in California for training. There the executives learned to drive a high-performance race car. The following is a sample of the number of laps completed in 10-minute runs by the five executives:

| 24 | 7  | 29 | 34 | 8  |
| 16 | 31 | 12 | 9  | 11 |

# PROBLEMS

|    |    |    |    |    |
|----|----|----|----|----|
| 9  | 6  | 19 | 35 | 10 |
| 32 | 27 | 36 | 14 | 33 |
| 34 | 11 | 10 | 7  | 32 |

In 1985, five more executives were sent to California. Here is a sample of the number of laps they completed in 10-minute runs:

|    |    |    |    |    |
|----|----|----|----|----|
| 16 | 20 | 15 | 30 | 18 |
| 22 | 17 | 24 | 16 | 21 |
| 19 | 17 | 16 | 20 | 19 |
| 15 | 21 | 22 | 15 | 15 |
| 23 | 19 | 14 | 8  | 14 |

**a)** Use the **SET** command to enter the 1984 data into **C1** and the 1985 data into **C2**.
**b)** Use the **DESCribe** command on **C1** and **C2**.
**c)** Obtain histograms and dot plots of **C1** and **C2**.
**d)** Examine the means of **C1** and **C2**. Which mean is a better representative of its respective data set?
**e)** Could we say that one group of executives was faster than the other? Why or why not?
**f)** Recalculate the histograms and dot plots for 1984 and 1985 specifying a starting point of 6 and an increment of 1. Redo with a starting point of 6 and an increment of 2. Redo again with a starting point of 6 and an increment of 4.
**g)** Plot both histograms and dot plots on the same scale using the **SAME** subcommand. Which figures do you feel give the best representation of the data?

# Probability and Sampling

**T**here are no new Minitab commands in this chapter.

In statistical analysis, we use inductive reasoning (that is, moving from a special case to a general statement) rather than deductive reasoning (that is, moving from a general statement to a special case). An example of deductive reasoning is

*General Statement*: College classes are a lot of work.
*General Statement*: Statistics is a college class.
*Special Case*: Statistics is a lot of work.

We can be absolutely certain that if the first two statements are true, then the special case is true. In statistics, however, we seldom enjoy this luxury. We are generally faced with inductive reasoning like the following:

*Special Case*: Statistics is a lot of work.
*General Statement*: Statistics is a college class.
*General Statement*: Some college classes are a lot of work.

Since we cannot be certain how many college classes are a lot of work, we would like to *estimate* a percentage. However, two problems arise in trying to estimate the correct percentage. First, it is important to understand probability theory. One must know, for example, the meaning of a 27% chance of something occurring in a given situation. Second, in almost all cases, we must collect a certain amount of data called a *sample*, and based upon that sample, we must make a probability estimate. How we collect a sample is critical to the entire process. *If proper procedures are not followed when the data is collected, then nothing effective can be done with the data later.*

Although it may be argued that statistics can be understood without understanding probability theory, we believe, and probably your instructor does too, that some understanding of probability will make the rest of this course much easier. Don't get discouraged! Whole courses are taught on probability theory, and whole classes full of aspiring mathematicians have had difficulty with it. In the rest of this chapter, we will first consider some general characteristics of probability theory. Afterwards, we will discuss some sampling procedures and highlight some typical problems in gathering data.

## 4-1 SOME GENERAL PROBABILITY CONCEPTS

### A Definition of Probability

Probability might be defined as likelihood. Thus, the probability that it will rain tomorrow is a numeric representation of the likelihood of that event. Over the years, mathematicians, gamblers, and theologians have tried to model phenomena that they could neither explain nor control, and in that effort they have developed what we now call probability theory. If you choose to study more statistics after this course, try to note the inventors of the various techniques and the original problems that they were trying to solve. The subject is fascinating.

### Types of Probability Assessment

There are three basic avenues for assessing probabilities, and each has its own usefulness.

▶ ▶ ▶ ▶ ▶ **Logical Probability** Logical probability is the ratio of possible successful events to total possible events. The probability that a toss of a coin will yield heads can be assessed logically. One side is heads and there are two sides, thus the ratio ½. As another example, consider the rolling of a die. There are six sides on a die, and together all six sides constitute the complete set of possibilities. In this case it is easy to define the set of possible outcomes. When we roll the die, there is a one-sixth chance of any given side being face up.

▶ ▶ ▶ ▶ ▶ **Relative Probability** Relative probability is the ratio of past successes to total attempts. If a football team has won 8 of its last 10 games, then we might assess the probability of a win next Saturday as the ratio 8/10. This is a common way to assess probabilities when there is no easily definable set of possible events.

▶ ▶ ▶ ▶ ▶ **Subjective Probability** This is the likelihood of an event, stated as a ratio, that is based on the experience and expertise of a decision maker. This method could be used to assess your probability of making an A in this course. The subjective method is probably the most widely used and the most maligned of the three techniques.

The discussion that follows is based on logical probabilities (for the sake of clarity and convenience), but you should remember that the axioms (or rules) presented here hold true regardless of the method used to assess the probabilities.

## Sets and Venn Diagrams

A good way to visualize events is by using a *Venn diagram*, as shown in Figure 4-1. This diagram depicts a *universal set* (*U*), which contains all the cases that are relevant to any particular problem. The universal set is also called the *population*. If in the Go-Fast Running Shoe Company we are interested in the ages of the employees, then the data set of all employee ages would be the universal set, or population. Within the Venn diagram shown in Figure 4-1, there are two collections of objects (called *sets*). One set has characteristic *A* and the other set lacks characteristic *A* (sometimes called *A complement*). For our Go-Fast example, for the universal set of ages, we might have characteristic *A* be all employees in their 20s. Then not *A*, or *A* complement, would be all those employees aged 19 and younger and all those aged 30 and older.

Another way to visualize this same information is with a plot. If we have all of the data on the Go-Fast employees entered into a Minitab worksheet with years of education in column 1 and ages in column 2, then we can plot `C1` versus `C2` and obtain a complete visual representation of the ages of all employees relative to their years of schooling. The output of this plot command is shown in Figure 4-2.

**FIGURE 4-1** Venn Diagram of Universal Set (*U*) with One Set (*A*)

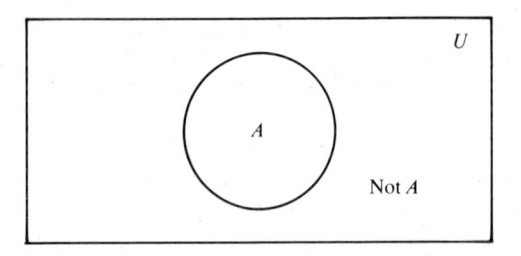

Each point represented by a * indicates a single individual of a given age with a specific level of education. In places where there are several individuals with the same age and same educational level, then the actual number of such employees is printed instead. For example, there are two employees who are 30 and have 12 years of education. This point is represented by a 2 at the intersection of 30 on the *X*-axis and 12 on the *Y*-axis.

Although this plot diagram is quite adequate for our needs, the years of education (the *Y*-axis) are shown in decimal form. Since we are going to use this particular plot figure extensively in the following discussion, it will be better to use a scale for the education axis that is in whole numbers. We can control the scales for each axis in Minitab by using a set of arguments similar to the arguments used for the `HISTogram` and `DOTPlot` commands. For the plot command, the arguments are

## SECTION 4·1 Some General Probability Concepts

```
         PLOT C vs. C;
         XINCrement = K;
         XSTArt at K;
         YINCrement = K;
         YSTArt at K.
```

**FIGURE 4·2** Default Minitab Plot of Education and Age of Go-Fast Employees

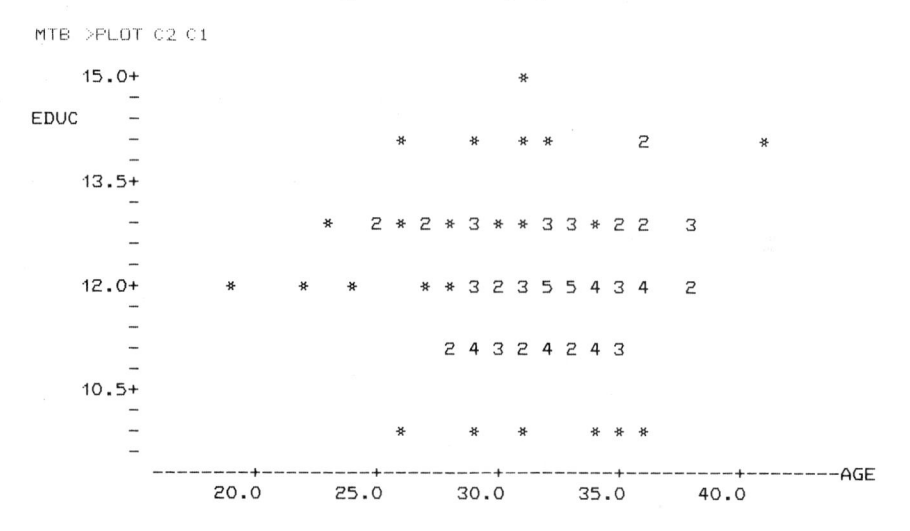

```
    MTB >PLOT C2 C1

       15.0+                              *
            -
EDUC        -
            -                  *    * *  *         2         *
            -
       13.5+
            -
            -           *   2 * 2 * 3 * * 3 3 * 2 2     3
            -
            -
       12.0+     *     *   *   * * 3 2 3 5 5 4 3 4     2
            -
            -
            -                   2 4 3 2 4 2 4 3
            -
       10.5+
            -
            -               *     *    *    * * *
            ---------+---------+---------+---------+---------+--------AGE
                   20.0      25.0      30.0      35.0      40.0
```

This set of arguments allows you to change the Minitab default scales. Remember that you are not changing the data but only the visual representation of the data. To provide us with a cleaner visual picture, the following plot command yields the results shown in Figure 4-3:

```
MTB >PLOT C1 C2;
SUBC> XINCrement = 1;
SUBC> XSTArt at 18;
SUBC> YINCrement = 1;
SUBC> YSTArt = 9.
```

We can now compare our plot in Figure 4-3 to the Venn diagram in the following manner. The area *A* in the Venn diagram of Figure 4-1 can represent the employees who are in their 20s. Employees not in their 20s are the rest of the population (*A* complement). The two groups together make up the entire population (the universal set).

Figure 4-4 is like Figure 4-3 except that we have drawn a circle around data for all those employees who are in their 20s. The employees inside the circled area of Figure 4-4 are symbolically represented by area *A* in the Venn diagram. The only difference is that in Figure 4-4 we can actually go in and count how many employees are in this particular classification. In our illustration, there are 27 employees between the ages of 20 and 29. With this information, we know the answer to the following logical probability question: "If we were to go to the personnel files and draw an employee at random, what is the probability that the employee drawn would be in his or her 20s?" Since there are exactly 100 employees, we have an intuitive feel that the answer is 27%.

In a similar analysis, we can examine the number of employees who have between 10 and 12 years of education. Figure 4-5 is similar to Figure 4-3 except that we have enclosed

**FIGURE 4-3** Plot of Education and Age of Go-Fast Employees with Axes Set by Arguments

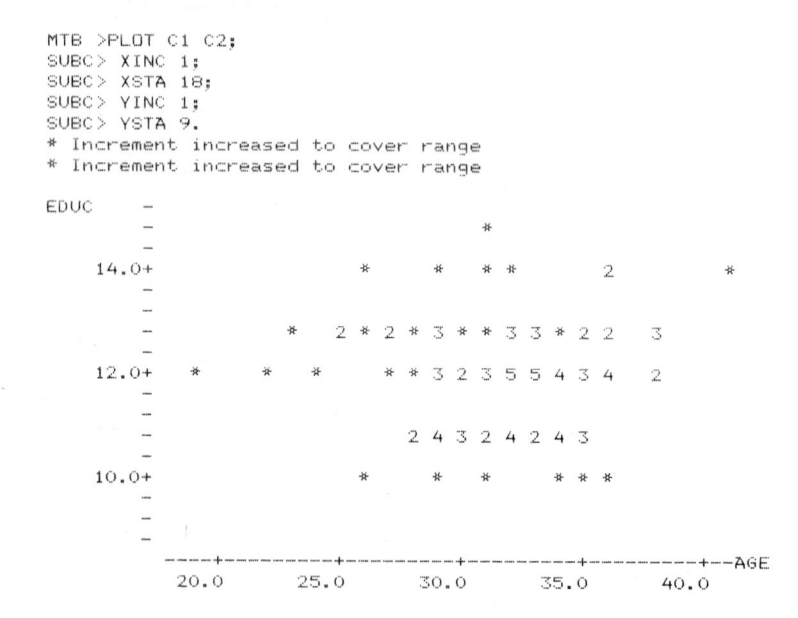

```
MTB >PLOT C1 C2;
SUBC> XINC 1;
SUBC> XSTA 18;
SUBC> YINC 1;
SUBC> YSTA 9.
* Increment increased to cover range
* Increment increased to cover range

EDUC    -
        -                              *
        -
   14.0+             *        *   * *        2           *
        -
        -        *  2 * 2 * 3 * * 3 3 * 2 2     3
        -
   12.0+  *      *      *      * * 3 2 3 5 5 4 3 4     2
        -
        -                        2 4 3 2 4 2 4 3
        -
   10.0+                  *       *     *       * * *
        -
        -
        ----+---------+---------+---------+---------+--AGE
          20.0      25.0      30.0      35.0      40.0
```

**FIGURE 4-4** Plot of Education and Age of Go-Fast Employees Indicating the Set of Employees Aged 20 to 29

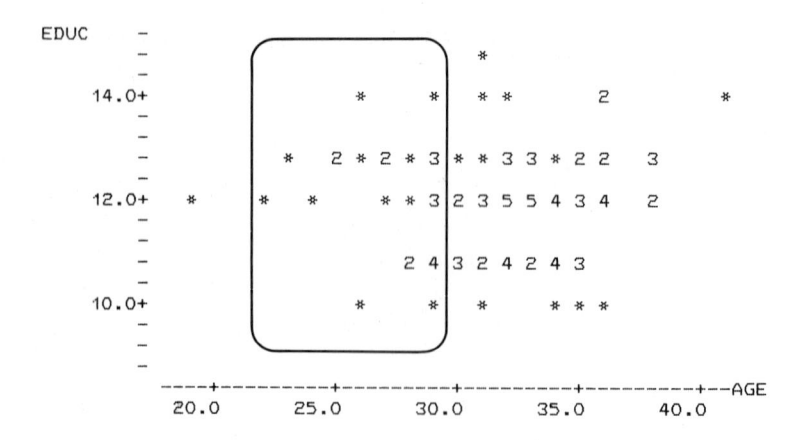

```
EDUC    -
        -                              *
        -
   14.0+             *        *   * *        2           *
        -
        -        *  2 * 2 * 3 * * 3 3 * 2 2     3
        -
   12.0+  *      *      *      * * 3 2 3 5 5 4 3 4     2
        -
        -                        2 4 3 2 4 2 4 3
        -
   10.0+                  *       *     *       * * *
        -
        -
        ----+---------+---------+---------+---------+--AGE
          20.0      25.0      30.0      35.0      40.0
```

an area extending from the Y-axis that includes individuals with between 10 and 12 years of education. By counting the number of employees within the enclosed area, we can determine that 66 employees are in that group. Symbolically we could have used the Venn diagram with all employees being the population, those with between 10 and 12 years of education being set A, and those with less or more education being not A.

We will very quickly learn to use formulas to calculate the various probability relationships shown in this illustration. For now, we hope that you are beginning to get a sense of probability relationships. Based on Figures 4-4 and 4-5, you should be able to determine

## SECTION 4·1 Some General Probability Concepts

**FIGURE 4·5** Plot of Education and Age of Go-Fast Employees Indicating the Set of Employees with 10 to 12 Years of Education

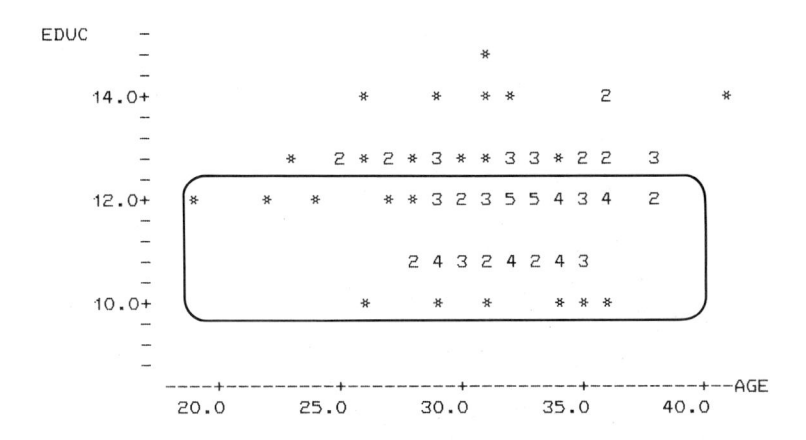

```
EDUC    -
        -
        -                           *
14.0+           *       *   * *         2           *
        -
        -       * 2 * 2 * 3 * * 3 3 * 2 2    3
        -   ┌─────────────────────────────────────┐
12.0+   -   │ *      *   *    * * 3 2 3 5 5 4 3 4    2 │
        -   │                                      │
        -   │                 2 4 3 2 4 2 4 3       │
        -   │                                      │
10.0+   -   │        *       *   *      * * *     │
        -   └─────────────────────────────────────┘
        -
        ----+---------+---------+---------+---------+--AGE
           20.0      25.0      30.0      35.0      40.0
```

the probability of an individual being between 20 and 29 (*A*) or not being between 20 and 29 (*A* complement or not *A*), and similarly for employees with between 10 and 12 years of education and those with either less or more than 10 to 12 years of education.

The next logical step is to ask, "What is the probability of someone being both between 20 and 29 and having between 10 and 12 years of schooling?" The formulas to calculate this will follow shortly, but now we can get a graphical solution. If the areas of our two previous figures are placed on the same graph, we can see the answer.

In Figure 4-6 the area that our two sets have in common contains individuals meeting both criteria. This *intersection* of employees with the right age characteristic and the right educational characteristic is the area of interest. By actually counting the observation points we can determine that there are 15 employees with both characteristics. In other words, if there are 100 employees and you go to the personnel files and draw one employee at random,

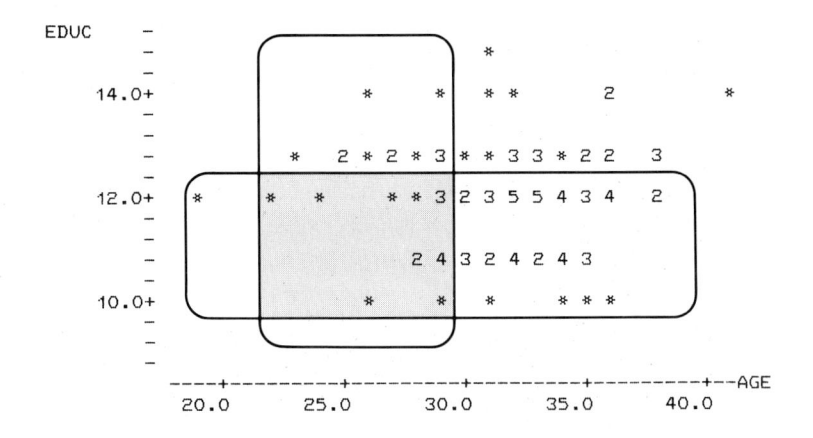

**FIGURE 4·6** Plot of Education and Age of Go-Fast Employees Indicating the Intersection of the Sets of Employees Aged 20 to 29 and of Employees with 10 to 12 Years of Education

the chances are 15 out of 100 that the employee drawn will be in his or her 20s and have between 10 and 12 years of education.

Before we look at more diagrams, we need to develop some formal logic for our relationships. If you do not understand the computation as we calculate various probability values, look at the appropriate plot diagram and count the data points to verify the accuracy of the calculation. At this point, our purpose is for you to have a sense of how the probability values are derived.

## Some Axioms of Probability Theory

If we know how many items are included in the universal set and how many items have characteristic A, we can compute the probability of A, notated as $P(A)$, as follows:

$$P(A) = \frac{\text{number of items with characteristic } A}{\text{number of items in the universal set}}$$

This is called a *simple probability*. If Go-Fast has 100 employees and 27 of them are in the age group between 20 and 29, then there is a 27% chance of randomly selecting an individual in that age group from all the employees. We can also compute the probability of not A as

$$P(\text{not } A) = 1 - P(A)$$

If 27 employees are in their 20s (set A), then not A will be $100 - 27$ or 73 employees. Consequently if there is a 27% chance (or .27) of selecting an employee from set A, then

$$P(\text{not } A) = 1 - .27$$
$$= .73$$

Thus, if a single employee is chosen at random, there is a 73% chance that the employee will not be in his or her 20s.

The Venn diagram in Figure 4-7 can be used to illustrate various possibilities in probability. Here we have pictured a situation with two distinguishing characteristics, A and B. Set A could be Go-Fast employees in their 20s and set B could be employees with between 10 to 12 years of education. The universal set would still be all Go-Fast employees. With this information, consider the possibilities if we draw a sample of a single employee and determine both the employee's age and years of schooling:

**(1)** $P(A)$ = probability of employee being in his or her 20s. The employee may or may not have between 10 and 12 years of schooling. (Also see Figure 4-4.)

**(2)** $P(B)$ = probability of employee having between 10 and 12 years of schooling. The employee may or may not be in his or her 20s. (Also see Figure 4-5.)

**(3)** $P(A \text{ and } B)$ = probability of employee being in his or her 20s and having between 10 and 12 years of schooling. This is called an *intersection*, because both characteristics represented in the Venn diagram are found in a single individual from this part of the diagram. This is shown in Figure 4-7 where sets A and B overlap. In Figure 4-6, the overlapping area is identical in concept to the intersection in the Venn diagram in Figure 4-7.

## SECTION 4·1 Some General Probability Concepts

**FIGURE 4-7** Venn Diagram of the Universal Set (U) and Two Intersecting Sets (A and B)

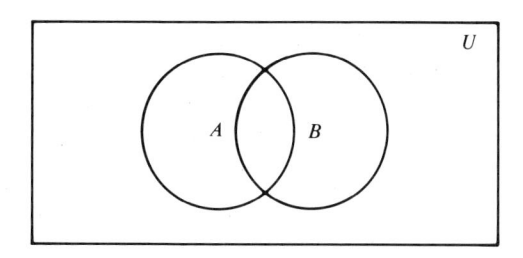

(4) P(A or B) = probability of possessing characteristic A or characteristic B or both. This is called a *union*. The condition is met when *either* or *both* characteristics are present. In Figure 4-8 the area that constitutes P(A or B) is shaded.

(5) P(A and not B) = probability of possessing characteristic A but not B. This case is illustrated in Figure 4-9, where the set of interest is all employees in their 20s who do not have between 10 and 12 years of schooling. Notice that in this case there are no employees with less than 10 years of schooling.

(6) P(not A and B) = probability of not possessing characteristic A but possessing characteristic B. Figure 4-10 illustrates this case, where the set of interest is all employees with 10 to 12 years of schooling who are not in their 20s. Notice in the figure that we have a few employees both below their 20s and above their 20s who have this amount of schooling.

(7) P(not A and not B) = probability of possessing neither characteristic. This probability is represented in our Venn diagram as the area inside the universal set but outside either set A or set B. Figure 4-11 illustrates this set. All employees in this group are either older than 29 or younger than 20 and have either less than 10 or more than 12 years of education.

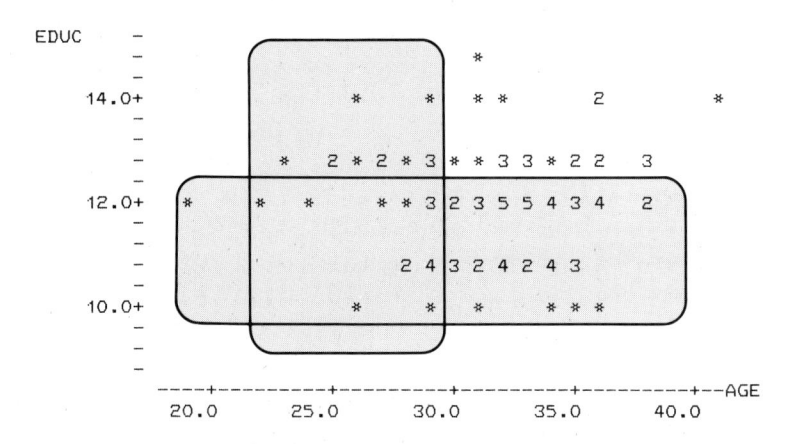

**FIGURE 4-8** Plot of Education and Age of Go-Fast Employees Indicating the Union of the Sets of Employees Aged 20 to 29 and of Employees with 10 to 12 Years of Education

**FIGURE 4-9** Plot of Education and Age of Go-Fast Employees Indicating All Employees Aged 20 to 29 Who Do Not Have from 10 to 12 Years of Education

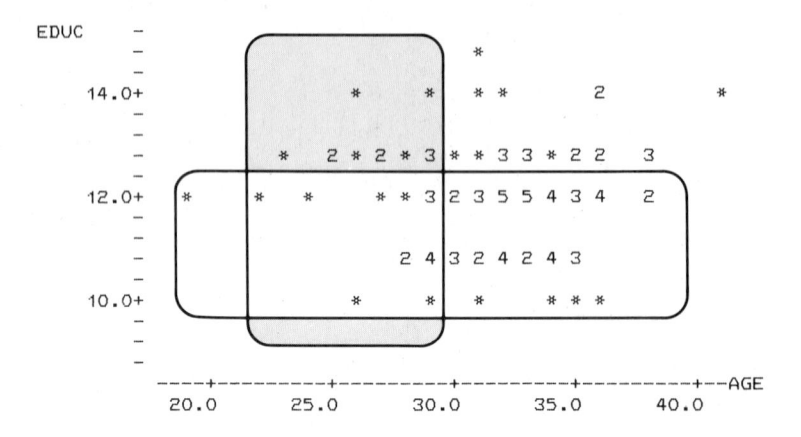

```
EDUC    -
        -
        -                                           *
14.0+                       *       *    * *     2           *
        -
        -           *    2 * 2 * 3 * * 3 3 * 2 2    3
        -
12.0+       *    *    *        * * 3 2 3 5 5 4 3 4    2
        -
        -                          2 4 3 2 4 2 4 3
        -
10.0+                  *          *    *         *  * *
        -
        ----+---------+---------+---------+---------+--AGE
          20.0      25.0      30.0      35.0      40.0
```

**FIGURE 4-10** Plot of Education and Age of Go-Fast Employees Indicating All Employees with 10 to 12 Years of Education Who Are Not Aged 20 to 29

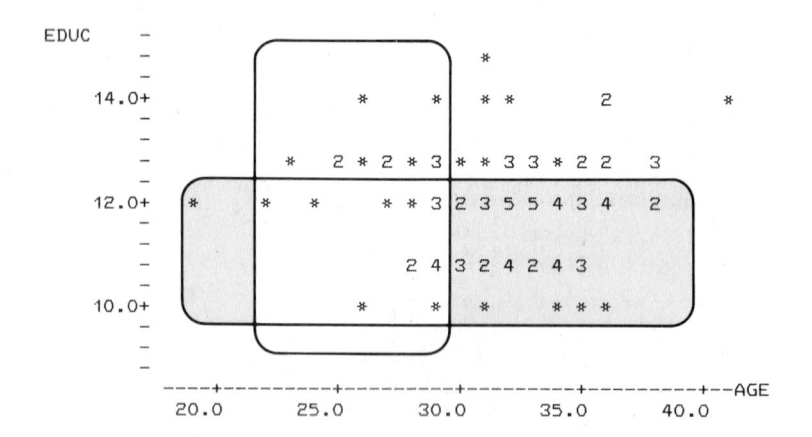

```
EDUC    -
        -
        -                                           *
14.0+                       *       *    * *     2           *
        -
        -           *    2 * 2 * 3 * * 3 3 * 2 2    3
        -
12.0+       *    *    *        * * 3 2 3 5 5 4 3 4    2
        -
        -                          2 4 3 2 4 2 4 3
        -
10.0+                  *          *    *         *  * *
        -
        ----+---------+---------+---------+---------+--AGE
          20.0      25.0      30.0      35.0      40.0
```

**(8)** $P(A \text{ or not } B)$ = probability of possessing characteristic $A$ or not possessing characteristic $B$. This is all the area of $A$ plus all the area outside $B$. Note that the area specified includes the intersection of $A$ and $B$. Figure 4-12 illustrates this probability.

**(9)** $P(\text{not } A \text{ or } B)$ = probability of possessing characteristic $B$ or not possessing characteristic $A$. This is all the area of $B$ plus all the area outside $A$. Note that the intersection of $A$ and $B$ is included. Figure 4-13 illustrates this probability.

**(10)** $P(\text{not } A \text{ or not } B)$ = probability of not possessing characteristic $A$ or not possessing characteristic $B$. This is the area outside the intersection of sets $A$ and $B$, as shown in Figure 4-14. Note in Figure 4-11 that the area that is not $A$ and not $B$ is shaded; the resulting probability is the area outside both sets. For the first portion of this stipulation, not $A$, all points not in $A$ qualify (including all points in $B$ that do not intersect set $A$). For the second portion of this stipulation, not $B$, all points not in $B$ (including all points in $A$ that do not

## SECTION 4·1 Some General Probability Concepts

**FIGURE 4·11** Plot of Education and Age of Go-Fast Employees Indicating All Employees Who Are Not Aged 20 to 29 and Who Do Not Have 10 to 12 Years of Education

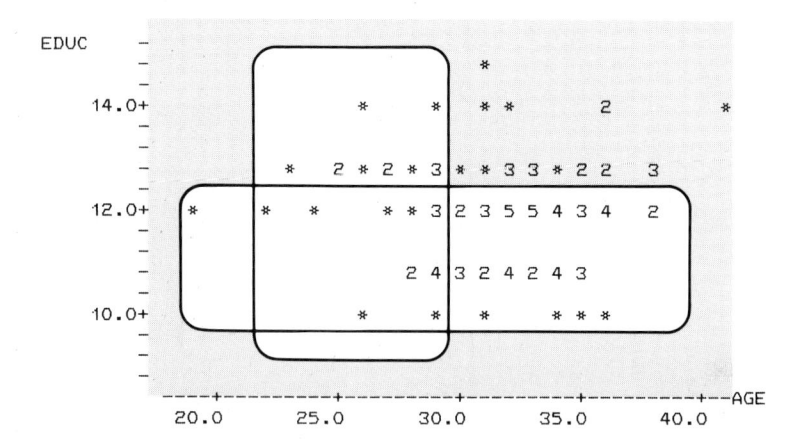

```
        EDUC    -
                -
         14.0+                          *
                -              *    *  * *        2
                -                                              *
                -      *   2 * 2 * 3 * * 3 3 * 2 2    3
                -
         12.0+    *      *   *       * * 3 2 3 5 5 4 3 4    2
                -
                -                       2 4 3 2 4 2 4 3
                -
         10.0+               *       *    *       * * *
                -
                -
                ----+---------+---------+---------+---------+--AGE
                   20.0      25.0      30.0      35.0      40.0
```

**FIGURE 4·12** Plot of Education and Age of Go-Fast Employees Indicating All Employees Who Are Aged 20 to 29 or Who Do Not Have 10 to 12 Years of Education

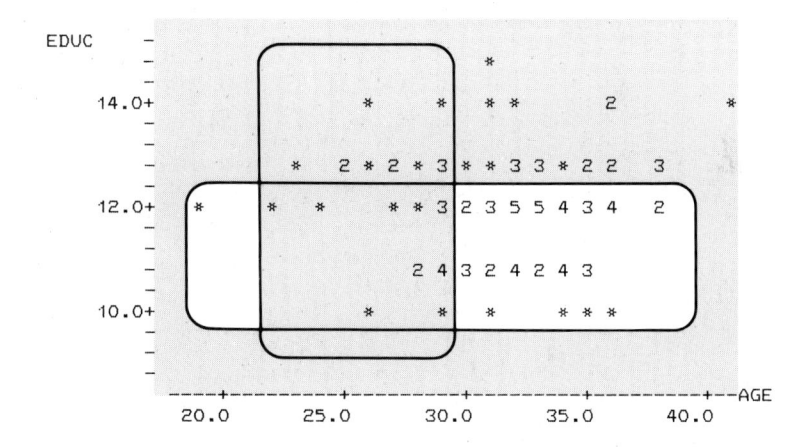

```
        EDUC    -
                -
         14.0+                          *
                -              *    *  * *        2
                -                                              *
                -      *   2 * 2 * 3 * * 3 3 * 2 2    3
                -
         12.0+    *      *   *       * * 3 2 3 5 5 4 3 4    2
                -
                -                       2 4 3 2 4 2 4 3
                -
         10.0+               *       *    *       * * *
                -
                -
                ----+---------+---------+---------+---------+--AGE
                   20.0      25.0      30.0      35.0      40.0
```

intersect $B$) qualify. The only points that do not qualify are those that are in both $A$ and $B$, namely, the intersection of the two sets.

**(11)** $P(A$ given $B) =$ probability of $A$ given the presence of characteristic $B$. Although this probability might seem like the others, it is very different. The difference is that we want to know the probability of $A$ given that we know that $B$ is already present. In other words, the universe from which we are drawing is not the universe of all possible employees (i.e., 100 employees) but the set $B$. Thus anyone not having characteristic $B$ is not part of the sample from which we are drawing.

In the Go-Fast data, this statement translates to the probability of an employee being of age 20 to 29 if he or she has between 10 to 12 years of education. The universal set here is the 66 employees with 10 to 12 years of education, and we are interested in drawing a random sample from that group of 66. In Figure 4-15, notice that we have eliminated all

**FIGURE 4-13** Plot of Education and Age of Go-Fast Employees Indicating All Employees Who Are Not Aged 20 to 29 or Who Have 10 to 12 Years of Education

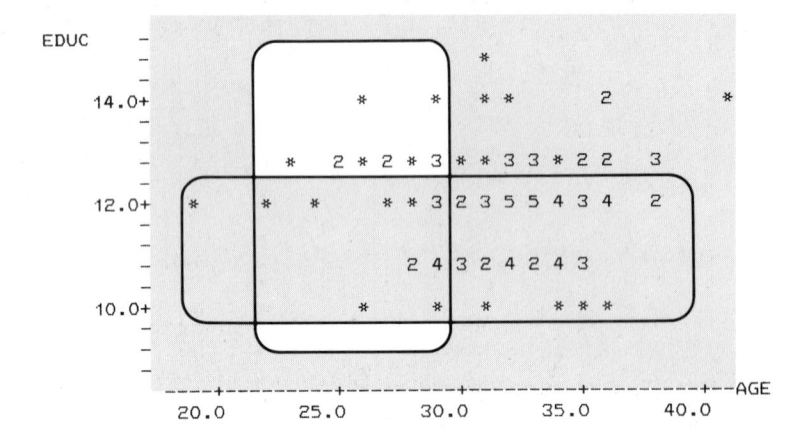

```
EDUC    -
        -
        -                                          *
  14.0+ -                *          *     * *              2                *
        -
        -                     * 2 * 2 * 3 * * 3 3 * 2 2     3
        -
  12.0+ -   *        *    *       * * 3 2 3 5 5 4 3 4      2
        -
        -                                2 4 3 2 4 2 4 3
        -
  10.0+ -                     *          *     *     * * *
        -
        -
        ----+---------+---------+---------+---------+--AGE
           20.0      25.0      30.0      35.0      40.0
```

**FIGURE 4-14** Plot of Education and Age of Go-Fast Employees Who Are Not Aged 20 to 29 or Who Do Not Have 10 to 12 Years of Education

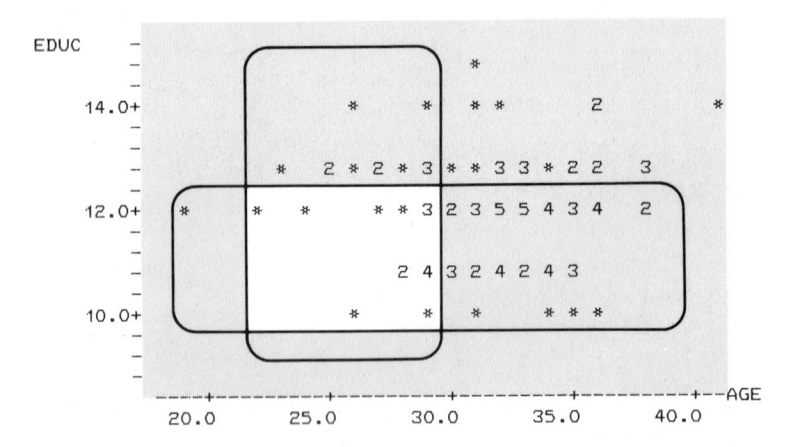

```
EDUC    -
        -
        -                                          *
  14.0+ -                *          *     * *              2                *
        -
        -                     * 2 * 2 * 3 * * 3 3 * 2 2     3
        -
  12.0+ -   *        *    *       * * 3 2 3 5 5 4 3 4      2
        -
        -                                2 4 3 2 4 2 4 3
        -
  10.0+ -                     *          *     *     * * *
        -
        -
        ----+---------+---------+---------+---------+--AGE
           20.0      25.0      30.0      35.0      40.0
```

observations not found in set *B* to start with. Instead of being 15 employees out of 100, the proper probability is 15 employees out of 66, or .23.

It is now time to go through the actual calculations of the various probabilities. We will continue to use the Go-Fast data; you can refer to the figures just presented to verify the values obtained from each formula.

## An Example of the Probability Axioms

There is a very important point to note before we begin the following example. Normally we do not have nice, neat populations of 100 in statistical analysis. Frequently our populations number in the thousands or more. However, if you understand how we obtain the probability values with a sample of 100, then exactly the same logic will allow you to apply

**SECTION 4·1** Some General Probability Concepts

**FIGURE 4-15** Plot of Go-Fast Employees with 10 to 12 Years of Education Who Are Aged 20 to 29

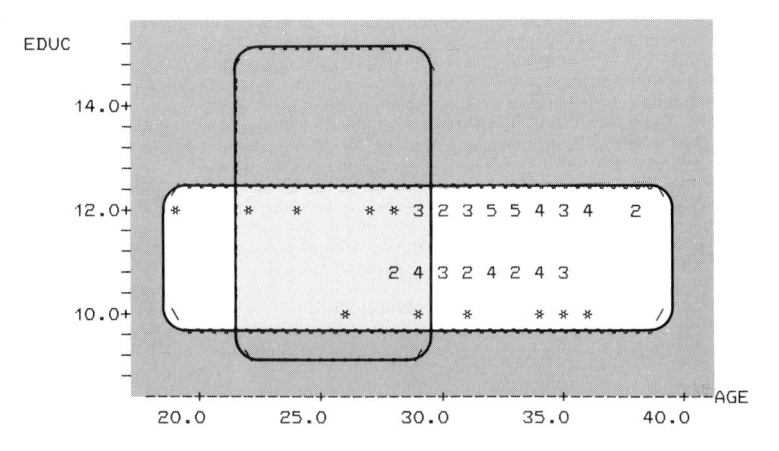

```
EDUC

 14.0+


 12.0+   *      *   *      * * 3 2 3 5 5 4 3 4    2

                              2 4 3 2 4 2 4 3
 10.0+ \           *         *    *     * * *    /
       +---------+---------+---------+---------+------AGE
      20.0      25.0      30.0      35.0      40.0
```

the same computational procedure to any large sample; you will not have to go and count the population to convince yourself that your calculations of the probability are correct.

We know the following about set A (Go-Fast employees aged 20 to 29) and set B (Go-Fast employees with 10 to 12 years of schooling):

$$P(A) = .27 \text{ (see Figure 4-4)}$$
$$P(B) = .66 \text{ (see Figure 4-5)}$$
$$P(A \text{ given } B) = .23 \text{ (see Figure 4-15)}$$

If we know $P(A$ given $B)$ as well as $P(A)$ and $P(B)$, then according to *Bayes' theorem* (named in honor of Rev. Thomas Bayes), we can compute $P(B$ given $A)$. Bayes' theorem states that the probability of $B$ given $A$ is equal to the probability of $B$ times the probability of $A$ given $B$ divided by the probability of $A$:

$$P(B \text{ given } A) = \frac{[P(B)] [P(A \text{ given } B)]}{P(A)}$$
$$= \frac{(.66)(.23)}{.27}$$
$$= .56$$

From our work so far, we also know the complements of $A$ and $B$, as follows:

$$P(\text{not } A) = 1 - P(A)$$
$$= 1 - .27$$
$$= .73$$
$$P(\text{not } B) = 1 - P(B)$$
$$= 1 - .66 = .34$$

Another axiom states that the probability of $A$ and $B$ is equal to the probability of $A$ times the probability of $B$ given $A$. This is called the *multiplication theorem*. The formula is as follows:

$$P(A \text{ and } B) = P(A)[P(B \text{ given } A)]$$
$$= \left(\frac{27}{100}\right)\left(\frac{15}{27}\right)$$
$$= (.27)(.56)$$
$$= .15$$

Another axiom states that the probability of A or B is equal to the probability of A plus the probability of B minus the probability of A and B. The reason for subtracting the area of A and B (the intersection) is that we have counted this portion of the probability space twice. If you look at Figure 4-6, the shaded area is counted in both the calculation of the probability of A and in the calculation of the probability of B. This rule is called the *addition theorem*. The formula for this computation is as follows:

$$P(A \text{ or } B) = P(A) + P(B) - P(A \text{ and } B)$$
$$= \frac{27}{100} + \frac{66}{100} - \frac{15}{100}$$
$$= .27 + .66 - .15$$
$$= .78$$

To compute the next axiom, we need to know $P(\text{not } B \text{ given } A)$, which is simply equal to $1 - P(B \text{ given } A)$. Since $P(B \text{ given } A) = 15/27$, then $P(\text{not } B \text{ given } A)$ is equal to $1 - 15/27$ or $12/27$. We now can compute the next axiom:

$$P(A \text{ and not } B) = P(A)[P(\text{not } B \text{ given } A)]$$
$$= \left(\frac{27}{100}\right)\left(\frac{12}{27}\right)$$
$$= (.27)(.44)$$
$$= .12$$

We present the remaining axioms without explanation:

$$P(\text{not } A \text{ and } B) = P(\text{not } A)[P(B \text{ given not } A)]$$
$$= \left(\frac{73}{100}\right)\left(\frac{51}{73}\right)$$
$$= (.73)(.70)$$
$$= .51$$

$$P(\text{not } A \text{ and not } B) = P(\text{not } A)[P(\text{not } B \text{ given not } A)]$$
$$= \left(\frac{73}{100}\right)\left(\frac{22}{73}\right)$$
$$= (.73)(.30)$$
$$= .22$$

$$P(A \text{ or not } B) = P(A) + P(\text{not } B) - P(A \text{ and not } B)$$
$$= \frac{27}{100} + \frac{34}{100} - \frac{12}{100}$$
$$= .27 + .34 - .12$$
$$= .49$$

# SECTION 4-1 Some General Probability Concepts

$$P(\text{not } A \text{ or } B) = P(\text{not } A) + P(B) - P(\text{not } A \text{ and } B)$$
$$= \frac{73}{100} + \frac{66}{100} - \frac{51}{100}$$
$$= .73 + .66 - .51$$
$$= .88$$

$$P(\text{not } A \text{ or not } B) = P(\text{not } A) + P(\text{not } B) - P(\text{not } A \text{ and not } B)$$
$$= \frac{73}{100} + \frac{34}{100} - \frac{22}{100}$$
$$= .73 + .34 - .22$$
$$= .85$$

## Statistical Independence

When working with statistical inference, we are often concerned with the issue of independence. This is the statistical correlate for the cause-and-effect phenomenon observed in science or logic. In statistical analysis, we cannot establish cause-and-effect relationships, but we can define the extent to which two or more phenomena occur together. In probability theory, such concurrences are defined by joint probabilities (intersections) and conditional probabilities.

One of the key questions concerns relationship versus independence. That is, "Is there a meaningful relationship between $A$ and $B$?" In terms of probability, the question is "Is $P(A)$ equal to $P(A \text{ given } B)$?" Thus, to determine whether $A$ and $B$ are independent, we calculate the probability of $A$, that is $P(A)$, and then the probability of $A$ given $B$, or $P(A \text{ given } B)$. If $P(A)$ does not equal $P(A \text{ given } B)$, then the two events are not statistically independent. If they are the same, then $A$ is said to be independent of $B$, and the two events are mutually exclusive. Figure 4-16 illustrates this case in a Venn diagram.

**FIGURE 4-16** Venn Diagram of the Universal Set (*U*) and Two Mutually Exclusive Events (*A* and *B*)

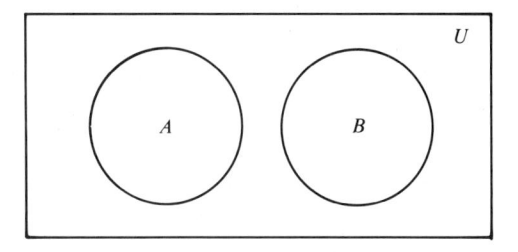

Suppose we wish to test the hypothesis that there is a relationship between sex and sales ability for sales personnel at Go-Fast. We would then take all the sales personnel (we will demonstrate later that we do not need to take all of them) and divide them into groups of successful and unsuccessful (according to some appropriate criterion) and into groups of male and female. A cross tabulation of this data produces Table 4-1.

**TABLE 4-1** Relationship between Sex and Sales Success for Go-Fast Employees

|        | Sales Ability |              |       |
|--------|---------------|--------------|-------|
| Sex    | Successful    | Unsuccessful | Total |
| Male   | 40%           | 30%          | 70%   |
| Female | 25            | 5            | 30    |
| Total  | 65%           | 35%          | 100%  |

We can now examine the data to see if there is a relationship between sales ability and sex by computing

$$P(\text{successful}) = 65\%$$

and

$$P(\text{successful given female}) = \frac{25}{30} = 83.33\%$$

Since these figures are not equal, sex and sales ability at Go-Fast are not statistically independent.

## Insufficient Reason

When revising probabilities because of some known event or characteristic, you must have sufficient reason.

Let us propose an example. You know that there are 4 aces in a deck of 52 playing cards. Thus the logical probability of selecting an ace randomly from a deck of cards is $4/52$, or $1/13$. Suppose that we remove half of the playing cards and discard them without ever looking at them. Now we have only 26 cards. What is the probability of selecting an ace? By the *principle of insufficient reason*, we would still assess the probability at $1/13$, because we do not have sufficient information to change it.

This example probably bothers you, so let's consider another. Suppose instead of discarding half the deck, you set half of the cards aside on the table and select from the remaining half. Clearly the probability of selecting an ace from the bottom half of the deck is the same as from the whole deck, $1/13$. Now simply ask yourself, "Does it make a difference if the top half of the deck is on the table or in the trash?" Of course it does not! From a managerial viewpoint, this observation has important implications. In making decisions, we must use the information available and not worry about what we do not know.

You can now calculate some basic probability sets and determine if two events are statistically independent. Although you cannot determine cause-and-effect based upon statistical inference, you have taken the first steps toward understanding the relationship between two variables. Now let's look at how management might use some concepts of probability in decision making.

## Decision Trees

We can make some practical applications of the basic concepts of probability theory. Frequently when one is working with probability concepts, a visual model of the relationships is very useful. One such visual model is called a *decision tree*. With such a model we can

## SECTION 4-1 Some General Probability Concepts

work on complex problems and keep track of how the various probability factors influence each other. Let's start with a very simple illustration.

Suppose your friend Joe wants to bet you about the outcome of a coin toss. We will assume the coin to be a fair one with a .5 probability of heads and .5 of tails. Joe's bet is as follows: If the outcome is heads, he pays you $1.00; if the outcome is tails, you pay him $1.00.

We can compute what is called the *expected monetary value* (EMV) of such a bet as follows: Multiply each outcome by the probability of that outcome. In this case the calculation is

$$P(\text{heads}) \times \text{outcome} + P(\text{tails}) \times \text{outcome} = \text{EMV}$$
$$(.5 \times \$1.00) + [.5 \times (-\$1.00)] = \$0.00$$

In this case, the EMV is zero. The EMV may take on any value from positive to negative, depending on the consequences of the bet. As a second example, let's assume you are rolling a die with Joe. If the die comes up 1, 2, 3, or 4, he pays you $1.00; however, if it comes up 5 or 6, you pay him $1.50. Since there are six possible outcomes, one way to calculate the EMV would be as follows:

$$\tfrac{1}{6} \times \$1.00 + \tfrac{1}{6} \times \$1.00 + \tfrac{1}{6} \times \$1.00$$
$$+ \tfrac{1}{6} \times \$1.00 + \tfrac{1}{6} \times (-\$1.50) + \tfrac{1}{6} \times (-\$1.50) = \$0.33$$

In this illustration, the EMV is $0.33. You can also see that if the bet was reversed—that is, you paid Joe $1.00 for a 1, 2, 3, or 4 and he paid you $1.50 for a 5 or 6—then the EMV would be −$0.33.

Notice that the two possible outcomes are gaining $1.00 or losing $1.50. You might well ask, "What does the $0.33 really represent?" The $0.33 is the expected result *per occurrence* if the bet was repeated a large number of times. If the bet was agreed upon and two rolls of the die made, the possible outcomes of two rolls would be for you to win $2.00, to lose $0.50, or to lose $3.00. The average outcome would be those amounts divided by 2. If you repeated this bet a large number of times, then the average result would be for you to win $0.33 per roll of the die. This is the EMV. Even though the theory requires a large number of repetitions of the experiment, we can apply the concept just as effectively to a single occurrence.

What the EMV tells us is that when the monetary values and probabilities of all the possible outcomes are taken into account, the net result is positive (or negative) by some amount. We can use this same logic to compare several outcomes. If two or more different choices are available, then the one with the highest EMV would be the most profitable. However, although the highest EMV is typically the best decision, this will not always be the case. We will examine such an example in just a little while.

Let's return to the coin toss bet with Joe and examine the example visually. We can do this with a decision tree, as shown in Figure 4-17. In a simple case with two possible outcomes, a decision tree is really not needed, but the concept can best be understood with a simple illustration. We will use a circle to illustrate a "chance mode." A chance mode is followed by the probabilities associated with it.

As you can see, this tree diagram represents both possible outcomes and their probabilities. Assuming that we have a fair and unbiased coin, we can go up the first leg of our tree, which is the probability of heads (.5) or down the second leg, which is the probability

**FIGURE 4-17**

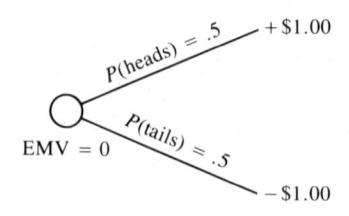

of tails (.5). At the end of each leg, we have the outcome of traveling along that path. If you multiply the probability on a given leg times the outcome along that leg, you obtain the value of that path. If you sum the monetary values from all possible paths on a given tree, you can obtain the EMV of that problem. As we have shown, the EMV of this tree is $0.00.

This simple decision tree can be extended to include decisions regarding strategies. For instance, we have the choice of either accepting Joe's wager or refusing to play. An expanded decision tree would then contain a "decision mode." A decision mode is represented by a rectangle. Our tree is now represented by Figure 4-18.

Having computed the expected return of the game, we can compare the EMV of each strategy. In this simple game, we would be indifferent to the two strategies. Thus, the determining factor would be our feelings about gambling and taking risks.

Recently while watching cartoons with his children on Saturday morning, one of us observed a game that comprises another good example. In each specially marked box of Crunchy Creatures cereal is a game card. If your game card says you are a winner, you get a neat plastic robot (worth about $10.00). If you lose, you have to eat the Crunchy Creatures cereal. Of course, to comply with the laws in many states, no purchase is necessary to play, so the makers of Crunchy Creatures will happily send you a game card if you mail them a stamped, self-addressed envelope. Just before returning to the show, the ad reveals that 1 out of every 35,000 cards is a winner.

Now, having been raised right, our author's kids wouldn't have touched Crunchy Creature cereal with a 10-foot pole, but they were interested in sending for a card game. Let's evaluate this problem. We need to add a "price tag" to our decision tree to indicate an action for which there is a specific price. In this case, there are two possible strategies. The first one is to do nothing and hope the kids forget by noon. The second is to send for a game

**FIGURE 4-18**

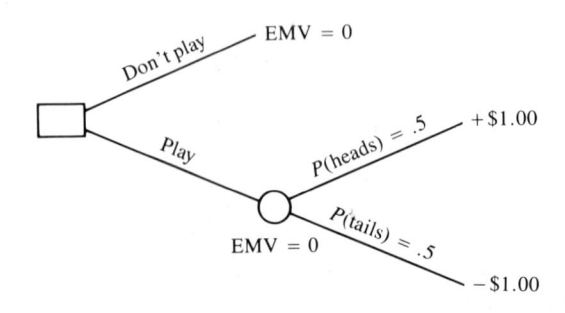

## SECTION 4·1 Some General Probability Concepts

**FIGURE 4-19**

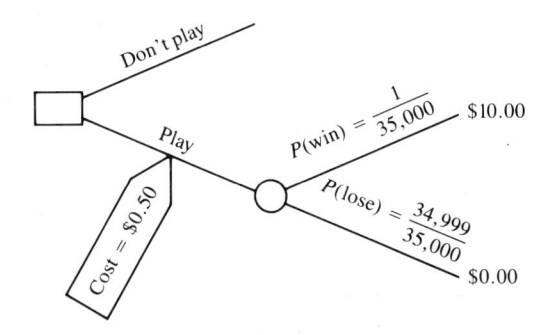

*[Decision tree: Don't play / Play (Cost = $0.50). Play leads to chance node with P(win) = 1/35,000 → $10.00 and P(lose) = 34,999/35,000 → $0.00]*

card. We will assume that sending for a game card costs $0.50. This cost is for two first-class stamps ($0.22 each) and two envelopes ($0.03 each). This cost is known with certainty since it will be incurred each time the kids send for a game card. Our decision tree is represented in Figure 4-19.

The payoff for the strategy of sending for a game card is

$$\frac{1}{35,000}(\$10) + \frac{34,999}{35,000}(\$0) = \frac{\$10}{35,000} = \$0.000286$$

We must now subtract the cost of playing the game from the expected return:

$$\$0.000285 - \$0.50 = -\$0.499715$$

Since the EMV is negative, it would be a poor decision to send for the game card. Our final decision outcomes are represented in Figure 4-20.

**FIGURE 4-20**

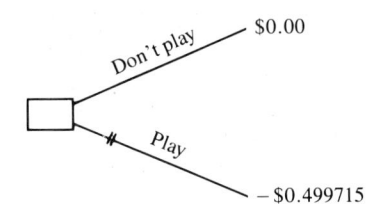

*[Decision tree: Don't play → $0.00; Play → −$0.499715]*

## Risk Aversion

In the real world of management decision making, one does not simply find the highest EMV and automatically assume that that decision path is the best one. For example, in our coin-flipping gamble, let's change the payoffs a bit and see what happens. Assume that you win $2.00 with heads and lose $1.00 with tails. The EMV is $0.50, and in general you will take the bet.

However, let's assume the bet is $201 if you win and $200 if you lose. The EMV is exactly the same, namely, $0.50. What if the bet were $20,001 if you win and $20,000 if you lose? The EMV would still be the same at $0.50. However, other concepts in management, such as risk, also come into play. While in a purely mathematical sense the EMV is positive and you should therefore take the bet, a manager might well decline the bet. A return of $1.00 on a $20,000 investment is not very attractive given the risk involved. Or a manager might consider the possibility of losing the first five coin tosses. Clearly, such considerations must also be taken into account.

Finally, there are some people and some companies that are simply more risk-averse than others. People who are very risk-averse would probably turn down the first bet. At the other extreme are individuals who seek risk. For them, the $20,000 bet might seem attractive.

The material just covered is basic to *decision theory*. Much more complicated decision trees with numerous branches and numerous possible outcomes per branch can be developed. In many cases, as the problems become more complex, a decision tree may be the only way to determine the possible outcomes and their consequences.

## 4-2 INITIAL CONSIDERATIONS OF SAMPLING

We must now turn our attention to sampling. *In the world of practical application, no probability estimate or EMV calculation is of value if the sample is improperly drawn.* In all of our discussions to this point, there has been an implicit assumption that every sample used is valid.

In inferential statistical analysis we always take a sample of some problem; from the results of analyzing the sample, we state some conclusions about the *population* from which the sample was drawn. By *population* we mean the complete set of possible occurrences having the characteristic that we wish to study. If we are talking about the employees of Go-Fast, then the population is all 100 employees and the sample that we have been using is the 20 selected for analysis. In different situations, the members of the population change. At a university, all students is one possible population; all part-time students is another. Other possible populations include all state residents, all night students, and all students over 30.

In statistical analysis, we will constantly be taking a sample of a population and then analyzing it. Two important factors must be kept in mind. First, you must be very sure about what group is being included in the population. What group do you really want to study? Second, in sampling there must be some type of representativeness within the sample of the members of the population.

The importance of representativeness cannot be overemphasized. If the sample is not representative of the population, then stop your analysis! It does not make any sense to proceed. No statistical techniques will be correct for a badly drawn sample. Nonrepresentative samples simply lead to inaccurate, erroneous, and misleading results. An old saying used by computer folks fits very well: "Garbage in, garbage out." The greatest care must be taken to ensure that a sample is representative.

Our basic strategy is simply to study a sample and then make some kind of statement about everyone in the population. For example, we might ask a group of students their

### SECTION 4-2 Initial Considerations of Sampling

opinion on a given issue and then, from analyzing their answers, draw a conclusion about the opinion of the general student body at the school. If such a conclusion is to be meaningful, the representativeness of the sample is essential.

In sampling, the fundamental sampling design is called a *simple random sample*. In a simple random sample, *each member of the population has an equal chance of being selected for the sample*. That requirement is normally very difficult to meet in actual practice. In our university example, if we stood outside the engineering building at noon and asked the first 50 students who walked by what they thought about a certain issue, would that survey sample be random? Our first question should be "What is the population?" If we are trying to sample all full-time students, then in all likelihood our sample is not random. Any student who did not pass in front of the engineering school at noon did not have a chance of being selected. For obvious reasons, our sample would probably have more engineering students. Would our sample be a random sample of engineering students? Probably not. Do all engineering students pass in front of the engineering building at noon? Some engineering students may have only morning classes and be gone by noon; others may have only afternoon classes. There are numerous reasons why all engineering students would not have a chance to be selected for our sample. Consequently, it is not a true random sample.

If you live in a medium-sized city such as Sacramento, California, and want to survey its population, how might you conduct such a survey? You probably cannot interview everyone in Sacramento. How would you take a simple random sample? In all honesty, you probably cannot do so. In theory, you would need a list of all residents; then, using a random numbers table or generator, you would select your sample. You would then find those people and ask them your question. Such a procedure would result in a pure random sample. But is such a complete list of all Sacramento residents available? A list that is complete and accurate on one day would not be so the next, because a few people would die, babies would be born, some people would move into the city, and some people would leave. Also, what constitutes the population of "the people of Sacramento"? In the previous sentence, we mentioned newborns. You obviously cannot ask them any questions (or at least they probably will not answer). Who, then, constitutes your population? People from 10 to 75? People from 21 to 100? People over 40? Each of these populations is different, and probably none of them can be sampled in a true simple random sample.

An alternative way to sample Sacramento is to use the telephone book, auto registrations, or similar sources as a list of Sacramento residents. This is a practical solution to a theoretical problem and is often done. However, it is important to realize that using the telephone book will systematically exclude two sections of the population. First, the very poor of a community do not have telephones and cannot be selected. Also, many people have unlisted numbers and they, too, cannot be selected. The absence of these groups will introduce *bias*, or distortion, into your sample. Thus you can use the telephone book and take a sample of "people listed in the telephone book" as your population, but such a sample is not a sample of the population of a city.

In addition, size is not a cure for bad sampling. A sample of 1,000 is not more accurate than a sample of 100 simply because it is bigger. A sample of 1,000 students in front of the engineering building at a large university is not a better sample of the "population of university students" than a sample of 100 students.

There are classic examples of very large samples and sophisticated techniques going astray. The best-known was the attempt to predict the 1936 election. *Literary Digest* magazine mailed out 10 million postcard ballots and received some 2 million ballots back. Based upon these 2 million returns (certainly a very large sample), it predicted that Roosevelt would lose the election. However, the poll had an error of 19 percentage points and Roosevelt won handily. Why? The *Digest* obtained its mailing lists from telephone books, auto registrations, and similar sources. In 1936, such sources typically reflected only upper-income voters. Although the poll properly reflected how that segment of society would vote, the poll was not representative of the general population of voters.

There is a second type of sampling error other than nonrepresentativeness that is also very common. This error concerns on what the opinions of respondents to a survey are based. For example, on May 3, 1983, the U.S. Catholic Bishops Conference passed a pastoral letter on nuclear weapons and war. The final document was approximately 175 pages in length. From May 15 to May 20 a research organization polled 602 adults in Nebraska by telephone on their opinion of the pastoral letter. A three-sentence introduction concluded with the question "Do you agree or disagree with the bishop's letter?"

Let us consider the results of this survey. First, the letter was so complex that a subcommittee was established to prepare a summary statement. In mid-June a 2,700-word summary of the 44,000-word letter was completed and approved by the bishops. That summary was distributed to Nebraskans on June 17, at which time the full text was still not generally available. Notice that these dates were about a month after the survey was taken. Unless you assume that the 602 adult Nebraskans who were polled all read the preliminary draft, then we must ask, "What question are the polled Nebraskans really answering when they respond to this survey?" Possible answers include that (1) they were responding to TV news reports, (2) they were responding to the three-sentence summary offered by the poll taker, or (3) they were responding to the full draft, which they read in the context of the discussion and dialogue held by the bishops. The news reports, of course, concentrated on the controversial and newsworthy portion of the pastoral letter. The poll taker had to make the question short enough to be understood in a telephone interview. Finally, the full draft of the pastoral letter was not available at the time of the survey.

With these issues in mind, what can you do with the results of such a survey? It would seem that, at best, respondents to the survey based their opinions on the publicity about the letter. If you assume that all such publicity was unbiased, then the results of the survey reflected accurate feelings about the letter. However, if you believe that the news reports concentrated on certain facets of the pastoral letter and excluded other facets, it seems most likely that the survey results indicated responses to the pastoral's "advertising" rather than to the pastoral letter itself. (Results of the survey were reported in a copyrighted *Omaha World-Herald* newspaper story, May 28, 1983, p. 1.)

A final problem with surveys concerns nonrespondents. If you send out 500 surveys and get 100 back, there is a tendency to analyze the 100 surveys received and then make conclusions about the population of your study. But what about the 80% who did not respond? They have in a sense told you something by not responding. Are there differences between those who did respond and those who did not? The response rate to any volunteer mail survey tends to be quite low. This should always be of concern. When magazines survey their readership and, on the basis of the results, try to conclude something about the general population, this same kind of question must be asked. Is the readership a legitimate sample of the general population? Would you expect the readership of *Mother Earth News* and *The American Rifleman* to respond the same to questions about military defense? Is the sexual behavior of the readership of *Playboy* or *Playgirl* representative of the sexual behavior of the general public? It seems obvious that the readers of a sexually oriented magazine would have a high interest in sex (probably higher than the general public). Similarly, it seems likely that the readership of *American Sportsman* has a greater interest in firearms and hunting than does the general population. Such interest is exactly why special interest magazines exist and prosper.

Another problem arises when surveying such interest-oriented groups. Not all *Playboy* readers return *Playboy* surveys, and not all *Psychology Today* readers return their surveys. Thus, even within such selected samples, there are still the biases of who does and who does not respond to a survey. In almost any area (but especially in personal areas such as love and sex), only the most interested of an already very interested group will take the time and energy to respond to a survey. Thus, if 100 people responded out of 500 surveys sent, there is a good possibility that those 100 responses cluster at one end of the spectrum. Not only are your responses from people with a high interest in the subject, but they are people with high interest among a readership with high interest. The danger is to take the results of such a survey and conclude that since this characteristic or that trait is found in 27% of the readership, then that same trait will be found in 27% of the general population. Such conclusions are ludicrous.

We will have a more detailed analysis of proper sampling procedures in a later chapter. For the present, it is important to be alert to the potential problems. When someone tells you that they have drawn a random sample and, based upon that sample, have concluded certain things, there is always reason to question the legitimacy of their sample. They may have done a very fine job, and their results may be accurate. However, if someone tells you that they stood on a street corner and asked the first 50 people who walked by what they thought, the results are likely to contain a significant error when applied to the general population.

## 4-3 SURVEY SAMPLING

So far we have limited our discussion of sampling to simple random sampling. Further, we have stated that when you use various subgroups (e.g., magazine readers) as a sample, there is the possibility of introducing great error into your data. However, sometimes we can use subgroups and reduce possible error by using more efficient sampling techniques. That is not to say that we will abandon random sampling, but that we will use it more intelligently.

### Cluster Sampling

Our first alternative sampling technique is *cluster sampling*. Let's assume that we have been assigned the task of gathering public opinion on a political issue. By its nature, the survey must be conducted in person. If we use simple random sampling (let's assume a sample size of 300), we might have to drive all over the city. In contrast, using cluster sampling, we could divide the city into blocks of 30 people each (our clusters) and then randomly select 10 of those blocks to interview. As long as we did not introduce a systematic bias in our selection of clusters, we could use the cluster sampling technique and reduce costs and the time involved.

## Stratified Sampling

*Stratified sampling* is probably one of the most popular methods used today. This second alternative sampling technique can also save cost and time as well as get information that better suits our needs. In our political survey, we might suspect that respondents will align themselves with other members of their political parties. If we have the following data on the distribution of party members:

| | |
|---|---|
| Republicans | 30% |
| Democrats | 35% |
| independents | 10% |
| not registered | 25% |

we might sample proportionately. Assuming that we want a sample of 300, we would randomly sample within each group until we had the following number of individuals in each group:

| | |
|---|---|
| Republicans | 90 |
| Democrats | 105 |
| independents | 30 |
| not registered | 75 |

If our theory of party alignment is right, we will eliminate a lot of variation from our sample. As we will see in subsequent chapters, a reduction in variation will allow us to make more accurate estimates. Thus, this technique allows us to use sample information more efficiently.

Whenever we have reason for wanting a sample to conform to other relationships, such as the percentages with certain political party affiliations or in certain age groups, a stratified sample might be effective. Because we get the appropriate number of individuals for a given group, we can skip interviewing any more members of that group and thus save time and money.

These are only two of many alternative ways to draw samples and still work within the concept of a random sample. There are much more sophisticated techniques such as multistage sampling, where you stratify by political party and then stratify within each political party by, say, age group. Further discussion, however, is beyond the scope of this book. Go to your library for further information on sampling theory. If this area is of special interest to you, consider a complete course on sampling theory. Clearly, our culture has become inundated with polls in which samples are drawn using numerous strata and cluster techniques.

## SUMMARY

In this chapter we have covered the basic theory of probability. An understanding of the basic concepts of probability is essential to function effectively in today's business environment. We can analyze numerous samples of products or manufacturing processes and then use data from such samples to draw conclusions about the production process. Further, numerous business products succeed or fail because of market research on potential cus-

# DISCUSSION QUESTIONS

1. What is the difference between inductive and deductive reasoning? Do you think statistics employs inductive reasoning, deductive reasoning, or both?
2. When taking a sample of a population, what must you consider?
3. What does a simple random sample entail?
4. Is a simple random sample from your city's telephone book a sample of the city's population? Why or why not?
5. Is a large sample drawn from the readership of a special interest magazine representative of the population of the United States? Why or why not?

# PROBLEMS

**4-1** First City Bank of Los Angeles found that of its customers, 60% are female and 52% make deposits. Of the females that come to First City Bank, 64% make deposits.

a) Given that a person is making a deposit, what is the probability that the person is female?

b) What is the probability that a customer is female and is making a deposit?

c) What is the probability that a customer is female or is making a deposit or both?

d) First City Bank is thinking of giving away free gifts to depositors. Should it use a gift that appeals to females or males? Why?

**4-2** Of people filling prescriptions at a local pharmacy, 72% are over 65 years old. Of the people over 65, 75% live within 10 miles of the pharmacy. Of all customers of the pharmacy, 82% live within 10 miles.

a) If a customer walks into the pharmacy, given that he lives within 10 miles, what is the probability that he is over 65?

b) If a second customer walks into the pharmacy, given that he lives within 10 miles, what is the probability that he is over 65?

c) What is the probability that a customer is over 65 and does not live within 10 miles of the pharmacy?

d) The owner of the pharmacy is considering making home deliveries to people over 65 who live within 10 miles. Out of 100 prescriptions, how many would not have to be delivered?

**4-3** James' Garden House has a 81% success rate when starting plants from clippings. If a clipping is first dipped in root starter, it has a 91% success rate. Often the Garden House runs out of root starter, so it is only able to use it 72% of the time.

a) Given that a clipping started to grow, what is the probability that it was dipped in root starter?
b) What is the probability that a clipping was not dipped in root starter or was successful?
c) Given that it was not dipped in root starter, what is the probability that a clipping started to grow?
d) Should James use root starter all the time? If root starter were used all the time, what would happen to the overall success rate?

4-4 Robb Good, a professional baseball player, hit a home run 59% of the time. Seventy-six percent of the time the team played a game, the sun was shining.
a) If hitting a home run and having the sun shine are independent events, what is the probability that Robb will hit a home run given that the sun is shining?
b) Explain the meaning of independent events.

4-5 Oysters produce pearls of various shapes and sizes. From a sample of 100, the following data was collected:

|       | Shape |     |       |
|-------|-------|-----|-------|
| Color | Round | Odd | Total |
| White | 62    | 0   | 62    |
| Blue  | 30    | 8   | 38    |
| Total | 92    | 8   | 100   |

a) What is the probability that a pearl is white?
b) What is the probability that a pearl is blue?
c) Given that a pearl is blue, what is the probability that it is oddly shaped?
d) If a jeweler has 575 blue pearls, how many should be oddly shaped?

4-6 Many Americans travel abroad to purchase European cars. Americans save 6% to 8% off the sticker price by buying a European car in Europe. Following is a sample of Americans who purchased cars in Europe and their respective annual income and age:

| American | Annual Income | Age |
|----------|---------------|-----|
| 1        | $ 30,000      | 60  |
| 2        | 42,000        | 55  |
| 3        | 64,000        | 48  |
| 4        | 66,000        | 38  |
| 5        | 55,000        | 46  |
| 6        | 39,000        | 41  |
| 7        | 92,000        | 52  |
| 8        | 46,000        | 49  |
| 9        | 54,000        | 41  |
| 10       | 61,000        | 56  |
| 11       | 88,000        | 52  |
| 12       | 60,000        | 49  |
| 13       | 61,000        | 51  |
| 14       | 140,000       | 57  |
| 15       | 68,000        | 46  |

*(continued on next page)*

## PROBLEMS

| American | Annual Income | Age |
|---|---|---|
| 16 | 62,000 | 48 |
| 17 | 61,000 | 42 |
| 18 | 200,000 | 53 |
| 19 | 69,000 | 43 |
| 20 | 72,000 | 60 |
| 21 | 81,000 | 59 |
| 22 | 44,000 | 36 |
| 23 | 46,000 | 41 |
| 24 | 65,000 | 47 |
| 25 | 31,000 | 32 |

**a)** What is the probability that an American purchasing a car in Europe earns between $60,000 and $69,000 a year?

**b)** What is the probability that the same American is between the ages of 40 and 49?

**c)** Given that the American earns between $60,000 and $69,000 a year, what is the probability that he or she is between 40 and 49 years of age?

**d)** Given that the American is between 40 and 49 years of age, what is the probability that he or she earns between $60,000 and $69,000?

**e)** What is the probability that the American is earning between $60,000 and $69,000 a year and is between the ages of 40 and 49?

**f)** A dealer would like to gear an advertising campaign to Americans who earn $60,000 or more and who are 40 or older. What percentage of Americans purchasing cars abroad fall into this category? Would the dealer appeal to a substantial number of Americans purchasing cars abroad?

**4-7** Recently managers of Farm Fresh Grocery noticed that there were a large number of cartons delivered that contained broken eggs. The managers wished to know which of the companies, Farmer Joe or Garden Grown, was sending more damaged cartons. Following are data on a sample of cartons received by Farm Fresh:

| Carton | Damaged | Company |
|---|---|---|
| 1 | Yes | Farmer Joe |
| 2 | No | Garden Grown |
| 3 | No | Garden Grown |
| 4 | No | Garden Grown |
| 5 | Yes | Garden Grown |
| 6 | Yes | Farmer Joe |
| 7 | No | Farmer Joe |
| 8 | No | Farmer Joe |
| 9 | No | Garden Grown |
| 10 | Yes | Farmer Joe |
| 11 | No | Garden Grown |
| 12 | No | Garden Grown |
| 13 | Yes | Farmer Joe |
| 14 | Yes | Farmer Joe |
| 15 | Yes | Garden Grown |

*(continued on next page)*

| Carton | Damaged | Company |
|--------|---------|---------|
| 16 | No | Garden Grown |
| 17 | Yes | Farmer Joe |
| 18 | Yes | Farmer Joe |
| 19 | Yes | Garden Grown |
| 20 | No | Garden Grown |

**a)** Given that Farm Fresh Grocery received a damaged carton of eggs, what is the probability it came from Garden Grown? From Farmer Joe?

**b)** Which company should Farm Fresh get rid of?

**4-8** Inflight Airlines sampled their customer reservations and found that 15% resulted in no-shows. Of the reservations sampled, 45% were booked through travel agents and 55% were booked by customers who called Inflight directly. Four percent of the reservations confirmed through travel agents resulted in no-shows, and 22% of the reservations confirmed by customers resulted in no-shows.

**a)** What is the probability that a reservation was from a travel agent, given that it resulted in a no-show?

**b)** What is the probability that a reservation was booked by an individual, given that it resulted in a no-show?

**c)** What is the probability that a reservation was both a no-show and booked through a travel agent?

**d)** What is the probability that a reservation was both a no-show and booked by an individual?

**e)** What is the probability that a reservation was booked by a travel agent and was not a no-show?

**f)** What is the probability that a reservation was booked by an individual and was not a no-show?

**g)** If Inflight wanted to reduce the number of no-shows, what should it do? Why?

**4-9** Refer to Problem 2-6.

**a)** For Doctor Kage's 9:00 class, what is the probability that a student is between the ages of 18 and 21 given that he received a test score of 40 or better? For the 12:00 class?

**b)** For the 9:00 class, what is the probability that a student received a score between 30 and 39 given that he is between 22 and 26 years old? For the 12:00 class?

**c)** From your answers above, what can you say about each of the classes?

**4-10** A random sample of 200 newborns was taken at St. Mary's Hospital to study eye color. Here are the results.

| Sex | Blue | Brown | Green | Total |
|-----|------|-------|-------|-------|
| Female | 26 | 65 | 9 | 100 |
| Male | 20 | 59 | 21 | 100 |
| Total | 46 | 124 | 30 | 200 |

**a)** What is the probability that a newborn is a girl? A boy?

**b)** Given that a newborn is a girl, what is the probability that she has blue eyes? Brown eyes? Green eyes?

# PROBLEMS

c) Given that a newborn is a boy, what is the probability that he has blue eyes? Brown eyes? Green eyes?

**4-11** The Senate voted on the 1985 budget. The results are as follows:

| Party | For | Against | Total |
|---|---|---|---|
| Democratic | 32 | 8 | 40 |
| Republican | 41 | 19 | 60 |
| Total | 73 | 27 | 100 |

a) If a senator is a Democrat, what is the probability that he voted in favor of the 1985 budget?
b) If a senator is a Republican, what is the probability that he voted in favor of the 1985 budget?
c) What is the probability that a senator voted against the budget and was not a Democrat?
d) What is the probability that a senator voted against the budget and was not a Republican?
e) If the Republicans and the Democrats each had needed a two-thirds majority within their party to pass the budget, would the budget have passed?

**4-12** The manager of College Grove Apartments is reluctant to rent to college freshmen because he believes that they are disruptive. Of the residents at College Grove, 22% are freshmen. Of the 22%, 77% have had at least one complaint filed against their behavior. The residents of College Grove have been found to be disruptive 29% of the time.
a) Given that a resident is disruptive, what is the probability that the individual is a freshman?
b) What is the probability that a resident is disruptive and is a freshman?
c) What is the probability that a resident is disruptive and is not a freshman?
d) Do you feel that the manager's bias is justified? Why?

**4-13** A sample of 300 bulbs was taken from the flower beds of a small park outside a city and then planted. The results follow:

|  | Tulip | Daffodil | Crocus | Total |
|---|---|---|---|---|
| Bloomed | 89 | 75 | 92 | 256 |
| Did not bloom | 11 | 25 | 8 | 44 |
| Total | 100 | 100 | 100 | 300 |

a) Given that you have a sample bulb, what is the probability it is a tulip? A daffodil? A crocus?
b) Given that a bulb did not bloom, what is the probability that it was a tulip? A daffodil? A crocus?
c) Given that a bulb bloomed, what is the probability that it was a tulip? A daffodil? A crocus?
d) Which bulb would a gardener prefer to plant?

**4-14** Delectables, a gourmet fast-food restaurant, obtained the following information from a random sample of its customers:

| Distance to Work (miles) | Clerical | Managerial | Professional | Total |
|---|---|---|---|---|
| Less than 6 | 112 | 3 | 8 | 123 |
| 6–10 | 8 | 57 | 12 | 77 |
| Total | 120 | 60 | 20 | 200 |

Occupation Type

**a)** Given that a customer works within 6 miles of the restaurant, what is the probability that the individual's occupation is clerical? Managerial? Professional?

**b)** Given that a customer works between 6 and 10 miles away, what is the probability that the individual's occupation is clerical? Managerial? Professional?

**c)** If Delectables wants to target the group that has the largest percentage working within 6 miles, on whom should it focus?

**4-15** A T-shirt manufacturer did a quality check on her products. She found that 12% were defective, and of these 12%, 94% resulted from threading the machine improperly. The machines were threaded incorrectly 18% of the time.

**a)** Given that a machine was improperly threaded, what is the probability that a garment was defective?

**b)** What is the probability that a machine was improperly threaded and that a garment was defective?

**c)** What is the difference between the probability asked for in (a) and the probability asked for in (b)?

**4-16** Kelly and Jon were left outside of the store while their mother went shopping. To amuse themselves, they decided to play a game. For every passerby, Kelly won $1.00 if it was a woman, and Jon won $2.00 if it was a man.

**a)** If 4 out of every 5 passersby were women, draw a decision tree for each child.

**b)** What was the EMV for Kelly?

**c)** What was the EMV for Jon?

**d)** Who had the advantage in this game?

**4-17** You are in an adventurous mood and are trying to decide whether to go hunting for old bottles. The probability of finding a bottle worth $100 is 1 in 500. The equipment and transportation costs associated with looking for bottles is $20.

**a)** Draw the decision tree.

**b)** What is your EMV for each decision?

**c)** Should you go?

**4-18** Randy needs help in deciding whether to attend the senior prom. If he goes, he must pay $100, which includes tuxedo rental and flowers for his date. Randy would be one of the attendants in the prom court. He has a 20% chance of being chosen prom king. If he is chosen, he receives $150 as a gift.

**a)** Draw Randy's decision tree.

**b)** What is the EMV for each decision?

**c)** Will Randy come out ahead if he attends?

**4-19** Donald loves to bet on horse races. If he goes to the racetrack and bets on a horse named Las Vegas, Las Vegas has a 35% chance of winning. If Las Vegas wins, Donald will win $200. If the horse loses, Donald will lose $100. Admission to the race track is $4.

**a)** Draw Donald's decision tree.

# PROBLEMS

**b)** What is his EMV for each decision?

**c)** What does the EMV tell you?

**4-20** A contractor must decide which of three projects to bid on. Following are the possible returns associated with each project and their respective probabilities.

| Project | Gain | Break-even | Loss |
|---|---|---|---|
| 1 | $45,000 (.45) | 0 (.35) | −$12,000 (.20) |
| 2 | 25,000 (.55) | 0 (.20) | − 7,000 (.25) |
| 3 | 75,000 (.65) | 0 (.02) | − 27,000 (.33) |

**a)** If the cost associated with each project is $5,000, draw the decision tree.

**b)** What is the EMV for each project?

**c)** Which project should the contractor bid on?

**4-21** You recently inherited a large sum of money from your rich aunt. Because you are attending college, you have decided to invest that money in real estate and to sell the property upon graduation. You've narrowed your choice of properties down to four. The following data shows the expected return and the probabilities for each property if sold after four years. Because of taxes and upkeep costs, it is possible to lose more than the original cost of the property.

| Property | Cost | Return Gain | Break-even | Loss |
|---|---|---|---|---|
| 1 | $12,000 | $76,000 (.38) | 0 (.42) | −$22,000 (.20) |
| 2 | 11,000 | 25,000 (.60) | 0 (.25) | − 16,000 (.15) |
| 3 | 16,000 | 52,000 (.45) | 0 (.03) | − 50,000 (.52) |
| 4 | 6,000 | 18,000 (.62) | 0 (.14) | − 4,000 (.24) |

**a)** Draw your decision tree.

**b)** Which property would you most likely purchase?

**c)** Given an unlimited amount of money, which properties would you purchase? Why?

**4-22** The owner of Go-Fast has recently found that the company has excess capacity. Because of this, he is considering the manufacture of three new products: bowling shoes, tennis shoes, and racquetball shoes. All three shoes would use the equipment presently in use and could thus be produced without any additional costs. A marketing study yielded the following probabilities for each type of shoe:

| Type of Shoe | Success | Break-even | Failure |
|---|---|---|---|
| Bowling | .5 | .2 | .3 |
| Tennis | .3 | .4 | .3 |
| Racquetball | .4 | .2 | .4 |

Given the costs of production and marketing, Go-Fast feels that the dollar amounts associated with each possible outcome would be as follows:

| Type of Shoe | Success | Break-even | Failure |
|---|---|---|---|
| Bowling | $60,000 | 0 | − $20,000 |
| Tennis | 80,000 | 0 | − 30,000 |
| Racquetball | 30,000 | 0 | − 5,000 |

a) Develop a decision tree for this problem. Include the option of doing nothing.
b) What is the EMV of each investment possibility?
c) If factory space and machine time were unlimited, which of the possible investments would you undertake?

4-23 In Problem 4-22, three options for investment were available. Assume that it will be necessary to spend $100,000 over 5 years, at the rate of $20,000 per year, to set up the equipment necessary to produce each type of shoe.
a) Do any of the investments remain attractive?
b) Which investment would you consider making? What other factors might you wish to consider besides a straight EMV analysis?

# **P**ROBABILITY DISTRIBUTIONS

*The following Minitab commands will be introduced in this chapter:*

```
RANDom K observations, put into C,...,C;
```
A command to generate a random set of numbers based on a specified distribution
```
PDF for values in C [put results in C];
```
A command to generate a probability density function based on a specified distribution
```
CDF for values in C [put results in C];
```
A command to generate a cumulative density function based on a specified distribution
Subcommands for each of the above commands:
```
  BINOmial with n = K and p = K.
  POISson with mean = K.
  BERNoulli with p = K.
  INTEgers with uniform on K to K.
  DISCrete with values in C, probabilities in C.
  NORMal with mu = K, sigma = K.
  UNIForm with continuous uniform on K to K.
  PARSum of data in C, put in C
```
A command to calculate the partial sums of a column
```
⎰STORe
⎱EXECute
```
A pair of commands for executing a repetitive series of commands

In Chapter 4, we examined the general principles of probability theory. We also discussed the necessity of drawing a valid sample for the laws of probability to apply. However, we discussed the probability calculations for drawing only one observation, such as the probability of drawing a single worker from Go-Fast who had between 10 and 12 years of education and was in his or her 20s. In this chapter, we will see how these principles apply to sampling more than one observation. In other words, if we draw a sample of 20 employees, what are the characteristics of that sample? Will that sample tend to follow any type of pattern? Because we know the probabilities for a specific event, it seems reasonable to assume that we can predict something about more than one observation. In the course of this brief survey, we will introduce several well-known and useful probability distributions as well as the central limit theorem, which is a basic tenet of statistical theory.

## 5-1 WHAT IS A PROBABILITY DISTRIBUTION?

Until now we have drawn individual samples from what is called a Bernoulli population. A *Bernoulli population* is a population divided into two groups. In such a population, each observation either has or does not have the characteristic of interest. For example, Go-Fast employees can be thought of as either being in their 20s or not being in their 20s. Thus, the same population can be treated as dichotomous for one statistical purpose but not for another. For example, we can also think of Go-Fast employees as being in their 20s, 30s, 40s, and so on. A binomial population is always dichotomous. Continuous distributions can have a wide variety of shapes such as normal, uniform, Poisson, and chi-square.

### Minitab Random Variate Generators

If we want to generate a sample from any type of distribution, the general command form in Minitab remains the same. Each generator is broken into two parts. In the first line, which is the same for all generators, you specify a numbers generator, how many observations you wish to generate, and in which column you wish to store the values generated. The second line specifies which generator to use and the characteristics required for that particular generator.

The command and subcommand for using the random number generator for a normal distribution is

```
RANDom K observations, put into C;
   NORMal with mu = K, sigma = K.
```

Be sure to use the semicolon at the end of the first line and the period at the end of the second. This command looks more complicated than it actually is. As an illustration, if we want a sample of size 30 to be placed in column 1 and we want that sample to be drawn from a normal population with mean (mu) equal to 100 and a standard deviation (sigma) equal to 10, the commands will be

```
MTB >RANDom 30 observations, put into C1;
SUBC> NORMal with mu = 100, sigma = 10.
```

This command can also be entered in shorter form as

```
MTB >RAND 30, C1;
SUBC> NORM 100, 10.
```

## SECTION 5-1 What Is a Probability Distribution?

Minitab will generate 30 observations from a uniform population with a range from 0 to 1 if you do not specify the subcommand (`NORM 100, 10.`) and just enter `RAND 30 C1` by itself. The reason for specifying the normal distribution in the subcommand is that we might instead want to generate a Poisson, binomial, or uniform distribution. To generate any of these other distributions, you use the same command format and simply change the subcommand line to reflect the distribution desired and the specific characteristics of that distribution.

Before returning to the specific statistical issue at hand, we should list all of the generators. As the discussion continues, you can refer to this list for the format used for each generator. These formats are as follows:

```
RANDom K observations, put into C,...,C;
  BINOmial with n = K and p = K.
  POISson with mean = K.
  BERNoulli with p = K.
  INTEgers with uniform on K to K.
  DISCrete with values in C, probabilities in C.
  NORMal with mu = K, sigma = K.
  UNIForm with continuous uniform on K to K.
```

We have already shown you the use of the normal distribution generator with the illustration

```
MTB >RANDom 30 observations, put into C1;
SUBC> NORMal with mu = 100, sigma = 10.
```

To place 30 observations in each of the first 20 columns, the command would be

```
MTB >RANDom 30 observations, put into C1-C20;
SUBC> NORMal with mu = 100, sigma = 10.
```

We will shortly cover many of the other distributions. However, to illustrate how the command set works, let's assume you wanted to generate 30 observations from a binomial distribution from which 5 objects were selected and the probability of an object belonging to a desired category was equal to .4. The command would be

```
MTB >RANDom 30 observations, put into C1;
SUBC> BINOmial with n = 5 and p = .4.
```

To create a sample from a Poisson distribution with mean equal to 8, the command would be

```
MTB >RANDom 30 observations, put into C1;
SUBC> POISson with mean = 8.
```

As before, this command group could be shortened to

```
MTB >RAND 30, C1;
SUBC> POIS 8.
```

To place samples from a Poisson distribution into each of the first 25 columns, the command would be

```
MTB >RANDom 30, C1-C25;
SUBC> POISson with mean = 8.
```

We will explain the meaning of each of these generators in the next few pages. Don't panic if you don't understand what each of them does. For now we just want to get the mechanics of Minitab out of the way. In summary, the general form of the Minitab generator is to specify that you are going to use the generator, how many observations you want, and in which column or columns you are going to place data. In the second line you specify which generator to use and the characteristics necessary for that generator.

## Probability Density Function (PDF)

There is another command that uses the exact same group of subcommands as the random variate generators. That command is the probability density function (PDF) command. For now we will briefly describe what the PDF means. Later in the chapter we will discuss the use and the interpretation of the PDF.

Assume that you have used the random generator along with the normal subcommand and have placed 500 values in column 1. Further assume that you had specified a normal population with a mean of 100 and a standard deviation of 5. As you might expect, there will be a lot of values close to the mean of your sample (100) and fewer values as you get further from the mean. In other words, the probability of a value close to 100 is higher—*in a predictable sequence*—than the probability of a value that is either much greater than 100 or much less than 100.

The probability of any given value occurring can be found by using the **PDF** command. Technically, the distribution created by a generator produces an output which, when plotted using either the **HISTogram** or **DOTPlot** commands, gives a visual representation of a curve. The height of that curve for any given value represents the probability of that value. The PDF determines this height, or probability.

As stated above, the **PDF** command uses the same subcommands as the **RANDom** command. Assuming that you have not yet panicked, remember that we are just going through the command structure at this time; a fuller explanation will be developed later in the chapter. The form of the **PDF** command is

```
PDF for values in C [put results in C];
```

and then one of the following subcommands:

```
BINOmial with n = K and p = K.
POISson with mean = K.
BERNoulli with p = K.
INTEgers with uniform on K to K.
DISCrete with values in C, probabilities in C.
NORMal with mu = K, sigma = K.
UNIForm with continuous uniform on K to K.
```

Thus, if you had generated a sample of 50 observations using the normal generator with mean 100 and standard deviation equal to 5 and had placed those observations in column 1, then the **PDF** command would be

```
MTB >PDF C1;
SUBC> NORMal with mu = 100, sigma = 5.
```

If you wanted to store the PDF values for the same illustration, the command group would be

```
MTB >PDF C1, put in C2;
SUBC> NORMal mu = 100, sigma = 5.
```

### SECTION 5-2 The Bernoulli Distribution

As you can see, you must first have a group of data in a column before using the `PDF` command. The subcommand informs the PDF calculator about the nature of the basic distribution—such as normal or Poisson—and the necessary characteristics. The command group is identical to the `RANDom` command except for the need to specify a column when calling the `PDF` command. This exception will be discussed later.

## Cumulative Density Function (CDF)

Another set of commands is used for the cumulative density function (CDF). This command is analogous to that for the PDF except that instead of producing the probability of a specific value, the CDF calculates the probability of obtaining that value *or a lower one*. Technically, with the `PDF` command you find the height of the curve at any point; with the `CDF` command you find the total area under the curve up to that value.

The form of the command is

```
CDF for values in C [put results in C];
```

and then one of the following subcommands:

```
BINOmial with n = K and p = K.
POISson with mean = K.
BERNoulli with p = K.
INTEgers with uniform on K to K.
DISCrete with values in C, probabilities in C.
NORMal with mu = K, sigma = K.
UNIForm with continuous uniform on K to K.
```

As with the `PDF`, you must have a group of values in a column before you call the `CDF` command. With the subcommand, the probability calculations are carried out on the type of distribution that you have specified.

With these three groups of Minitab commands available, it is time to look at the various probability distributions. To start, we will consider the consequences of sampling a dichotomous population. Then we will discuss the consequences of sampling continuous populations. As the discussion progresses, several important factors will be shown to be common to both types of distribution.

## 5-2 THE BERNOULLI DISTRIBUTION

The *Bernoulli distribution* is a dichotomous population. Activities like coin tossing yield Bernoulli distributions. Also, rolling a die and determining the number of 4's obtained would produce a Bernoulli distribution, where the result of the die roll is either "4" or "not a 4." As can be seen, everything can conceivably be divided into dichotomous, or Bernoulli, distributions. This leads us to "Barth's distinction," a wry view of statisticians: "There are two types of people in the world, those who divide people into groups and those who don't."

If we determined that 40% of the people in the world divide people into groups and 60% do not, then we might ask, "What is the probability of choosing a 'grouper' at random from the world's population?" The answer is intuitively 40%. What if we draw a sample of 30 people? How many would we expect to be groupers? The Bernoulli distribution and

Minitab will allow us to experiment with this question using any of the random number generators of the Minitab system. To use the Bernoulli distribution, we must be able to break the population into two groups and to assign a probability to each group.

In coin tossing, the outcomes are heads and tails, and the probability of a head is .5. In rolling a die, the probability of obtaining a 4 is ⅙. In our grouper illustration, the probability of a person being a grouper is .4. The final stipulation necessary to use the Bernoulli generator is that the outcome of one trial be independent from all preceding trials. In tossing a coin, rolling a die, or sampling from an extremely large distribution, the outcome of any given observation is independent of any other observation. In coin tossing, getting a head on your first toss does not in any way affect the probability of the outcome of your second toss.

Notice, however, that sampling from an extremely large distribution also meets the requirement of independence. For the moment, assume that we had only four cards from a standard deck of playing cards. Let's say we had the four kings. The probability of drawing the king of hearts would clearly be 1 in 4, or ¼. If we then drew a card and did not replace it, only three cards would remain. Now the probability of drawing a particular card of the three would be 1 in 3, or ⅓. Clearly, the first draw would have an effect on the probabilities associated with the second draw, since the sample was very small. However, if we had 10,000,000 cards and drew a first card and then a second, the effect of drawing the first card would be so small on the probabilities associated with the second draw that we would be willing to say that there was no practical effect.

## Simulating Bernoulli Trials ▼▲▼▲▼▲▼▲▼▲▼▲▼▲▼▲▼▲▼▲▼▲▼▲▼▲▼▲

The Minitab command for the Bernoulli random number generator is to call the `RANDom` generator and then to specify the `BERNoulli` subcommand. In our example of groupers, we want to generate a sample of 30 observations, each with a .4 chance of occurrence. Since the probability of someone being a grouper is 40%, the input factor in the subcommand for *p* is `.4`. The command would be

```
MTB >RANDom 30 observations, put into C1;
SUBC> BERNoulli with p = .4.
```

The `.4.` value in the subcommand might look a bit awkward. Remember that it specifies the 40%. The final period in the subcommand line is required to tell Minitab that you have ended the command group. Of course, this command could be shortened to

```
MTB >RAND 30, C1;
SUBC> BERN .4.
```

When using the random number generators, the samples generated are not printed out. However, if you use the `PRINt` command, the values obtained when we drew our sample will appear as shown in Figure 5-1. Remember, your values will be slightly different, because although you are sampling from the same population, you are not drawing the same samples.

In Figure 5-1, we see a string of zeros and ones. Each of the ones represents an occasion when we drew a grouper. In this particular experiment, we have 10 groupers. Since 40% of the population are groupers, we would have theoretically expected 12 (.4 × 30).

## SECTION 5-2 The Bernoulli Distribution

**FIGURE 5-1** Bernoulli Distribution for Groupers Example with $n = 30$, $p = .4$

```
MTB >RAND 30;
SUBC> BERN .4.

   30 BERNOULLI TRIALS WITH P =  .4000
 0.    0.    1.    1.    1.    0.    0.    0.    1.    0.
 1.    1.    1.    0.    1.    1.    0.    0.    1.    0.
 0.    0.    0.    0.    0.    0.    0.    0.    0.    0.
```

However, since this is a random sampling of the entire population, the fact that we did not get exactly 12 is not surprising. When you repeat this command yourself, you should get about 12 ones. Next time you are on your Minitab system, try this command several times with a sample size of 30. You should find that the number of ones varies between slightly above and slightly below 12.

We can now demonstrate an important principle by expanding our present example. In your sample of 30 observations, the number of ones may or may not have been exactly 12. In fact, there might have been a good bit of variation in the number of ones if you had repeated the experiment several times. What would you expect to happen if you generated a noticeably larger sample, such as 300? Repeated samples of size 300 would probably have less variation and a proportion of ones consistently closer to 40% than samples of size 30. But why?

## The Law of Large Numbers

A good way to illustrate the law of large numbers is to find the mean of a large sample. For our present illustration, we will draw a sample of size 300. If we add up all the zeros and ones, the mean should be close to 0.4. Also, we could calculate a running average. In a running average, we would add the first two values of our sample of 300 and find their average; then we would add the third value to the first two and find the average of these three samples; then we would add the fourth value to the first three and find the mean of these four samples, and so on. This process would continue until we had added the means of all 300 samples together and found the average of all 300. The value of this running average might jump around wildly at first when we have only a few values, but as more and more trials are added, the running average should approach the mean of the population and become quite stable.

In Figure 5-1, if we add the first two values together (going horizontally), our total is 0 and thus our mean is 0. When we add the third value, our total becomes 1 and our mean becomes ⅓. With the fourth value added, the total becomes 2 and the mean becomes ½. As more and more values are included, the mean value becomes much more constant.

We can illustrate this quite easily using several Minitab commands. However, before we actually list the command lines for such an experiment, we need one new command. It is possible to add the values in a column *one at a time* and create a new column using the `PARSum` command. This command creates a running total of any other column. The command is

```
PARSum of data in C, put in C
```

If our original data was in column 1, then we might execute the following command:

`MTB >PARSum data in C1, put in C2`

This command could also be entered in the form

`MTB >LET C2 = PARSum(C1)`

In C2 the first value would be the same as the first value in C1, but the second value in C2 would be the sum of the first and second values of C1. The third value of C2 would be the sum of the first three values in C1, and so on.

With this new command, let us return to our larger experiment. What we will want to do is the following:

**(1)** Generate 300 observations from our Bernoulli distribution with $p = .4$ and put them in column 1.

**(2)** Add each observation together and obtain a running total; put these values in column 2.

**(3)** Divide each value in the running total column by the proper denominator, namely, the first two values by 2, the first three values by 3, and so on; put the results into column 4. To do this, we will create a column of numbers from 1 to 300 in consecutive order in column 3.

**(4)** We then simply print the final running average values found in step 3.

Examine the following Minitab command group as it executes the set of instructions just discussed:

```
MTB >RANDom 300 observations, put into C1;
SUBC> BERNoulli with p = .4.
MTB >PARSum of data in C1, put in C2
MTB >SET data into C3
DATA>1:300
DATA>LET C4 = C2/C3
MTB >PRINt C4
```

With these commands, we obtain the results shown in Figure 5-2.

**FIGURE 5-2** Running Average Calculation for a Bernoulli Distribution with $n = 300$, $p = .4$

```
MTB >PRIN C4

COLUMN         C4
COUNT         300
 1.00000     .50000     .33333     .25000     .20000     .16667
  .28571     .25000     .22222     .30000     .36364     .33333
  .38462     .35714     .33333     .31250     .35294     .33333
  .31579     .30000     .28571     .27273     .30435     .29167
  .32000     .34615     .37037     .39286     .37931     .40000
  .41935     .40625     .42424     .41176     .40000     .38889
  .40541     .39474     .41026     .40000     .39024     .40476
  .39535     .40909     .42222     .43478     .44681     .43750
  .42857     .42000     .41176     .42308     .43396     .42593
  .43636     .42857     .42105     .41379     .42373     .41667
  .42623     .43548     .44444     .45313     .46154     .46970
  .47761     .48529     .47826     .48571     .49296     .48611
  .47945     .47297     .46667     .46053     .45455     .44872
  .45570     .45000     .45679     .46341     .46988     .47619
  .47059     .46512     .47126     .46591     .47191     .47778
  .47253     .46739     .46237     .45745     .46316     .46875
  .46392     .46939     .46465     .47000     .46535     .46078
```

*(continued on next page)*

**FIGURE 5-2** (*Continued*)

| | | | | | |
|---|---|---|---|---|---|
| .45631 | .46154 | .45714 | .46226 | .45794 | .45370 |
| .44954 | .44545 | .45045 | .45536 | .45133 | .44737 |
| .45217 | .45690 | .45299 | .45763 | .45378 | .45833 |
| .46281 | .45902 | .45528 | .45161 | .45600 | .46032 |
| .45669 | .46094 | .45736 | .45385 | .45038 | .44697 |
| .44361 | .44030 | .43704 | .43382 | .43796 | .44203 |
| .43885 | .43571 | .43262 | .43662 | .43357 | .43056 |
| .43448 | .43151 | .42857 | .42568 | .42282 | .42667 |
| .42384 | .42105 | .42484 | .42208 | .41935 | .42308 |
| .42675 | .42405 | .42767 | .42500 | .42857 | .43210 |
| .42945 | .42683 | .43030 | .42771 | .42515 | .42262 |
| .42604 | .42353 | .42690 | .43023 | .42775 | .42529 |
| .42286 | .42045 | .42373 | .42135 | .41899 | .41667 |
| .41989 | .41758 | .42077 | .42391 | .42162 | .41935 |
| .41711 | .42021 | .41799 | .41579 | .41885 | .42188 |
| .42487 | .42268 | .42051 | .41837 | .42132 | .41919 |
| .42211 | .42500 | .42786 | .42574 | .42857 | .43137 |
| .42927 | .42718 | .42512 | .42308 | .42105 | .42381 |
| .42180 | .42453 | .42254 | .42056 | .42326 | .42593 |
| .42857 | .43119 | .42922 | .42727 | .42534 | .42342 |
| .42152 | .41964 | .41778 | .42035 | .41850 | .41667 |
| .41485 | .41739 | .41558 | .41379 | .41631 | .41453 |
| .41277 | .41102 | .40928 | .41176 | .41423 | .41250 |
| .41079 | .41322 | .41152 | .41393 | .41224 | .41463 |
| .41700 | .41532 | .41767 | .41600 | .41434 | .41667 |
| .41502 | .41339 | .41569 | .41406 | .41245 | .41473 |
| .41699 | .41538 | .41379 | .41221 | .41065 | .40909 |
| .40755 | .40602 | .40824 | .41045 | .41264 | .41111 |
| .40959 | .40809 | .40659 | .40876 | .40727 | .40942 |
| .40794 | .40647 | .40860 | .40925 | .40780 | .40780 |
| .40636 | .40493 | .40351 | .40210 | .40070 | .39931 |
| .40138 | .40345 | .40206 | .40411 | .40614 | .40476 |
| .40678 | .40541 | .40404 | .40268 | .40468 | .40667 |

In the sample generated by our Minitab command, we have 122 ones; a theoretically ideal sample would have contained 120. We are certainly close, and any sample you generate will probably vary slightly from a total of 120 ones. As you can see, the first few running mean values jump around significantly. However, by the time we have accumulated 30 observations in our sample, the mean value is 0.40000 (out of sheer luck did it hit 0.4 exactly), and it remains reasonably close for the rest of the sample. This very important fact is an illustration of the *law of large numbers,* which states:

> As a sample becomes larger and larger, if it is randomly drawn, then the mean of the sample becomes closer and closer to the population mean.

## 5-3 THE BINOMIAL DISTRIBUTION

It is sometimes helpful to look at the possible sequences of outcomes when drawing a sample. In the previous section, we simply took samples of sizes 30 and 300 and then analyzed the results. However, we can also look at the probabilities of drawing a particular sequence of samples, such as 3 groupers or 2 groupers and then 1 nongrouper. Such a probability analysis can be diagrammed in what is called a probability tree, such as the one shown in Figure 5-3. This chart depicts the possible outcomes of all samples of size $n = 3$ and their respective probabilities.

**CHAPTER 5** Probability Distributions

**FIGURE 5-3** Three-Level Probability Tree for Groupers and Nongroupers

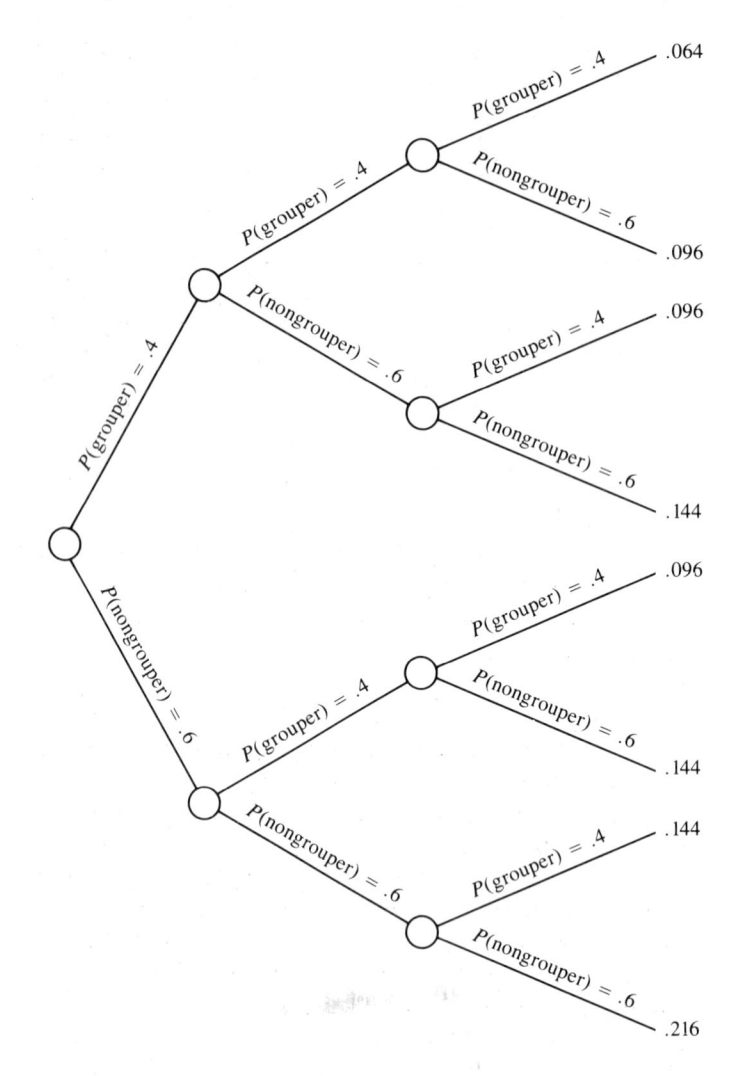

*[Three-level probability tree diagram showing branches at each level with P(grouper) = .4 and P(nongrouper) = .6, with final joint probabilities: .064, .096, .096, .144, .096, .144, .144, .216]*

A probability tree is read from left to right. As you can see, the first possible outcome is either a grouper with a .4 probability or a nongrouper with a .6 probability. This is the first branch of our probability tree. The second set of branches follows from the first. If the first individual drawn was a grouper, then we would follow the upper branch of the probability tree to our second intersection. At that point we again have the possibility of either a grouper or a nongrouper. However, if the first individual was a nongrouper, then we would follow the lower branch of the probability tree to the second intersection, where we would again have the probabilities of a grouper or nongrouper on the second draw.

You will note that since each subsequent observation is independent, we can compute the following probabilities for the first two levels of our probability tree with $n = 2$:

## SECTION 5-3 The Binomial Distribution

$P$(grouper, grouper) = (.4)(.4) = .16
$P$(grouper, nongrouper) = (.4)(.6) = .24
$P$(nongrouper, grouper) = (.6)(.4) = .24
$P$(nongrouper, nongrouper) = (.6)(.6) = .36

With this background, you should be able to develop a similar table for the probability tree for all possible samples of size $n = 3$. The set of calculations is shown below:

$P$(grouper, grouper, grouper) = (.4)(.4)(.4) = .064
$P$(grouper, grouper, nongrouper) = (.4)(.4)(.6) = .096
$P$(grouper, nongrouper, grouper) = (.4)(.6)(.4) = .096
$P$(grouper, nongrouper, nongrouper) = (.4)(.6)(.6) = .144
$P$(nongrouper, grouper, grouper) = (.6)(.4)(.4) = .096
$P$(nongrouper, grouper, nongrouper) = (.6)(.4)(.6) = .144
$P$(nongrouper, nongrouper, grouper) = (.6)(.6)(.4) = .144
$P$(nongrouper, nongrouper, nongrouper) = (.6)(.6)(.6) = .216

Take a minute to be sure that you can calculate each of these values before you continue.

We have determined the probabilities for every possible outcome for a sample size of 3. We can also determine the probability of a particular number of groupers when drawing a particular sample size. For example, if we wanted to know the probability of drawing 2 groupers in a sample size of 3, all we would have to do is add together the probabilities for each branch of our probability tree that contained 2 groupers. In our case, there are three ways to get exactly 2 groupers in a sample of size 3. The probability for each tree branch of 2 groupers is .096, and since there are 3 such branches, there is a (3)(.096) = .288 chance of drawing 2 groupers.

Be careful of how any such question is stated. There is a big difference between the probability of *exactly* 2 groupers in a sample of size 3 and the probability of 2 or more groupers in a sample of size 3. In the latter case, not only do we have to account for the case of 2 groupers but also for the case of 3. Thus the probability of 2 *or more* groupers is .288 + .064 = .352.

## Simulating Binomial Trials ▼▲▼▲▼▲▼▲▼▲▼▲▼▲▼▲▼▲▼▲▼▲▼▲▼▲▼

There is a Minitab subcommand that allows us to generate numerous samples of size 3 and to determine the number of groupers and nongroupers in each sample. This subcommand is **BINOmial**. In the **BERNoulli** subcommand, we simulated a single experiment with any given number of observations in that experiment. However, if we wanted to simulate 100 experiments of 3 samples each, then instead of using the **BERNoulli** subcommand 100 times, we could use the **BINOmial** subcommand just once. If we wanted to continue our illustration and simulate 100 trials of size 3 with a probability equal to .4, the command set would be

```
MTB >RANDom 100 observations, put into C1;
SUBC> BINOmial with n = 3 and p = .4.
```

If you are in doubt about this command set, refer back to the discussion of commands at the start of this chapter. As you will note, for the **BINOmial** subcommand you must

**FIGURE 5-4** Binomial Distribution for Groupers with $n = 3$, $p = .4$

```
MTB >RAND 100 C1;
SUBC> BINO 3 .4.

100 BINOMIAL EXPERIMENTS WITH N =   3   AND P =   .4000

    0.    1.    1.    1.    1.    2.    3.    1.    1.    0.
    1.    1.    1.    2.    1.    2.    2.    2.    0.    1.
    1.    1.    0.    1.    1.    0.    1.    2.    2.    1.
    0.    0.    2.    1.    0.    1.    1.    1.    0.    1.
    2.    2.    1.    1.    0.    1.    1.    1.    1.    2.
    2.    0.    2.    2.    1.    2.    1.    2.    0.    2.
    2.    0.    0.    1.    1.    1.    2.    2.    3.    3.
    2.    0.    1.    0.    1.    0.    2.    2.    1.    1.
    2.    2.    1.    1.    2.    1.    1.    2.    0.    0.
    2.    0.    2.    2.    2.    1.    1.    1.    2.    1.

    C1           N = 100       MEAN =      1.1800      ST.DEV. =       .783
```

specify both $n$ and $p$ values. The value $n$ is the sample size, and $p$ is the probability of an occurrence per sample. After completing this command set, the results can be printed; the results from using the `PRINt` command on `C1` are shown in Figure 5-4.

In this figure we are counting the number of successes—the number of groupers in a sample of size 3. Therefore, for each individual sample, a `0` is interpreted as no groupers, a `1` as one grouper, and so on. Using either a histogram or stem-and-leaf plot, we can see that in 20 of the experiments there were no groupers (value of 0). Notice that from our calculations above, we would have expected 21.6% of our samples to have contained no groupers. Our experiment of 100 samples seems to be reasonably close.

The mean of the binomial distribution is the product of the sample size $n$ and the probability $p$. Thus we know that the mean of our population of sample size $n = 3$ is $(3)(.4) = 1.2$. In Figure 5-4 we have also included a portion of the `DESCribe` command, which shows our mean to be 1.18. What is the mean of your sample? It should also be near 1.2.

## Calculating the Binomial Probability Distribution Function ▼▲▼▲

In Appendix B, Table 5 is a binomial probability table. You might wonder where all those numbers in the table actually come from. Well, it is not magic, nor does Minitab simply store large banks of numbers for every possible binomial distribution. These values are the results of calculations. By looking at the formulas by which the binomial distribution is calculated, we hope that you can gain a better grasp of the logic of this particular distribution. Since there may be occasions when you do not have a computer or any statistical tables to use, it will be useful to know how to calculate binomial distribution values by hand.

▶▶▶▶▶**The Binomial Probability Density Function** The general formula for the binomial distribution is

$$P(x, n - x) = \frac{n!}{x!(n - x)!} p^x (1 - p)^{n-x}$$

## SECTION 5-3 The Binomial Distribution

where

$x$ = number in a sample with some given characteristic
$n$ = number in a sample
$p$ = proportion of the universal set containing the given characteristic

For those who might have forgotten factorial notation, the form

$$x! = x(x - 1)(x - 2) \cdots (1)$$

Also, by definition,

$$0! = 1$$

Let's now take an example and put values into the formula to determine the binomial probability. This probability can then be compared with the value in the table and the value computed by Minitab.

Suppose that the recruit class of a major city's police academy contains 10 rookies, 9 of them white and 1 black. The city has a population that is 40% black. What is the probability that this class was taken from the city population without bias (intentional or unintentional)?

In this case

$$x = 1$$
$$n = 10$$
$$p = .4$$

We can thus substitute into our equation as follows:

$$P(x, n - x) = \frac{n!}{x!\,(n - x)!}\,p^x(1 - p)^{n-x}$$

$$P(1, 9) = \frac{10!}{(1!)(9!)}(.4)^1(.6)^9$$

$$= \frac{(10)(9)(8)(7)(6)(5)(4)(3)(2)(1)}{(1)(9)(8)(7)(6)(5)(4)(3)(2)(1)}(.4)(.0100777)$$

$$= 10(.4)(.0100777)$$

$$= .0403108$$

In other words, there is about a 4% chance that this sample of 9 whites and 1 black would occur at random from a population that is 40% black.

We might check our work by looking up the appropriate value in a binomial probability table, such as Table 5 in Appendix B. Remember that our basic data is $p = .4$, $n = 10$, and $x = 1$. The value for $p$ is located across the top of the table, and the values for $n$ and $x$ are located down the left-hand margin. If you go down the left-hand margin to $n = 10$ and then down one more row to $x = 1$, you will find the appropriate value about halfway over in the table under $p = .40$. That value is .040. Thus, the value calculated above is accurate.

Note that using the formula offers a few advantages. For example, the table in Appendix B has values for only a few $p$-values. What if the black population of this particular city were 27%? Table 5 could not be used (unless you approximated the answer by using $p = .30$). Of course, tables with more divisions of the *p*-values are available. Hand calculations also let us control the degree of accuracy. Our hand calculation found seven places of accuracy. In some scientific applications that degree of accuracy may be needed.

▶▶▶▶▶▶**Using the PDF Command** Finally, we can cross-check the accuracy of our hand calculation by using Minitab. Minitab can calculate any portion of a normal binomial table and provide both the exact value for a given occurrence and the cumulative value for that occurrence (cumulatives will be discussed in the next section). The command to produce a binomial table is part of the **PDF** command. As already noted, the **PDF** command is performed on an existing column of numbers. However, if you do not specify a column, the command creates a normal binomial table for the *n* and *p* specified.

Such a command on the police academy data is given below.

```
MTB >PDF;
SUBC> BINOmial n = 10, p = .4.
```

Notice that the command line does not specify a column. Minitab prints the entire portion of the binomial table that is relevant; then you can simply look up the binomial value. This command produces Figure 5-5. As can be seen, the value found by the hand calculation is again verified with .0403 in the second column. (Note that Minitab uses **K** for the number of occurrences.) Depending on your computer, there will be limits on how large you can specify *n* and *p*. Typical limits are 250 for *n* and .5 for *p*.

**FIGURE 5-5**  Minitab Printout for the Command **PDF** for a Binomial Population with $n = 10, p = .4$

```
MTB >PDF;
SUBC> BINO 10 .4.

BINOMIAL WITH N =  10   P = 0.400000
    K            P( X = K)
    0             0.0060
    1             0.0403
    2             0.1209
    3             0.2150
    4             0.2508
    5             0.2007
    6             0.1115
    7             0.0425
    8             0.0106
    9             0.0016
   10             0.0001
```

▶▶▶▶▶▶**The Binomial Cumulative Density Function** In the police academy example, we calculated the exact probability of a sample being drawn without bias from a given universe. If we were really testing for systematic bias, we would have to use a *cumulative binomial distribution*. At issue is the probability of 1 or *fewer* black police recruits in a group of 10. Cumulative binomial probabilities simply require that you calculate the specific probabilities for each possible occurrence and then add them together.

In general, cumulative binomial probabilities can be calculated as either some occurrence and less than that occurrence or as some occurrence and greater than that occurrence. For example, we might wonder about the probability of 4 or fewer black recruits or 3 or

## SECTION 5-3 The Binomial Distribution  ▶119

more black recruits. In both cases the mechanics are the same. For "fewer"-type problems the calculations run from the given point down to 0. For "greater"-type problems, the calculations run from that point up until the probability calculation results in a value of 0. All of the various values are then summed to determine the cumulative total.

In our police example, we know the exact probability of 1 black recruit. The probability of none is

$$P(0, 10) = \frac{10!}{0!\,10!}(.4)^0(.6)^{10}$$
$$= (1)(1)(.0060467)$$
$$= .0060467$$

We can now calculate the total probability that this recruit class was selected without bias (with respect to race) as

$$.0403108 + .0060467 = .0463575$$

▶▶▶▶▶ **Using the CDF Command**  If we return to Table 5 of Appendix B, we see that adding .040 and .006 yields a total of .046, which is the same value we found by hand calculations (when they are rounded off).

We can also generate this value with Minitab using the **CDF** command. Just as with the **PDF** command, if you do not specify a column, the general portion of the cumulative probability table will be provided for the distribution specified. In this case, the commands would be

```
MTB >CDF;
SUBC> BINOmial with n = 10 and p = .4.
```

This command creates the output shown in Figure 5-6.

**FIGURE 5-6**  Minitab Printout for the Command CDF for a Binomial Population with $n = 10$, $p = .4$

```
MTB >CDF;
SUBC> BINO 10 .4.
        BINOMIAL WITH N = 10  P = 0.400000
          K    P( X LESS OR = K)
          0         0.0060
          1         0.0464
          2         0.1673
          3         0.3823
          4         0.6331
          5         0.8338
          6         0.9452
          7         0.9877
          8         0.9983
          9         0.9999
         10         1.0000
```

In Figure 5-6, the second column is the cumulative value. For the value of **K** = 1, the cumulative binomial probability is .464, which agrees with our hand-calculated value after rounding. This illustration should give you some insights into the possibilities for hand calculations; we will not develop the hand calculations for the Poisson distribution and the normal distribution.

## 5-4 THE POISSON DISTRIBUTION

Another distribution that describes many business situations is the *Poisson* distribution. This distribution is frequently used to describe rare occurrences. In such situations, the number of successes (*p*) is quite small, but the number of opportunities (*n*) to succeed is quite large. For example, automobiles coming off an assembly line might have a few flaws in their paint. We might know that the population mean is 4 flaws per car. This does not mean that each car has 4 flaws, but rather that as the sample gets large, there will be an average of 4 flaws per car.

### Simulating Poisson Trials

We can generate Poisson samples much as we used the `BINOmial` subcommand to generate binomial experiments. The command set for the Poisson generator has the same format as that for previous generators, but in this case we must specify the mean number of items of interest (in this case, defects per car). If we wanted to simulate 100 cars coming off the assembly line with a mean number of 4 paint defects per car, the command would be

```
MTB >RANDom 100 observations, put into C1;
SUBC> POISson with mean = 4.
```

Figure 5-7 contains the results of this simulation. Column 1 was printed, and each value in the upper portion of Figure 5-7 is the number of defects found on each car. The two graphs have the general shape of a Poisson distribution: a rapidly rising set of values at the lower end that trails off as the probability value for a higher number of defects becomes smaller and smaller. We expect the average number of defects to be about 4, and that is borne out by the `DESCribe` command. Do not forget that these observations occur randomly. Your results will not be exactly the same as ours or as those of other students in your class, but the distributions should be quite similar.

### Calculating the Poisson PDF

The Poisson PDF can be determined using the `PDF` command with the `POISson` subcommand in Minitab. In much the same way that we can simulate binomial probabilities and ask ourself questions like "What is the probability of 2 groupers being in a sample of size 3?", we can determine similar probabilities with the `POISson` subcommand. If we know that the distribution tends to follow a Poisson pattern, we can determine the probability that a given car will have no defects, or 3 defects, or fewer than 5 defects. All we must specify is the mean number of defects (or whatever else we are studying) per observation. For the present illustration, the command would be

```
MTB >PDF;
SUBC> POISson with mean = 4.
```

This command generates the output shown in Figure 5-8. The first column indicates the number of defects per car, and the second indicates the probability of exactly that number of defects.

## SECTION 5-4 The Poisson Distribution

**FIGURE 5-7** Results of Random Experiments, the DESCribe Command, the Histogram, and the Dot Plot for a Poisson Distribution with Mean = 4

```
MTB >RAND 100 C1;
SUBC> POIS 4.
MTB >PRIN C1

C1
   4    2    4    6    5    1    8    1    7    1    7
   8    4    1    5    4    4    8    5    5    3    7
   3    2    3    4    7    2    3   12    2    5    4
   5    2    3    3    4    6    7    6    5    3    2
   6    4    1    8    3    5    2    4    5    3    3
   4    3    2    2    7    1   10    4    6    2    2
   2    2    5    1    2    4    2    7    4    1    2
   4    2    7    3    9    5    3    2    2    3    5
   2    2    3    3    3    5    2    3    5    2    7
   3

MTB >DESC C1

              N       MEAN     MEDIAN    TRMEAN     STDEV    SEMEAN
C1          100      3.970      3.500     3.833     2.218     0.222

            MIN        MAX         Q1        Q3
C1        1.000     12.000      2.000     5.000

MTB >HIST C1

Histogram of C1   N = 100

Midpoint   Count
     1        8   ********
     2       23   ***********************
     3       19   *******************
     4       15   ***************
     5       14   **************
     6        5   *****
     7        9   *********
     8        4   ****
     9        1   *
    10        1   *
    11        0
    12        1   *

MTB >DOTP C1
```

                      .
                      :
                      :        .
                      :        :
                      :        :
                      :        :    .
                      :        :    :
                      :        :    :
                      :        :    :    :
                 :    :        :    :    :
                 :    :    :   :    :    :
                 :    :    :   :    :    :     .
                 :    :    :   :    :    :     :     .             .
         +---------+---------+---------+---------+---------+--C1
       0.00      2.50      5.00      7.50     10.00     12.50

One other very important fact can be seen in this figure. Notice that in column 2, the individual probabilities create an interesting pattern. The first two values are relatively small, then values 2 through 6 are relatively large; the rest of the values become smaller and smaller. In other words, the probabilities start out small, quickly rise, and then trail off.

**FIGURE 5-8**  Poisson Probabilities Using the PDF Command with Mean = 4

```
MTB >PDF;
SUBC> POIS 4.

    POISSON WITH MEAN =    4.000
       K             P( X = K )
       0               0.0183
       1               0.0733
       2               0.1465
       3               0.1954
       4               0.1954
       5               0.1563
       6               0.1042
       7               0.0595
       8               0.0298
       9               0.0132
      10               0.0053
      11               0.0019
      12               0.0006
      13               0.0002
      14               0.0001
      15               0.0000
```

Unlike the normal distribution, there is not an equal number of small probabilities at both ends of the curve. This is a unique property of the Poisson curve: It does not tend to have an equal number of observations at each end. The Poisson distribution is particularly useful when modeling the arrivals of individuals or items in waiting lines. You will see it again in operations research and management science.

## Calculating the Poisson CDF

As in the previous section, we might want to know the probability of 3 or fewer defects. For this question the `CDF` command is required. The command group would be

```
MTB >CDF;
SUBC> POISson with mean = 4.
```

The results of this command set are shown in Figure 5-9.

## 5-5 THE NORMAL DISTRIBUTION

Now we come to the most important distribution in statistical theory. It is called the *normal distribution*, and it is governed by a very important fact known as the central limit theorem. Roughly, it states that whenever you draw repeated *random* samples from *any* distribution, *the distribution of the means of those samples will be normal*. We will discuss this idea in greater detail at the end of this section.

In the paint defects example, the original population followed a Poisson distribution. However, according to the central limit theorem, the distribution of the means of repeated samples of that Poisson distribution will be normal. Let us see if we can actually generate normal distributions from repeated samplings of a nonnormal population.

## SECTION 5·5 The Normal Distribution

**FIGURE 5-9** Cumulative Poisson Probabilities Using the CDF Command with Mean = 4

```
MTB >CDF;
SUBC> POIS 4.

       POISSON WITH MEAN =    4.000
        K    P( X LESS OR = K)
        0              0.0183
        1              0.0916
        2              0.2381
        3              0.4335
        4              0.6288
        5              0.7851
        6              0.8893
        7              0.9489
        8              0.9786
        9              0.9919
       10              0.9972
       11              0.9991
       12              0.9997
       13              0.9999
       14              1.0000
```

First, however, we need to define a normal distribution. What does such a distribution look like? How does it behave? What are its characteristics? Since the normal distribution is the most important statistical concept, you should first get some idea of its general characteristics. To do so, try the following experiment. Get out your popcorn popper. (Wash it if necessary!) Now, following the manufacturer's instructions, make some popcorn. Do *not* go to the refrigerator for a drink while the popper is running. Try to count the number of kernels that pop in each 10-second interval, and write these numbers down. At first the figures will be low, but they will increase rapidly until you can no longer count the number of pops and can only guess at the proper number. After that the rate will slow to only a few pops per time interval. Now take your popcorn (we like ours with Parmesan cheese), a drink, and your pad back to your desk. If you now sketch the number of pops in each time interval, it will probably look something like Figure 5-10.

This is a *normal curve*: The number of observations is low at first and then increases to a much greater number of observations. It then decreases at approximately the same rate at which it increased, creating a symmetric curve. There are many types of data that will occur in such a distribution, including weights, heights, body temperatures, production quotas, and grade-point averages.

**FIGURE 5-10** A Normal Curve

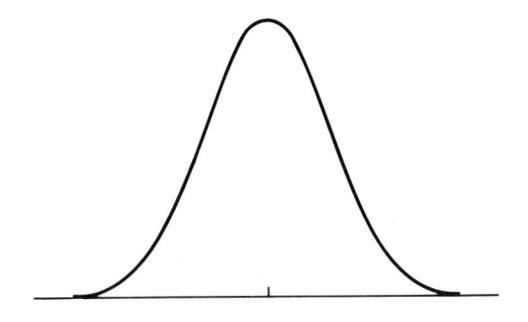

The normal distribution is completely described by two measures: the mean and the standard deviation. The mean is at the center of the normal distribution, and the standard deviation determines the relative height of the distribution. If the standard deviation is relatively small, then the values in the distribution will be tightly clustered about the mean, and the distribution will be "tall." If the standard deviation is relatively large, then the values in the distribution will not be tightly clustered, and the distribution will appear spread out or flat.

For example, if you have a load of sand delivered to your house and the dump truck simply unloads the sand in your front yard, you will get a pile of sand that looks like a normal distribution. However, to describe the location of the pile of sand, you will need to specify the two measures just noted: the location of the center of the pile and the degree of spread from the center. As you can imagine, the same amount of sand can be very high and have little spread or very low and have a great deal of spread.

Figure 5-11 shows two normal distributions with the same population mean but different standard deviations. We will discuss the characteristics of the normal distribution in Chapter 6.

**FIGURE 5-11**  Two Normal Curves with the Same Mean Value But Different Standard Deviations

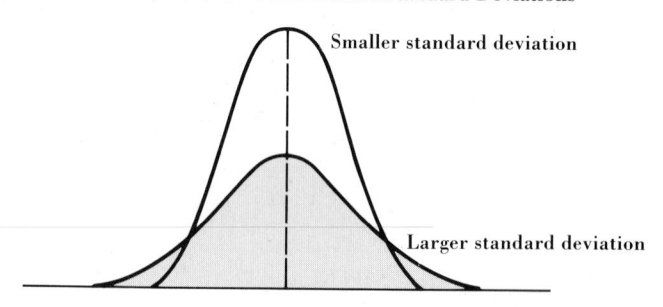

All normal distributions are symmetric; that is, they have the same rate of decrease on each side of the mean. Because of this symmetry, we can work with just one-half of the curve and know what is happening in the other half. This relationship between "upper" and "lower" halves will become important in Chapter 6 and later. In the real world, normal distributions will be very close to symmetric. If you use the `HISTogram` command on the data columns for the Go-Fast Company, you will see that several, but not all, of the columns are normally distributed. However, this is not a sufficient reason for saying that the normal distribution is the most important distribution. That justification must still be developed.

## 5-6  THE CENTRAL LIMIT THEOREM

According to the central limit theorem, the means of repeated samples from *any* population will form a normal distribution; and if we know we have a normal distribution, then we know the characteristics of that population.

Now let us see the formal statement of the central limit theorem.

▶ ▶ ▶ ▶ ▶ **Central Limit Theorem**  For large simple random samples drawn from any parent population, the distribution of sample means (sampling distribution) will be normal, the mean of the sampling distribution will be equal to the mean of

## SECTION 5-6 The Central Limit Theorem

the parent population, and the standard deviation of the sampling distribution will be equal to the standard deviation of the parent population divided by the square root of the sample size.

The key points are that (1) the repeated samples must each be randomly drawn, (2) we must take large samples, and (3) the distribution of the sample means will be approximately normally distributed in all cases.

Let us first consider the binomial distribution in light of the central limit theorem. We will use a binomial distribution, generate samples from that distribution, and compute their means.

## A Binomial Population

In our first example we will use a binomial population of an equal number of zeros and ones. Remember that we *know with certainty* that the original population is made up of equal numbers of zeros and ones. The mean of the population is 0.5, and there are no tails trailing out from the mean. All values are either zero or one.

The **BINOmial** subcommand allows us to generate samples from a binomial distribution. Let us start by drawing 5,000 samples of size $n = 5$ with a probability of $p = .5$. If we place those 5,000 samples in column 1 and then divide each total by 5 and place the results in column 2, then column 2 will contain the mean value for each sample drawn.

The following Minitab command group executes the first command set, namely, taking samples of size 5 and placing the mean value for each sample in column 2. Included are **DESCribe**, **HISTogram**, and **DOTPlot** commands on column 2 to determine the nature of the distribution found in that column.

```
MTB >RANDom 5000 observations put into C1;
SUBC> BINOmial with n = 5 and p = .5.
MTB >LET C2 = C1/5
MTB >HISTogram of data in C2
MTB >DOTPlot of data in C2
MTB >DESCribe data in C2
```

**FIGURE 5-12**  Summary Analysis and Histogram of 5,000 Samples from a Binomial Number Generator with $n = 5$, $p = .5$

```
MTB >RAND 5000, C1;
SUBC> BINO 5 .5.
MTB >LET C2 = C1/5
MTB >HIST C2

Histogram of C2    N = 5000
Each * represents 35 obs.

Midpoint    Count
    0.0      146   *****
    0.1        0
    0.2      764   **********************
    0.3        0
    0.4     1516   *******************************************
    0.5        0
    0.6     1621   **********************************************
    0.7        0
    0.8      805   ***********************
    0.9        0
    1.0      148   *****
```

*(continued on next page)*

**FIGURE 5-12** (*Continued*)

```
MTB >DOTP C2

Each dot represents 101 points

                    :       :
                    :       :
                    :       :
                    :       :       :
              :     :       :       :       :
              :     :       :       :       :
        :     :     :       :       :       :
        +---------+---------+---------+---------+---------+--C2
      0.00      0.20      0.40      0.60      0.80      1.00

MTB >DESC C2

              N      MEAN    MEDIAN    TRMEAN     STDEV    SEMEAN
C2         5000   0.50476   0.60000   0.50520   0.22182   0.00314

            MIN       MAX        Q1        Q3
C2      0.00000   1.00000   0.40000   0.60000
```

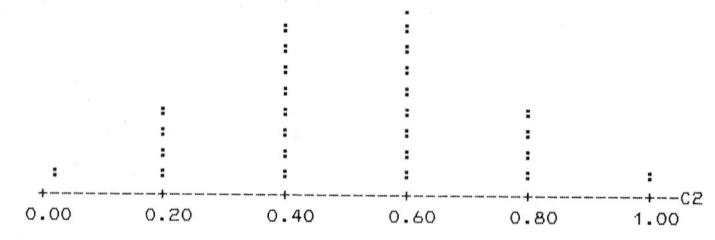

Our result is shown in Figure 5-12.

Now, we can repeat this instruction set but this time using a sample of size 15:

```
MTB >RANDom 5000 observations, put into C1;
SUBC> BINOmial with n = 15 and p = .5.
MTB >LET C2 = C1/15
MTB >HISTogram of data in C2
MTB >DOTPlot of data in C2
MTB >DESCribe data in C2
```

The results of this command set are shown in Figure 5-13.

**FIGURE 5-13** Summary Analysis and Histogram of 5,000 Samples from a Binomial Number Generator with $n = 15$, $p = .5$

```
MTB >RAND 5000, C1;
SUBC> BINO 15 .5.
MTB >LET C2 = C1/15
MTB >HIST C2

Histogram of C2    N = 5000
Each * represents 20 obs.

Midpoint   Count
  0.05        3   *
  0.10        0
  0.15       17   *
  0.20       80   ****
  0.25      194   **********
  0.30        0
  0.35      455   ***********************
  0.40      788   ****************************************
  0.45      946   ************************************************
  0.50        0
  0.55      942   ************************************************
  0.60      776   ***************************************
  0.65      480   ************************
  0.70        0
  0.75      232   ************
  0.80       64   ****
  0.85       21   **
  0.90        0
  0.95        2   *
```

*(continued on next page)*

**SECTION 5-6** The Central Limit Theorem

**FIGURE 5-13** (*Continued*)

```
MTB >DOTP C2
Each dot represents 59 points

                               .
                           : :   : :
                           : :   : :
                           : :   : :
                           : :   : :  .
                       : : : :   : :
                       : : : :   : :   :
                     : : : : :   : : : :
                   : : : : : :   : : : : :
               .   : : : : : :   : : : : :     .    .
       +---------+---------+---------+---------+---------+--C2
      0.00      0.20      0.40      0.60      0.80      1.00

MTB >DESC C2

           N      MEAN    MEDIAN   TRMEAN    STDEV    SEMEAN
C2       5000  0.50159  0.53333  0.50193  0.13108   0.00185

           MIN      MAX       Q1       Q3
C2     0.06667  0.93333  0.40000  0.60000
```

**FIGURE 5-14** Summary Analysis and Histogram of 5,000 Samples from a Binomial Number Generator with $n = 30$, $p = .5$

```
MTB >RAND 5000 C1;
SUBC> BINO 30 .5.
MTB >LET C2 = C1/30
MTB >HIST C2

Histogram of C2    N = 5000
Each * represents 25 obs.

Midpoint    Count
   0.20        2   *
   0.25       47   **
   0.30       67   ***
   0.35      400   ****************
   0.40      403   *****************
   0.45     1223   *************************************************
   0.50      725   *******************************
   0.55     1219   *************************************************
   0.60      382   ****************
   0.65      411   *****************
   0.70       72   ***
   0.75       41   **
   0.80        7   *
   0.85        1   *

MTB >DOTP C2

Each dot represents 45 points

                        .
                      : : :
                      : : :
                      : : :    : .
                    . : : :    : : .
                      : : :    : : :
                      : : : :  : : : :
                      : : : :  : : : : :
                    . : : : :  : : : : :  .
            . . . : : : : : :  : : : : :  : . . .    .
      ---+---------+---------+---------+---------+---------+---C2
        0.15      0.30      0.45      0.60      0.75      0.90
```

*(continued on next page)*

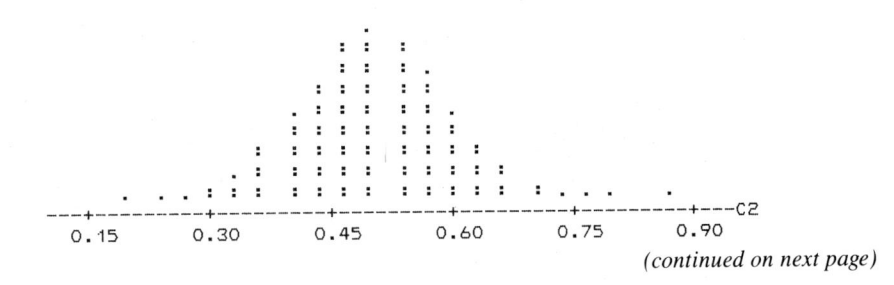

**FIGURE 5-14** *(Continued)*

```
MTB >DESC C2

            N      MEAN    MEDIAN    TRMEAN     STDEV    SEMEAN
C2       5000   0.50027   0.50000   0.50013   0.09275   0.00131

           MIN       MAX        Q1        Q3
C2     0.20000   0.86667   0.43333   0.56667
```

Notice that we are approaching a more normal distribution. Figures 5-14 and 5-15 show the results for samples of size $n = 30$ and $n = 50$. Notice that for each histogram and dot plot, the curve of mean values becomes more and more symmetric. The means of our figures approach 0.5.

**FIGURE 5-15** Summary Analysis and Histogram of 5,000 Samples from a Binomial Number Generator with $n = 50$, $p = .5$

```
MTB >RAND 5000 C1;
SUBC> BINO 50 .5.
MTB >LET C2 = C1/50
MTB >HIST C2

Histogram of C2    N = 5000
Each * represents 35 obs.

Midpoint    Count
    0.25        2   *
    0.30       38   **
    0.35      111   ****
    0.40      619   ******************
    0.45      854   *************************
    0.50     1639   ***********************************************
    0.55      909   ***************************
    0.60      647   *******************
    0.65      137   ****
    0.70       40   **
    0.75        4   *

MTB >DOTP C2

Each dot represents 35 points

                              .
                           : : :
                           : : :           
                           : : : : : .
                         : : : : : : :
                       . : : : : : : : .
                       : : : : : : : : :
                     . : : : : : : : : : : .
                     : : : : : : : : : : : :
                   . : : : : : : : : : : : : .
             . . . : : : : : : : : : : : : : : . .   .   .
          ---+---------+---------+---------+---------+---------+---C2
          0.30      0.40      0.50      0.60      0.70      0.80

MTB >DESC C2

            N      MEAN    MEDIAN    TRMEAN     STDEV    SEMEAN
C2       5000   0.50210   0.50000   0.50217   0.07085   0.00100

           MIN       MAX        Q1        Q3
C2     0.24000   0.74000   0.46000   0.56000
```

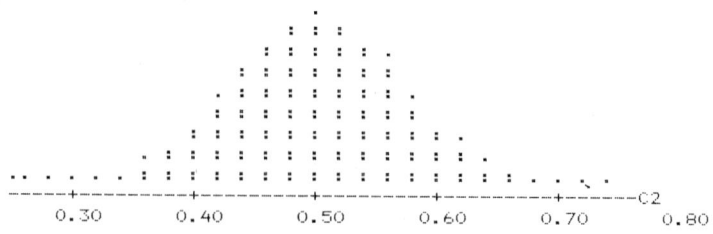

## SECTION 5-6 The Central Limit Theorem

If we change the value of $p$ from .5 to .3 we will obviously change the original population, and the population mean will change from 0.5 to 0.3. However, will we change the shape of the distribution of sample means? The following Minitab commands generate 5,000 samples of $n = 50$ with $p = .3$:

```
MTB >RANDom 5000 observations, put into C1;
SUBC> BINOmial with n = 50 and p = .3.
MTB >LET C2 = C1/50
MTB >HISTogram of data in C2
MTB >DOTPlot of data in C2
MTB >DESCribe data in C2
```

This set of commands generates Figure 5-16. A probability of $p = .1$ produces a distribution like that shown in Figure 5-17.

**FIGURE 5-16** Summary Analysis and Histogram of 5,000 Samples from a Binomial Number Generator with $n = 50$, $p = .3$

```
MTB >RAND 5000 C1;
SUBC> BINO 50 .3.
MTB >LET C2 = C1/50
MTB >HIST C2

Histogram of C2    N = 5000
Each * represents 40 obs.

Midpoint    Count
    0.10      12   *
    0.15      90   ***
    0.20     583   ***************
    0.25     928   ************************
    0.30    1771   ********************************************
    0.35     865   **********************
    0.40     616   ****************
    0.45     105   ***
    0.50      27   *
    0.55       2   *
    0.60       1   *

MTB >DOTP C2

Each dot represents 39 points


                              .
                           . : .
                         : : : : .
                       . : : : : :
                       : : : : : : :
                     : : : : : : : : :
                   . : : : : : : : : : :
                 . : : : : : : : : : : : :
             . . . : : : : : : : : : : : : : . . . . .     .
           -+---------+---------+---------+---------+---------+--C2
          0.10      0.20      0.30      0.40      0.50      0.60

MTB >DESC C2

              N      MEAN    MEDIAN    TRMEAN     STDEV    SEMEAN
C2         5000   0.30124   0.30000   0.30077   0.06575   0.00093

              MIN       MAX        Q1        Q3
C2        0.08000   0.58000   0.26000   0.34000
```

**FIGURE 5-17** Summary Analysis and Histogram of 5,000 Samples from a Binomial Number Generator with $n = 50$, $p = .1$

```
MTB >RAND 5000 C1;
SUBC> BINO 50 .1.
MTB >LET C2 = C1/50
MTB >HIST C2

Histogram of C2    N = 5000
Each * represents 20 obs.

Midpoint    Count
    0.00       28   **
    0.02      128   *******
    0.04      389   ********************
    0.06      674   **********************************
    0.08      902   **********************************************
    0.10      939   ***********************************************
    0.12      760   **************************************
    0.14      548   ****************************
    0.16      335   *****************
    0.18      158   ********
    0.20       82   *****
    0.22       39   **
    0.24       14   *
    0.26        1   *
    0.28        3   *

MTB >DOTP C2

Each dot represents 58 points


                                  .       .
                              :       :
                          :   :   :   :
                          :   :   :   :   :
                      .   :   :   :   :   :
                      :   :   :   :   :   :
                      :   :   :   :   :   :   .
                  .   :   :   :   :   :   :   :   .
                . :   :   :   :   :   :   :   :   .   .   .      .
          +---------+---------+---------+---------+---------+--C2
        0.000     0.060     0.120     0.180     0.240     0.300

MTB >DESC C2

              N      MEAN    MEDIAN    TRMEAN     STDEV    SEMEAN
C2         5000   0.10080   0.10000   0.09962   0.04269   0.00060

              MIN       MAX        Q1        Q3
C2        0.00000   0.28000   0.08000   0.12000
```

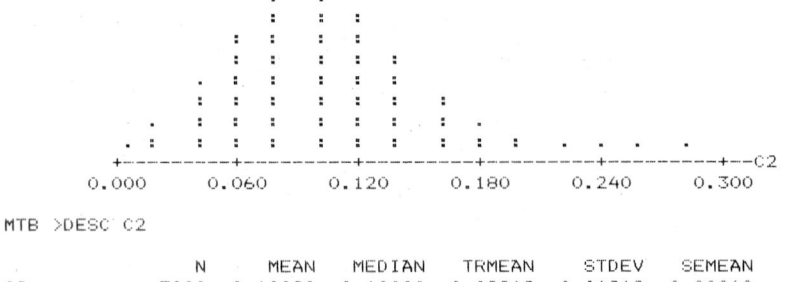

Thus, although you can change the sample size and the probability value, the resulting distribution will still be normal. In general, for a binomial population, as long as $n$ is greater than 50 and the product $np$ is greater than 5, the distribution of sample means will conform to a normal distribution.

## A Poisson Population

Would we obtain the same result if we used a Poisson distribution? Let's assume that we have a Poisson distribution with a mean of 1 defect per sample. The following Minitab command set will allow us to answer this question:

## SECTION 5-6 The Central Limit Theorem

```
MTB >RANDom 5000 observations, put into C1;
SUBC> POISson n = 1.
MTB >LET C2 = C1/1
MTB >HISTogram of data in C2
MTB >DOTPlot of data in C2
MTB >DESCribe data in C2
```

Figure 5-18 shows the results of our simulation of 5,000 observations from a Poisson distribution with a mean of 1. (In this illustration we are dividing by 1 because of the value of the mean. In other examples where the mean is other than 1, this step will not be superfluous as it is here.)

**FIGURE 5-18** Summary Analysis and Histogram of 5,000 Samples from a Poisson Number Generator with Mean = 1

```
MTB >RAND 5000 C1;
SUBC> POIS 1.
MTB >LET C2 = C1/1
MTB >HIST C2

Histogram of C2   N = 5000
Each * represents 40 obs.

Midpoint   Count
   0        1846    ******************************************************
   1        1819    *****************************************************
   2         922    ************************
   3         321    *********
   4          73    **
   5          15    *
   6           3    *
   7           1    *

MTB >DOTP C2

Each dot represents 115 points


        :       :
        :       :
        :       :
        :       :       .
        :       :       :
        :       :       :
        :       :       :
        :       :       :       .       .       .       .
        +-------+-------+-------+-------+-------+---C2
       0.00    1.50    3.00    4.50    6.00    7.50

MTB >DESC C2

          N      MEAN    MEDIAN   TRMEAN   STDEV   SEMEAN
C2      5000    1.0036   1.0000   0.9227   1.0061  0.0142

          MIN     MAX      Q1       Q3
C2      0.0000  7.0000   0.0000   2.0000
```

Figure 5-19 is a similar figure of a Poisson distribution with a mean of 5. Since we have increased the possible sampling area by five times, we must divide each sample by 5 to reduce the defects per sample to a distribution comparable to our sample with a mean of 1. The commands for this second Poisson distribution follow:

**132** ◀ **CHAPTER 5** Probability Distributions

```
MTB >RANDom 5000 observations, put into C1;
SUBC> POISson with n = 5.
MTB >LET C2 = C1/5
MTB >HISTogram of data in C2
MTB >DOTPlot of data in C2
MTB >DESCribe data in C2
```

Notice how quickly we have begun to approach a normal curve.

We can now continue this process in exactly the same way we did for the binomial distribution with samples of 5,000 and mean numbers of defects rising from $n = 15$, to $n = 30$, and finally to $n = 50$. The results of these simulations are shown in Figures 5-20, 5-21, and 5-22, respectively.

**FIGURE 5-19** Summary Analysis and Histogram of 5,000 Samples from a Poisson Number Generator with Mean = 5

```
MTB >RAND 5000 C1;
SUBC> POIS 5.
MTB >LET C2 = C1/5
MTB >HIST C2

Histogram of C2    N = 5000
Each * represents 20 obs.

Midpoint    Count
   0.0        23   **
   0.2       189   **********
   0.4       412   *********************
   0.6       711   ************************************
   0.8       892   *********************************************
   1.0       865   *******************************************
   1.2       689   ***********************************
   1.4       516   **************************
   1.6       346   ******************
   1.8       180   *********
   2.0        98   *****
   2.2        45   ***
   2.4        24   **
   2.6         9   *
   2.8         1   *

MTB >DOTP C2

Each dot represents 55 points







                                                            .
                   :  :
                :  :  :  .
                :  :  :  :
             :  :  :  :  :
             :  :  :  :  :  .
          :  :  :  :  :  :  :
       .  :  :  :  :  :  :  :  .   .   .         .
     +---------+---------+---------+---------+---------+--C2
    0.00      0.60      1.20      1.80      2.40      3.00

MTB >DESC C2

              N      MEAN    MEDIAN    TRMEAN    STDEV    SEMEAN
C2         5000    1.0027    1.0000    0.9889   0.4534    0.0064

              MIN       MAX       Q1        Q3
C2         0.0000    2.8000   0.6000    1.2000
```

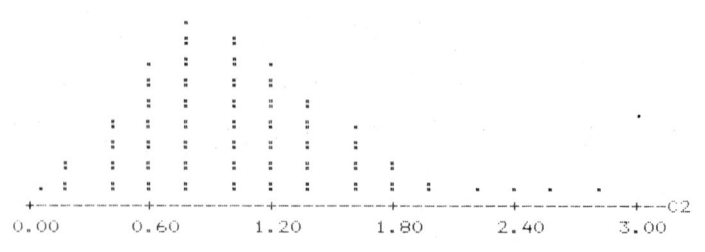

## SECTION 5-6 The Central Limit Theorem

**FIGURE 5-20** Summary Analysis and Histogram of 5,000 Samples from a Poisson Number Generator with Mean = 15

```
MTB >RAND 5000 C1;
SUBC> POIS 15.
MTB >LET C2 = C1/15
MTB >HIST C2

Histogram of C2    N = 5000
Each * represents 35 obs.

Midpoint    Count
    0.0         1   *
    0.2         6   *
    0.4        95   ***
    0.6       538   ****************
    0.8      1207   ***********************************
    1.0      1509   *******************************************
    1.2      1017   *******************************
    1.4       464   **************
    1.6       130   ****
    1.8        28   *
    2.0         3   *
    2.2         2   *

MTB >DOTP C2

Each dot represents 32 points


                            ..
                          . ::.
                         .: ::: .
                        .:: ::: ::
                      . ::: ::: ::.
                      : ::: ::: :::  .
                     :: ::: ::: :::  :.
                    ::: ::: ::: :::  :::
     .   .. ..: ::: ::: ::: ::: ::. ... ..   .
    +---------+---------+---------+---------+---------+--C2
  0.00      0.50      1.00      1.50      2.00      2.50

MTB >DESC C2

            N      MEAN    MEDIAN    TRMEAN     STDEV    SEMEAN
C2       5000   0.99489   1.00000   0.99050   0.26279   0.00372

              MIN       MAX        Q1        Q3
C2        0.06667   2.13333   0.80000   1.18333
```

Clearly, as the sample size increases, the distribution of sample means becomes normal. Generally, a sampling distribution is normally distributed when the sample size is greater than 30 to 50. This fact, the central limit theorem, accounts for the primary importance of the normal distribution. We can use the normal distribution for any case where we are drawing large samples *regardless* of the shape of the original population. Of course, this is only true if we have met the proper conditions of good sampling described in Chapter 4. If the sample is random, the central limit theorem holds. If the samples are poorly drawn, then nothing, including the central limit theorem, will make any difference. Our results will be garbage.

**FIGURE 5-21** Summary Analysis and Histogram of 5,000 Samples from a Poisson Number Generator with Mean = 30

```
MTB >RAND 5000 C1;
SUBC> POIS 30.
MTB >LET C2 = C1/30
MTB >HIST C2

Histogram of C2   N = 5000
Each * represents 25 obs.

Midpoint   Count
   0.4        1   *
   0.5       31   **
   0.6      104   *****
   0.7      296   ************
   0.8      659   ***************************
   0.9      969   ****************************************
   1.0     1033   ******************************************
   1.1      861   ***********************************
   1.2      596   ************************
   1.3      280   ************
   1.4      115   *****
   1.5       42   **
   1.6       11   *
   1.7        2   *

MTB >DOTP C2

Each dot represents 22 points
                        ...:::  ..
                       .::::::  ::
                       :::::::  ::..
                      .:::::::  ::::.
                    . :::::::::  ::::::
                    : :::::::::: :::::::.
                   .::: :::::::::: :::::::. .
                . ...:::::: :::::::::: :::::::: :::......  ..
           +---------+---------+---------+---------+---------+--C2
          0.30      0.60      0.90      1.20      1.50      1.80

MTB >DESC C2

              N       MEAN    MEDIAN    TRMEAN     STDEV    SEMEAN
C2         5000    0.99749   1.00000   0.99573   0.18736   0.00265

              MIN      MAX        Q1        Q3
C2        0.43333  1.73333   0.86667   1.13333
```

## A Uniform Population

Just to make sure that you fully accept the central limit theorem, let's try one more experiment using uniform distributions. In a *uniform distribution*, each value occurs the same number of times. For example, a distribution that includes the values of 1, 2, 3, 4, and 5 where each value occurs three times would constitute a uniform distribution. Such a distribution looks like a rectangle with the height equal to the number of occurrences of each value. Figure 5-23 illustrates such a distribution.

We can take samples from such a distribution using the **DISCrete** subcommand in Minitab. In the present example, we have five values from 1 to 5, and each has an equal chance of occurrence. To use the **DISCrete** subcommand, we must first enter the actual

## SECTION 5-6 The Central Limit Theorem

**FIGURE 5-22** Summary Analysis and Histogram of 5,000 Samples from a Poisson Number Generator with Mean = 50

```
MTB >RAND 5000 C1;
SUBC> POIS 50.
MTB >LET C2 = C1/50
MTB >HIST C2

Histogram of C2   N = 5000
Each * represents 30 obs.

Midpoint    Count
    0.5        3   *
    0.6       26   *
    0.7      179   ******
    0.8      575   ********************
    0.9     1037   ***********************************
    1.0     1356   **********************************************
    1.1     1055   ************************************
    1.2      531   ******************
    1.3      177   ******
    1.4       52   **
    1.5        8   *
    1.6        1   *

MTB >DOTP C2

Each dot represents 32 points


                                :   :
                                :   :
                            .   :   :
                            :  :.:.:    :
                           :...::::::.  :
                          .:::::::::::::
                         :::::::::::::::::::::
                   . ......::::::::::::::::::::::.:..... ..
                 ---+---------+---------+---------+---------+-------C2
                   0.50      0.75      1.00      1.25      1.50     1.75

MTB >DESC C2

              N      MEAN    MEDIAN   TRMEAN    STDEV   SEMEAN
C2         5000    1.0011    1.0000   0.9999   0.1479   0.0021

              MIN       MAX        Q1       Q3
C2         0.5000    1.5600    0.9000   1.1000
```

**FIGURE 5-23** A Uniform Distribution Curve

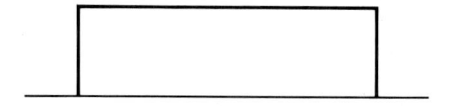

values of the distribution into column 1 and the probabilities of their occurrence into column 2. The Minitab instructions would look as follows:

```
MTB >READ C1 C2
DATA>1 .2
DATA>2 .2
DATA>3 .2
DATA>4 .2
DATA>5 .2
```

It is critical that the total of your probabilities always equals 1.0 for such a distribution. Thus, this example could be repeated with any group of numbers with any set of probabilities provided that the probabilities totaled to 1.0. If you wish to repeat this illustration with the numbers 0, 1, 2, 3, 4, and 5 and the probabilities .3, .2, .1, .1, .1, and .2, you are encouraged to do so. You can then compare those results with the results we obtain in this example.

In the **DISCrete** subcommand you must specify the column from which you wish to draw the sample and the column in which the probabilities for each population value are found. The general form of the **DISCrete** subcommand is as follows:

```
RANDom K observations, put into C;
    DISCrete with values in C, probabilities in C.
```

To draw a sample of 50 observations from column 1 (*with replacement*, since there are only five possible values), we would use the following command set to put our output into column 3:

```
MTB >RANDom 50 observations, put into C3;
SUBC> DISCrete with values in C1, probabilities in C2.
```

This command will draw a sample of 50 observations and place those values into column 3.

We now want to find the mean of the sample. We then want to repeat this process, taking a predetermined number of samples and finding the mean of each. Finally, we want to build a new column that contains the means of all these samples. To create this column of sample means, we must be able to insert values into a column one row at a time. We do this by using a form of the **LET** command. For example, if we wanted to place a specific value (say 46) into the first row of a column (say column 15), we would use the command

```
MTB >LET C15(1) = 46
```

To place the value of 99 in the second row of column 15, we would use

```
MTB >LET C15(2) = 99
```

Finally, to place the value of 57 in the 40th row of column 15, the command would be

```
MTB >LET C15(40) = 57
```

To create the column of sample means, we will calculate the mean of column 3 and designate this value as **K1**. Then we will insert that mean into a column (say column 4) of mean values. Then we will generate a new random sample of 50, put it into column 3 (thereby deleting the values of the previous sample), and repeat the process. Examine the following command set:

```
MTB >RANDom 50 observations, put into C3;
SUBC> DISCrete with values in C1, probabilities in C2.
MTB >MEAN data in C3, put in K1
MTB >LET C4(1) = K1
MTB >RANDom 50 observations, put into C3;
SUBC> DISCrete with values in C1, probabilities in C2.
MTB >MEAN data in C3, put in K1
MTB >LET C4(2) = K1
```

### SECTION 5-6 The Central Limit Theorem ▶ 137

What we have done with this command set is to take a sample of size 50 and put the mean of that sample in column 4, row 1, and then repeat the process, placing the second mean in column 4, row 2. If we typed in this command set 500 or 1,000 times, we would build in column 4 the means of 500 or 1,000 samples of size 50. Although the result is OK, the means leave a bit to be desired!

To issue such a command system would be a great deal of work. However, Minitab has a system that allows you to execute a set of commands more than once using the command group **STORe** and **EXECute**. To use this system, you first enter the **STORe** command and then enter all the commands necessary for your problem. Once that is finished, you enter the command **END**. You can then enter the **EXECute** command followed by the number of times you want the command group to be executed. In general, the command group looks as follows:

```
MTB >STORe
  :
MTB >END
MTB >EXECute 500 times
```

In our present example we have an additional problem. Notice that in each set of commands the *row* assigned advances by 1. To use the **EXECute** command we need to advance the row assignment 1 unit each time Minitab cycles through the command group. The following set of commands will allow us to accomplish this:

```
MTB >LET K2 = 1
MTB >STORe
STOR>RANDom 50 observations, put into C3;
STOR> DISCrete with values in C1, probabilities in C2.
STOR>MEAN data in C3, put in K1
STOR>LET C4(K2) = K1
STOR>LET K2 = K2 + 1
STOR>END
MTB >EXECute 1000 times
```

Notice that we use the value **K2** to tell Minitab where to place the mean in **C4**. By just setting the value of **K2** to 1 and then increasing its value by 1 with each execution, we can use it as a counter. If you have done any computer programming, you will recognize this concept.

The set of commands above would generate 1,000 samples of size 50 and place the mean values of those 1,000 samples in column 4. Remember that the original distribution was uniform—a nice, neat rectangle. However, if we now execute **HISTogram** and **DOTPlot** commands on column 4, the results, shown in Figure 5-24, indicate that again we have a distribution of sample means that is approximately normally distributed. (This example has one minor flaw: The computer prints each of the 1,000 mean values every time it calculates a new mean! Because this will result in a waste of computer paper, be certain that your printer is *off*. Your instructor will show you how.)

**FIGURE 5-24** Histogram and Dot Plot of Mean Values of 1,000 Samples of Size 50 Drawn from a Uniform Distribution of Values 1 through 5

```
MTB >HIST C4

Histogram of C4    N = 1000
Each * represents 5 obs.

Midpoint   Count
    2.4       3   *
    2.5       9   **
    2.6      27   ******
    2.7      53   ***********
    2.8     117   ************************
    2.9     183   *************************************
    3.0     199   ****************************************
    3.1     180   ************************************
    3.2     118   ************************
    3.3      64   *************
    3.4      36   ********
    3.5       8   **
    3.6       2   *
    3.7       1   *

MTB >DOTP C4

Each dot represents 5 points

                          .
                        : :
                      : : : .
                    . : :.: : .
                  : :.: :::::::.::      :
                . : .::::::::::::::..:
              . ....:.:.::::::::::::::::::::... .
           ------+---------+---------+---------+---------+---------C4
              2.50      2.75      3.00      3.25      3.50      3.75

MTB >DESC C4

              N      MEAN    MEDIAN    TRMEAN     STDEV    SEMEAN
C4         1000    3.0060    3.0000    3.0063    0.1973    0.0062

              MIN       MAX        Q1        Q3
C4         2.3800    3.6600    2.8800    3.1400
```

# SUMMARY

For any distribution from which we take a large random sample, we can be reasonably sure that the mean of the sample approaches the mean of the population from which the sample was drawn (the law of large numbers). If we take repeated random samples from *any* population, the distribution of the means of those random samples will form a normal distribution (the central limit theorem).

This chapter has laid the conceptual foundation necessary for much of statistical analysis. Armed with an understanding of the central limit theorem, you now can enter the world of inferential statistics. We have concluded that there are three possible types of statistical cases. First is when the parent population is normal. Here we can take a reasonably small sample and (cautiously) draw some statistical inferences. The second case is when the

# PROBLEMS

parent population is not normal. If we take much larger samples than when the population is normal, we can still draw statistical inferences because the resulting samples will follow a normal distribution (the central limit theorem). The third case is when the sample size is small and the parent population is not normal. Here we cannot rely on the usual tools of statistical analysis.

## DISCUSSION QUESTIONS

1. What is the relationship between a Bernoulli distribution and a binomial distribution?
2. What is a Poisson distribution?
3. Explain the central limit theorem.
4. What is the law of large numbers?

## PROBLEMS

*Warning:* If you are using Minitab on a dual-disk-drive PC, some of these problems will take hours to compute, especially those requiring the generation of 2,000 or more samples. In such cases, the fundamental point can be illustrated with samples of smaller size. On mainframe computer systems, the time required for these large samples is not significant.

**5-1** In the classified ads there is a probability of .40 that any particular ad is placed by a car dealer.
**a)** You would like to buy a car, but you feel that car dealers ask for a higher price than do private individuals. If you were to use the Bernoulli random generator to generate 50 observations representing classified ads for autos, how many ads would not have been placed by car dealers?
**b)** Using the Bernoulli random number generator to generate a sample of 50 observations representing classified ads for autos, how many ads would not have been placed by car dealers?
**c)** Was your answer in (a) correct? If not, why was it different?

**5-2** Forty percent of all automobile classified ads are placed by car dealers. Assume that out of 15 ads, you found approximately 45% to be placed by car dealers.
**a)** If a friend remarked that the probability of having an ad placed by a car dealer must be 45% rather than 40%, what would you say?
**b)** Generate 200 observations using the Bernoulli random number generator with $p = .40$. Compute the running average. Does the average support your answer in (a)?
**c)** Do a `TSPLot` of the means to see how the means converge as the sample size increases.

**5-3** In Europe many forests are being destroyed by acid rain. Assume that the probability of a tree dying is .45.
**a)** Use the Bernoulli random number generator to generate a sample of 15 observations. Compute the running average.
**b)** Use the `RANDom` command to generate a sample of 2,000 Bernoulli observations. Compute the running average.
**c)** Compare the last value of the running average in (a) to that in (b). What do you notice?

**5-4** On most airline flights, a round of drinks must be served to passengers, and each passenger must choose between an alcoholic drink and a nonalcoholic drink. Each row on the plane has 5 seats. Glenn, a flight attendant, dislikes serving alcohol because he has to collect money for these beverages.

**a)** If 30% of the passengers choose alcoholic drinks and 70% choose nonalcoholic drinks, develop a table for a probability tree for all possible samples of a size $n = 5$.

**b)** Glenn made a bet with another flight attendant. If 2 or fewer of the passengers seated in a row of 5 ordered an alcoholic drink, Glenn would lose and have to serve the round of drinks on the next flight. What is the probability of serving 2 or fewer passengers alcoholic drinks? Does Glenn have a good chance of winning?

**c)** To compute the probability of a sequence of outcomes, what must be true of the observations?

**5-5** Glenn knows that there are 100 rows with 5 seats in a row on a 747.

**a)** If the probability of serving alcoholic drinks is .30 for a sample of size 5, what is the mean of the population?

**b)** Use the binomial random generator to generate 100 samples of a sample size of 5.

**c)** Refer to Problem 5-4(b). Does the percentage of 2 or fewer drinks coincide with the outcomes found in the summary table just created?

**5-6** At a local grocery store, a sales clerk is giving away promotional items. To every male passerby, she gives a razor, and to every female a pair of nylon stockings. In a 15-minute period, there are 6 passersby.

**a)** Assume that 30% of the passersby are male. Use the binomial random generator to generate 100 samples of size 6.

**b)** Review the summary data. If the sales clerk determined the number of razors needed for every 15 minutes by the highest frequency found in the summary table, approximately how many razors would she need in an 8-hour day?

**5-7** A sewing shop manager found the mean number of flaws in a bolt of material to be 10.

**a)** Use the **POISson** command to generate output with a mean equal to 10.

**b)** If the manager only accepted bolts of material with 12 flaws or fewer, what percentage of bolts would he accept?

**5-8** The sewing shop manager believes that the percentage of the number of bolts that he can accept is quite low.

**a)** To confirm the answer from Problem 5-7, use the Poisson random number generator to generate a sample of 1,000 bolts with a mean number of flaws equal to 10.

**b)** Create a histogram for the data just generated.

**c)** From the histogram, manually count the number of bolts that would be accepted and divide it by the sample size. Does this percentage coincide with the answer given in Problem 5-7(b)?

**5-9** Print It, Inc., publishes a large number of romance books. When printing these books, the company tries to keep its overhead low by using a slightly older method for reproducing texts. However, this method results in a few pages with lightly printed copy.

**a)** Use the Poisson random number generator to generate 150 romance books with a mean number of lightly printed pages equal to 6.

**b)** What will the histogram for the number of lightly printed pages look like?

**c)** Create the histogram.

# PROBLEMS

**5-10** A wholesaler stores large quantities of record albums for distribution to record stores within the city. Eight percent of the records distributed have scratches and are returned to the wholesaler.

**a)** Draw a rough sketch of a histogram representing a single large sample for the wholesaler.
**b)** Use the RANDom command and BINOmial subcommand to generate 2,000 samples with 50 observations in each sample. Compute the mean value for each sample.
**c)** Create a histogram of the means.
**d)** Do the distributions in (a) and (c) differ? If so, why?

**5-11** After careful study, the wholesaler realized that the number of records that were returned because of scratches increased from 8% to 12%.

**a)** Use the RANDom command and BINOmial subcommand to generate 5,000 samples with 50 observations in each sample. Compute the mean value for each sample.
**b)** Will the histogram of the means be any different from the one created in Problem 5-10 because of the increase in the number of scratched records? Why or why not?
**c)** Create a histogram of the means.

**5-12** A manufacturing firm produces clear plastic handles for various items like umbrellas and suitcases. The company knows that the mean number of air bubbles that occur in the handles is equal to 5. You are the production manager. When you reported to the president of the company, you presented him with a graph of a sample of the number of air bubbles in handles. The figure resembled a Poisson distribution. The president was confused since he had a graph showing the mean number of air bubbles to be normally distributed.

**a)** What sort of explanation would you give to the president?
**b)** To support your explanation, use the Poisson random number generator to generate 5,000 samples with a mean equal to 5. Divide the values by the mean.
**c)** Create a histogram of the means.

**5-13** The president does not understand your explanation of Problem 5-12.

**a)** Use the Poisson generator to generate 5,000 samples with a mean of 15. Divide by the mean and create a histogram of the means.
**b)** Do the same as in (a) except use a mean of 25.
**c)** Use the two histograms just created to support your explanation. What would you tell the president?

**5-14** The distribution for the number of chocolate chips found in a cookie from Chips, Inc., appears as follows:

| No. of chocolate chips | 5 | 6 | 7 | 8 | 9 |
|---|---|---|---|---|---|
| Probability | .10 | .15 | .20 | .35 | .20 |

**a)** You, as the chief cookie maker, realize that the distribution is not normal. However, you want to show your boss that after taking a large number of samples, the means of the sample are normally distributed. Use the RANDom command to generate 100 observations using the distribution described. Find the mean of each of 5,000 samples of size 100.
**b)** Create a histogram of the sample means. Use the DESCribe command on the data.
**c)** Does the sample appear to be normal?

**5-15** You quit your job as chief cookie maker at Chips, Inc. After doing some job hunting, you were hired as chief cookie maker at Kookies, Inc. The distribution for the number of chocolate chips found in cookies from Kookies is as follows:

| No. of chocolate chips | 5 | 6 | 7 | 8 | 9 |
|---|---|---|---|---|---|
| Probability | .20 | .35 | .20 | .15 | .10 |

a) Use the RANDom command to generate 100 observations using the data developed. Find the mean of each of 5,000 samples of size 100.
b) Create a histogram of the means.
c) Compare the histogram just created to the one in Problem 5-14. What are your observations concerning the histograms and the distributions of their respective populations?

# 6

# STATISTICAL ESTIMATION

**T**he following Minitab commands will be introduced in this chapter:

`ZINTerval, confidence level = K, sigma = K, data in C`
A command to calculate the confidence interval for a normal population with a mean of mu

`RN C,...,C, put in C`
A command to count the number of entries (actual numbers) in a row and to store that number

`RNMIss C,...,C, put in C`
A command to count the number of missing values in a row and to store that number

`RMEAn C,...,C, put in C`
A command to compute the row mean and to store that number

`RMEDian C,...,C, put in C`
A command to compute the row median and to store that number

`RSTDev C,...,C, put in C`
A command to compute the row standard deviation and to store that number

`RMAX C,...,C, put in C`
A command to compute the row maximum and to store that number

`RMIN C,...,C, put in C`
A command to compute the row minimum and to store that number

```
RSUM C,...,C, put in C
```
A command to compute the row sum and to store that number
```
RSSQ C,...,C, put in C
```
A command to compute the row sum of squares and to store that number
```
RCOUnt C,...,C, put in C
```
A command to count the number of observations in a row and to store that number

Until now we have been developing statistical tools and concepts. Now we will undertake our first building project, using these tools and concepts to estimate population parameters. In this chapter we will learn how to estimate points and intervals. We will also introduce the standard normal (Z) distribution and its use.

## 6-1 POINT ESTIMATES

Suppose that you wanted to learn the average age of all workers in all shoe factories in the United States. (This might be important for negotiating fringe benefit packages and group life and health insurance policies.) One way to approach this problem would be to obtain a list of all the shoe factory workers in the country and compute the mean age. Although this method would probably accomplish the desired task, there would be some apparent problems. As mentioned in Chapter 4, obtaining an accurate list would be difficult because new employees would be hired, old employees would retire, individuals would be fired, and so on. Even if we could generate an accurate list of all workers, the cost in time and money would most likely be prohibitive.

Fortunately, we have ways to deal with this type of problem. You should recall from the law of large numbers that the mean of a sample will approximate the mean of the population from which that sample is drawn if the sample is sufficiently large and is randomly drawn. Now, if we could assume that our work force at Go-Fast was representative of all shoe factory workers in the country, then we could assume that the average age of the Go-Fast work force is the same as the average age of the population. So let's make this assumption and use our employees as a sample. We can then compute the mean of our workers and use this as an estimate of the population mean.

If the ages of the Go-Fast workers has been entered into column 1, the command and result would be

```
MTB >MEAN C1
    MEAN = 31.72
```

With this information, we would estimate the national average of shoe factory workers to be 31.72 years old. Is this a reasonable technique? With a little caution, it is! Assuming that there is not some obvious bias (such as if the factory was located in a retirement community and employed only retirees over 65 years old), we can proceed. However, we would certainly not assume that the national average is exactly 31.72 years old. Instead, we would say that it is approximately that age. The use of a sample mean as an estimate of the population mean is a form of *point estimation*. If we are forced to estimate a population parameter with a single value, then we use a point estimate. However, you should already have a somewhat uneasy feeling about the accuracy of this single value. It is probably close to the actual population mean, but we can hardly be 100% certain of its precision. (In fact, the probability of the sample mean exactly equaling the population mean is 1/∞. If you have taken a course in integral calculus, you may recall that you can only calculate the area under a portion of a continuous curve using an interval along its base. The area defined by a single point on the base of a continuous function is approximately 0.)

## 6-2 CHARACTERISTICS OF THE NORMAL CURVE

We said before that we could estimate the population's average age to be approximately 31.72 years. What do we mean by "approximately"? Do we mean between 31 years and 32 years? Do we mean between 30 and 34 years? The answer is difficult to know, but we can control and predict how much variation there is in our point estimates.

As an illustration, let's perform an experiment. We will select a population with a known mean of 100 and a standard deviation of 10. Remember, we know with *certainty* the population's mean and standard deviation. We will draw a sample from that population and compare the sample with the parent population. For this task we can use the Minitab normal random data simulator. Using the commands available in Minitab, we can specify the number of observations to draw, the population mean (mu), the population standard deviation (sigma), and the column in which to store the data values. (When we refer to the population standard deviation, we will use sigma [$\sigma$]; to the sample standard deviation, $S$; to the population mean, mu [$\mu$]; to the sample mean, "x bar" [$\bar{x}$]. This system of notation is fairly common in statistics books.)

### Mean of a Single Sample

We can generate the desired sample with the RANDom command and the NORMal subcommand discussed in Chapter 5. The command group is

```
MTB >RANDom 30 observations, put into C1;
SUBC> NORMal with mu = 100, sigma = 10.
```

Be sure to put the semicolon at the end of the first command line and a period at the end of the subcommand. We can then determine the mean of our sample with the command

```
MTB >MEAN C1
```

When we performed the above commands, the sample mean was computed to be 98.688. If you execute these commands and the student next to you does also, your answers will probably be slightly different. Remember, the RANDom command takes a sample from a population and then estimates the mean.

### Mean of Sample Means

Let's try to get a better feel for how these various estimates behave. We have just found the mean of a single sample of size 30 to be 98.688. That is reasonably close to 100 but certainly not exact. In fact, the probability of the mean of a sample of size 30 being exactly 100 is very small. What if we took a second sample of size 30 and found the mean of that sample? We would probably get another figure very close to but not exactly on 100.

## CHAPTER 6 Statistical Estimation

If we repeated this process 100 times, we would collect 100 sample means. Since all 100 samples would have come from the same population, how those samples *clustered* would assist us in understanding estimation. (The mechanics of drawing 100 samples of size 30 are not particularly complex, but they detract from our discussion here. In an appendix to this chapter, we describe how to draw 100 samples quickly.) Assume that the work has been done and that the 100 sample means, each determined for a sample of size 30 from a population with a mean of 100 and a standard deviation of 10, have been stored in column 1.

To obtain a better sense of the values in column 1, we will put them in order from lowest to highest. This will allow us to determine the extreme values much faster. The **SORT** command will rearrange the values in any column and place them in a new column, as discussed on page 28. The command takes the form

```
SORT data in C, put in C
```

For this example, we want to sort the values and then print the list. The commands are

```
MTB >SORT data in C1, put in C2
MTB >PRINt C2
```

The results of these commands for our data are shown in Figure 6-1. We have also printed the results of **DESCribe**, **HISTogram**, **DOTPlot**, and **STEM-and-leaf** commands on the same data.

The pattern shown in the histogram is just what we expect. Our modal value is 100 and the great majority of our values are between 97 and 103. Outside of this range, the number of values falls off quite quickly. Of course, when you run these commands, your results will be slightly different.

**FIGURE 6-1**   Results of Analyzing 100 Samples of Size 30 Drawn from a Population with Mean = 100, Standard Deviation = 10

```
MTB >SORT C1 C2
MTB >PRIN C2

C2
    95.968    96.374    96.696    96.792    96.893    97.080    97.500
    97.569    97.668    97.677    97.795    97.863    97.993    98.065
    98.252    98.344    98.452    98.483    98.494    98.592    98.769
    98.795    98.875    98.915    98.918    98.970    99.098    99.165
    99.200    99.207    99.213    99.287    99.400    99.442    99.492
    99.538    99.586    99.608    99.643    99.655    99.657    99.740
    99.779    99.781    99.804    99.821    99.879    99.886    99.936
    99.989    99.992   100.082   100.108   100.124   100.143   100.164
   100.183   100.282   100.282   100.301   100.420   100.541   100.543
   100.743   100.796   100.839   100.896   100.930   100.938   100.956
   101.028   101.142   101.146   101.203   101.290   101.292   101.358
   101.403   101.451   101.468   101.489   101.495   101.570   101.662
   101.783   101.890   101.891   101.970   101.981   101.982   102.018
   102.245   102.373   102.411   102.905   103.210   103.223   103.382
   103.734   104.565

MTB >DESC C2

              N       MEAN    MEDIAN    TRMEAN    STDEV    SEMEAN
C2          100     100.05     99.99    100.05     1.70      0.17

              MIN       MAX       Q1        Q3
C2          95.97    104.56    98.93    101.29
```

*(continued on next page)*

**FIGURE 6-1** (*Continued*)

```
MTB >HIST C2

Histogram of C2    N = 100

Midpoint   Count
      96       2  **
      97       4  ****
      98      13  *************
      99      16  ****************
     100      26  **************************
     101      21  *********************
     102      12  ************
     103       4  ****
     104       1  *
     105       1  *
MTB >DOTP C2


                                    .: :        . . .
                        .       . . : :::.:.. .: :: :
. . ....  .:.:..::::::::::::::::: :::::.:.:    . :. .           .
-+---------+---------+---------+---------+---------+------C2
96.00     97.60     99.20     100.80    102.40    104.00

MTB >STEM C2

Stem-and-leaf of C2       N = 100
Leaf Unit = 0.10


     1    95  9
     5    96  3678
    13    97  05566789
    26    98  0234445778999
   (25)   99  0122223445566667778888999
    49   100  0111112234557788999
    30   101  011222344444456788999
    10   102  02349
     5   103  2237
     1   104  5
```

The distribution of these sample means can be used to illustrate an important principle. You will notice that the lowest value in Figure 6-1 is 95.968. Remember, to obtain that value, we randomly drew 30 observations from a population with a mean of 100 and a standard deviation of 10. In this particular case, our estimate of the true population mean is not particularly good. So when we use the age of Go-Fast employees to estimate the age of employees in the entire shoe industry, how do we know if our estimate is one of the central values (near the true population mean) or one of the values at an extreme? In fact, we really do not know which case is true! All we do know is that *all random samples will tend to behave in exactly the same way as the 100 samples we just drew*. Most of the time our estimates will be very close to the population mean, but once in a while they will not be close.

# The Normal Curve

If we were to take all possible samples of size 30 from a population with mean equal to 100 and standard deviation equal to 10, a pattern very similar to the one obtained in the graphs in Figure 6-1 would emerge. The distribution of *all* possible samples would look like that shown in Figure 6-2. This fact follows from the central limit theorem.

**148** ◀  **CHAPTER 6** Statistical Estimation

**FIGURE 6-2** A Normal Curve of the Means of All Possible Samples of Size 30 from a Normal Population with Mean = 100, Standard Deviation = 10

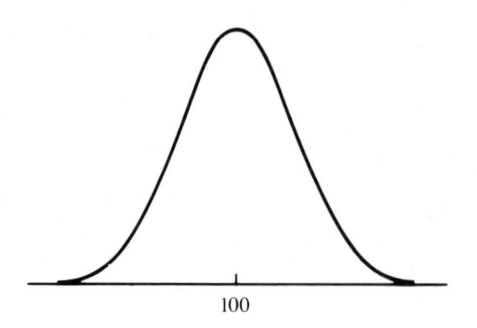

If we eliminate the 5 lowest and the 5 highest of our 100 sample means, what are we left with? We have 90 means that should bracket our estimate of the population mean 90% of the time. Similarly, if we had all possible samples and eliminated the bottom 5% and the top 5% of those sample means, the remainder would bracket the true population mean. This 90% bracket would be our *interval estimate*. We could also construct interval estimates for 95%, 99%, or any other percentage of sample means drawn from our target population. Figure 6-3 illustrates the most commonly used intervals.

**FIGURE 6-3** Normal Curves with Various Interval Estimates

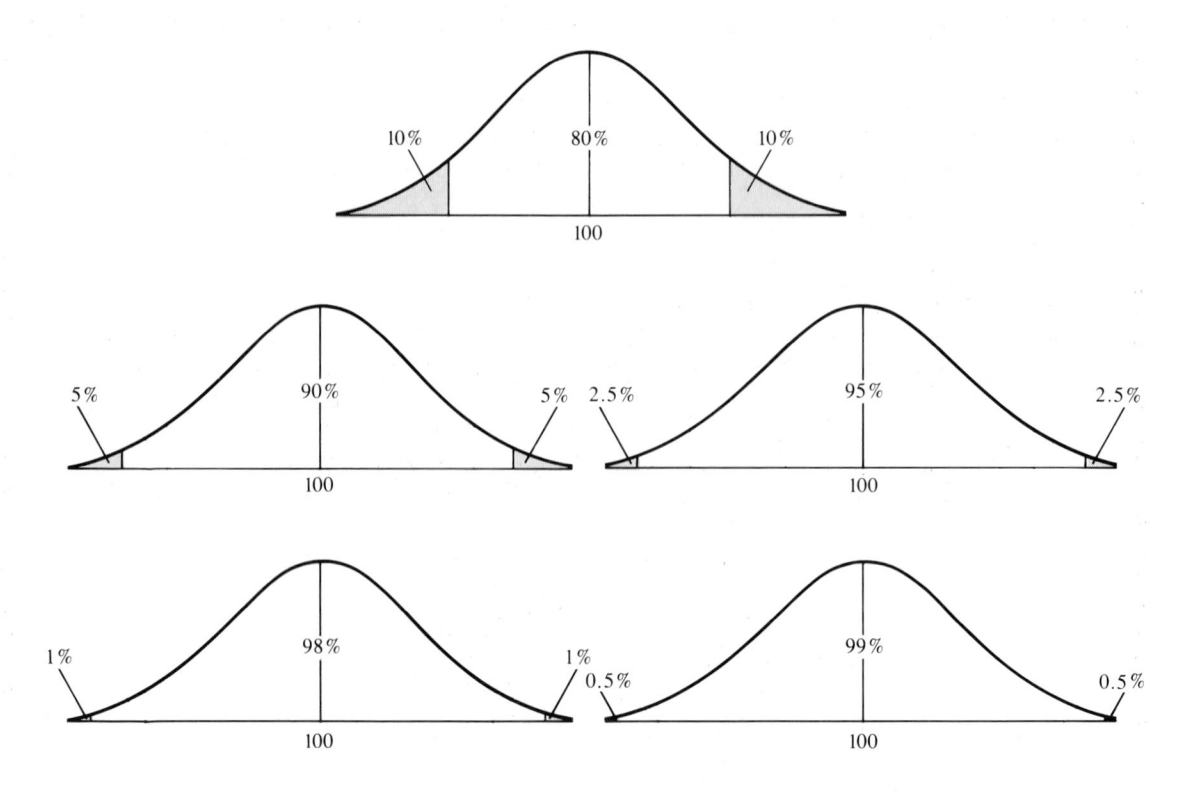

SECTION 6-3 The Standard Normal Curve

**FIGURE 6-4** 90% Confidence Interval Estimate for a Normal Curve with Mean = 100, Standard Deviation = 10

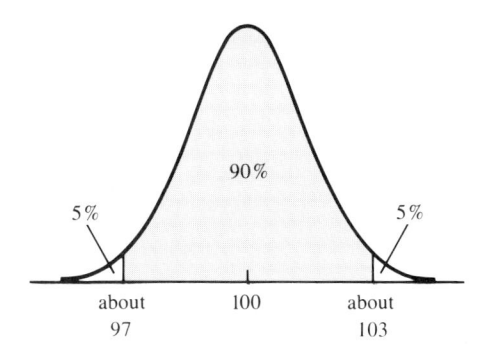

Let's review what we've done so far. We have taken 100 separate samples of size 30 from a normal distribution. No sample mean hit 100 exactly; in fact, they were spread out over a range of values. However, *the pattern of how those means spread out is regular and systematic*. From looking at the data, we know that about 90% of the time, the value of the sample mean was between approximately 97 and approximately 103. This range of 97 to 103, which is a 90% interval estimate, simply means that if the actual population has a mean of 100 and a standard deviation of 10, then 90% of the time our estimate of that mean will fall between approximately 97 and 103. Also, 10% of the time our estimate will be outside of this interval, as illustrated in Figure 6-4.

## 6-3 THE STANDARD NORMAL CURVE

We are using a population with a mean of 100 and a standard deviation of 10. The normal curve with a mean equal to 0 and a standard deviation of 1 is known as the *standard normal curve*. The standard normal curve is the reference point for developing confidence intervals and testing hypotheses. Almost all of our statistics refer to the standard normal curve. As we develop the material in this chapter, the importance of the standard normal curve will become apparent.

### The Z Statistic

The *population* of Go-Fast workers is normally distributed, and we do not need to compute all possible sample means or eliminate the end percentages to look at the intervals for the means. Because normal distributions all have the same shape, we can make interval estimates based on the Z score table. A complete Z table is given in Appendix B, Table 1.

The Z *statistic* is a statistic that allows us to standardize the area under a normal curve. The Z statistic itself refers to the number of standard deviations a particular value happens to fall from the mean. Basically, 1 standard deviation on either side of the mean is 1 Z-value; 2 standard deviations on either side of the mean is 2 Z-values. Because of the unique shape of the standard normal curve, we know that 1 Z on both sides of the mean always includes 68.27% of the area under the curve. This means that plus and

**FIGURE 6-5** Areas under the Normal Curve for 1, 2, and 3 Standard Deviations

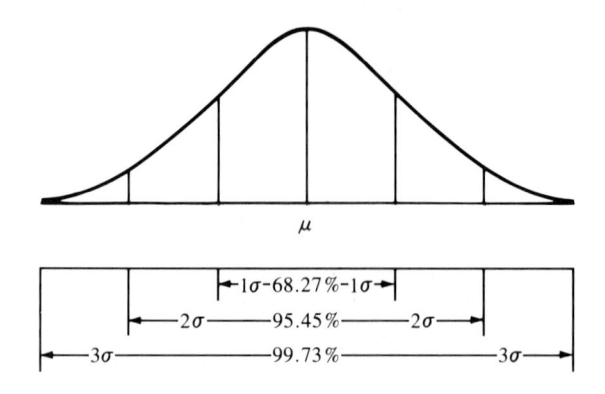

minus 1 Z score (or 1 standard deviation) includes about 68% of all observations in a normal distribution.

Plus and minus 2 standard deviations (or 2 Z-values) always encompass 95.45% of the area under the curve, and plus and minus 3 standard deviations (3 Z-values) encompass 99.73% of the area. These ranges are illustrated in Figure 6-5.

With this concept, we can answer the question "What is the necessary range if we want to capture 90% of the area under the curve?", or "What is the necessary value if we want to include 90% of the possible mean values around a population mean?"

## One- and Two-Tail Z Statistics

The Z statistic allows us to calculate the interval estimate for *any* normal population. Before building interval estimates, we must determine what percentage of the possible range of values we wish to include in our estimate. We have been discussing a 90% range, that is, dropping 5% from each end. We could also talk about a 90% range with all 10% dropped off only one end (let's say the lower end). In this case our estimate would have a lower limit but no upper bound. In our example, such an estimate would be a 90% confidence interval that the true mean value was 97.6 or above. Both cases are illustrated in Figure 6-6.

**FIGURE 6-6** 90% One-Tail and Two-Tail Confidence Intervals for a Normal Curve with Mean = 100, Standard Deviation = 10

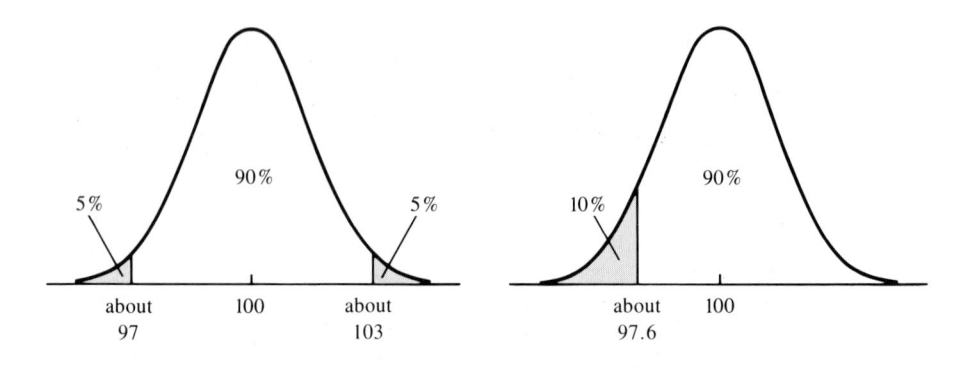

Thus, it is important to consider if your estimate is going to be a *two-tail* estimate (dropping values off each end of the curve) or a *one-tail* estimate (dropping values off only one end of the curve), because the Z-value will change. For example, as you will quickly see, the Z-value for a 90% one-tail test is 1.28, but the Z-value for a 90% two-tail test is 1.65.

## 6-4 THE CONFIDENCE INTERVAL

Working with one-tail and two-tail tests will probably be awkward at first. The problem is that for a 90% two-tail interval estimate (dropping values from both ends of the range), you do *not* use the Z-value for dropping off 10%. Instead you drop 5% off each end of the distribution for a total of 10%. The two-tail test is more common, and you will quickly become used to looking up the 5% value for a 90% interval, the 2.5% value for a 95% interval, and so on.

Before we examine the form for the estimate of the confidence interval, we should make certain points clear. In the first portion of the discussion below, we will be working with a known population. From that population, we will draw samples and examine how well they reflect the true population parameters. In some cases, the sample will reflect the population from which it was drawn quite closely; in other cases, it will not.

Let's assume that Go-Fast has just received a shipment of shoe string fabric for its running shoes. Further, let's assume that there are 10,000 spools of shoe string material in the shipment, each 100 inches long with a standard deviation of 10 inches. Remember, these are *known* values of the population. We will then take a sample from that population, and based on that sample, we will develop a method for estimating the mean within a certain range (called the *confidence interval estimate*). This concept is extremely important. If we take a single sample of size 100 from a population of 10,000 and find the sample mean, intuition should tell us that the probability of that calculated mean being exactly the same as that of the population is very remote. However, our intuition should also tell us that there is a good chance that the calculated sample mean will be close to the true population mean. Instead of using a single point estimate of the mean (the single calculated value), we will attempt to determine a reasonable *range* around the sample mean within which the population mean is most likely to be found.

Next we will see how well an interval around a given population mean predicts sample means. In other words, given the population mean and standard deviation and the size of randomly drawn samples from that population, we will predict the range within which sample means will most likely fall. In the following illustration, we will show that if the mean of the population is 100 and the standard deviation is 10 and we draw samples of size 30, then the values of 90% of the sample means obtained will fall between 96.9 and 103.0.

In the practical world of statistics, we almost never know the true population mean and standard deviation. If we did, there would be no need to sample that population. Instead, we almost always have a sample from which we try to estimate the characteristics of the population. Our final step is thus to provide a practical tool for estimating a population mean based on sample data and inductive logic. The logic behind this tool will be the reverse of the logic described in the previous paragraph, where we established an interval estimate of 96.9 through 103.0 for a *given* population mean. Instead, we will establish an interval *around a sample mean* and try to estimate the probability that the true population mean falls within that interval.

With this background, let us now return to the concept of the confidence interval and the assumption of a known population from which we have drawn one or more samples. The probability that a confidence interval estimate contains a population mean in the two-tail case is

$$P(\text{lower limit} < \text{population mean} < \text{upper limit}) = \text{confidence level}$$

The actual calculations for the confidence interval are not particularly complicated: We subtract from the population mean a factor for the lower limit and add to the mean the same factor to obtain the upper limit. For confidence intervals, we take the appropriate Z-value and multiply it by the population standard deviation divided by the square root of the sample size. Thus, to determine the lower limit of the interval, we use the formula

$$\text{lower limit} = \text{population mean} - (Z\text{-value})\left(\frac{\text{population standard deviation}}{\text{square root of sample size}}\right)$$

The upper limit is found in exactly the same way except that the right-hand product is added to the population mean.

For our example, we have reprinted the most frequently used values of Table 1 in Appendix B in Table 6-1.

**TABLE 6-1**  Critical Values of Z Statistic

| Alpha | One Tail | Two Tail |
|---|---|---|
| .01 | 2.33 | 2.58 |
| .02 | 1.76 | 2.33 |
| .05 | 1.65 | 1.96 |
| .10 | 1.28 | 1.65 |

The alpha value refers to the percentage we wish to drop from the confidence interval. A 90% confidence interval will have an alpha of .10 for a one-tail test and an alpha of .05 for each tail of a two-tail test. We can now calculate the specific values for our interval estimate. With a known population mean of 100, a known standard deviation of 10, and a sample size of 30, the limits of our interval estimate of the distribution of sample means are as follows:

$$\text{lower limit} = 100 - \frac{(1.65)(10)}{\sqrt{30}}$$
$$= 96.99$$

$$\text{upper limit} = 100 + \frac{(1.65)(10)}{\sqrt{30}}$$
$$= 103.01$$

In other words, if our true population mean is 100 and the standard deviation is 10, then 90% of our sample means of size 30 will fall between 96.99 and 103.01. The remaining 10% will fall outside that interval. Figure 6-7 shows both one-tail and two-tail tests for a

## SECTION 6-4 The Confidence Interval

**FIGURE 6-7** Z-Values for 90% One-Tail and Two-Tail Confidence Intervals for a Normal Curve with Mean = 100, Standard Deviation = 10

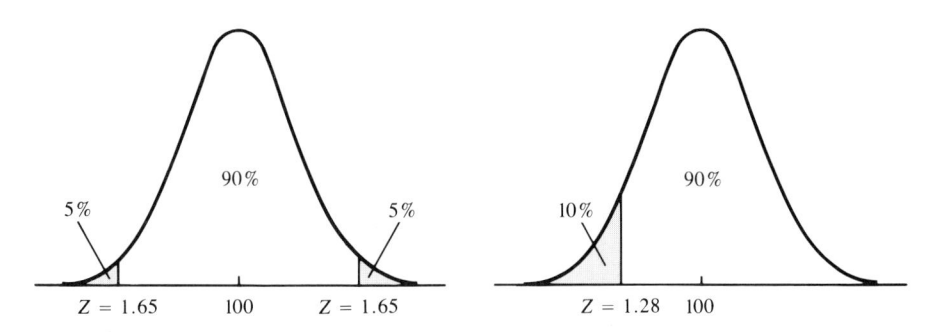

90% interval. In a two-tail 90% interval, we use the Z-value of 1.65 (or 5% per tail). In a one-tail 90% interval, we use a Z-value of 1.28 (or 10% in only one tail).

Likewise, we could compute interval estimates that would include 95%, 98%, or 99% of the sample means. You will note that our sample does not match these computed intervals exactly, but it is very close. We did not take *all* possible samples but only 100. Your analysis will probably yield similar results.

## The Z Table

At this point we should take a minute and explain how to look up a Z-value in Table 1 of Appendix B. Simply go inside the major portion of the table and find the value that represents the area you want to include in your interval estimate. Since the normal distribution is symmetric, the values in the table go only to 50%. For example, if you wanted to build a 98% two-tail confidence interval, you would include 49% above your estimate and 49% below your estimate. That would leave 1% excluded in each tail for a total of 2%. You would first find the value of .4900 (or as close to that value as possible) in the body of Table 1 of Appendix B. In the 24th row we have the values of

.4893   .4896   .4898   .4901   . . .

The value of .4901 is closest to our desired value. You then find the first two digits for the Z-value in the left-hand column (in this case, 2.3) and the final digit for the Z-value in the top row (0.03). The last digit is added to the first two for a total value of 2.33; this is your Z-value for the 98% two-tail confidence interval.

For a one-tail confidence interval at the 98% level, you would look up .4800. Remember, if 98% is to be inside our confidence interval in a one-tail estimate, the remaining 2% must be together on one side of the distribution. Thus we need to eliminate 2% of the possible values from the same tail (50% − 2% = 48%, or .4800). The Z-value in this case is about 2.05. This value would either be added *or* subtracted from the mean, depending on which tail we wanted to exclude.

Now that we have calculated the confidence interval for a known population, let's work with our sample values and develop sample interval estimates for the unknown mean of a population.

## Minitab Calculations

You might be asking yourself, "Why would I want to estimate sample means?" After all, we can compute sample means directly. In fact, that is the only thing we can compute. What we really want to do is use our sample mean to estimate the *population mean*. Generally, we will take a sample and then determine the confidence interval *around that sample mean*. This interval estimate will tell us something about the population mean. Fortunately, the process of calculating the interval estimate around a sample mean is exactly the same as calculating it around a population mean. The only difference is which mean we use.

Interval estimates made from Z-values are normally called Z intervals. (In Chapter 8 we will calculate intervals using *t*-values, which are called *t* intervals). We can compute the Z interval in Minitab with the following command:

```
ZINTerval, confidence level = K, sigma = K, data in C
```

If we had a random sample of data stored in column 11, the command for a 90% confidence interval with an assumed standard deviation of 10 would be

```
MTB >ZINTerval, confidence level = 90, sigma = 10, data in C11
```

With this command we can easily create estimates of 90%, 95%, or 99% confidence intervals. We have the Go-Fast data for ages entered in column 1. If we assume that the population standard deviation of the ages for all shoe industry employees is 4, then we might enter the following command:

```
MTB >ZINTerval, confidence level = 90, sigma = 4, data in C1
```

This command can be shortened to

```
MTB >ZINT 90, 4, C1
```

The execution of this command creates the printout shown in Figure 6-8. The results indicate a 90% Z interval of

$$31.061 < \text{population mean} < 32.379$$

If you calculate the 90%, 95%, and 99% Z interval values, you will note that the 95% interval is wider than the 90% interval, and the 99% interval is wider than the 95% interval. Does that make sense? Yes, it does. In a 90% estimate, we are willing to be wrong 10% of the time. In a 95% estimate, we are willing to be wrong only 5% of the time. To decrease our chances of being wrong, we must increase the width of the confidence interval. In Figure 6-9, the estimates of the 95% and 99% confidence intervals, as well as the 90%, have been calculated. Examine the width of each interval.

**FIGURE 6-8** 90% Z Interval for Age of Go-Fast Employees

```
MTB >ZINT 90 4 C1

THE ASSUMED SIGMA =4.00

            N       MEAN    STDEV   SE MEAN   90.0 PERCENT C.I.
C1         100     31.720   3.798    0.400    ( 31.061, 32.379)
```

## SECTION 6-4 The Confidence Interval

**FIGURE 6-9**  90%, 95% and 99% Z Intervals for Age of Go-Fast Employees

### 90% Z Interval

```
MTB >ZINT 90 4 C1

THE ASSUMED SIGMA =4.00

            N      MEAN     STDEV    SE MEAN    90.0 PERCENT C.I.
C1         100    31.720    3.798     0.400    ( 31.061,  32.379)
```

### 95% Z Interval

```
MTB >ZINT 95 4 C1

THE ASSUMED SIGMA =4.00

            N      MEAN     STDEV    SE MEAN    95.0 PERCENT C.I.
C1         100    31.720    3.798     0.400    ( 30.935,  32.505)
```

### 99% Z Interval

```
MTB >ZINT 99 4 C1

THE ASSUMED SIGMA =4.00

            N      MEAN     STDEV    SE MEAN    99.0 PERCENT C.I.
C1         100    31.720    3.798     0.400    ( 30.688,  32.752)
```

A word of warning! The 90% interval does not include 90% of the possible values of the population mean. The population mean is a parameter; it is a set constant that has only one possible value. What we can say is that 90% of the intervals computed in this manner would contain the true population mean. That is, 90% of the time we will correctly predict that the population mean lies within our computed range—and 10% of the time we will be wrong. This is a major idea.

Let's perform another experiment. If we sample our original population with known mean of 100 and standard deviation of 10 and build a confidence interval based on the sample, how often will that confidence interval encompass the known population mean? Based upon our work above, we would say that the answer is about 90 out of 100 times for a 90% confidence interval. We can simulate the issue quite easily with Minitab and examine the empirical results. Study the following command sequence:

```
MTB >STORe
MTB >RANDom 30 observations, put into C1;
SUBC> NORMal with mu = 100, sigma = 10.
MTB >ZINT confidence level = 90, sigma = 10, data in C1
MTB >END
```

Now send your output to the printer (or work on a printing terminal) and execute this command set 100 times with the command

```
MTB >EXECute 100
```

What you are doing is drawing repeated samples of size 30 from a known population and then using each individual sample to build a confidence interval. How many times does your interval include the true population mean of 100? It should be about 90 times.

In Figure 6-10, the results of running this experiment one time are shown. (Your results will be different because your simulations will be different.) In the figure the intervals have been plotted and connected to show the range. Since we know the actual mean to be 100, you can see those estimates that do not include the value of 100.

You should also rerun this experiment using 95% and 99% confidence levels. Notice how the intervals widen, thus increasing the number of intervals that include 100. Also repeat the experiment with samples larger than size 30. The intervals will narrow as the sample size increases. (Look at the formula for confidence intervals again and notice that we divide by the square root of the sample size. Thus as the sample size increases, the width of the interval decreases.) However, notice that as the sample size increases, the percentage of confidence intervals that do not contain the population mean remains about the same.

**FIGURE 6-10** 90% Confidence Intervals Determined from 100 Samples of Size 30 from a Population with Mean = 100, Standard Deviation = 10

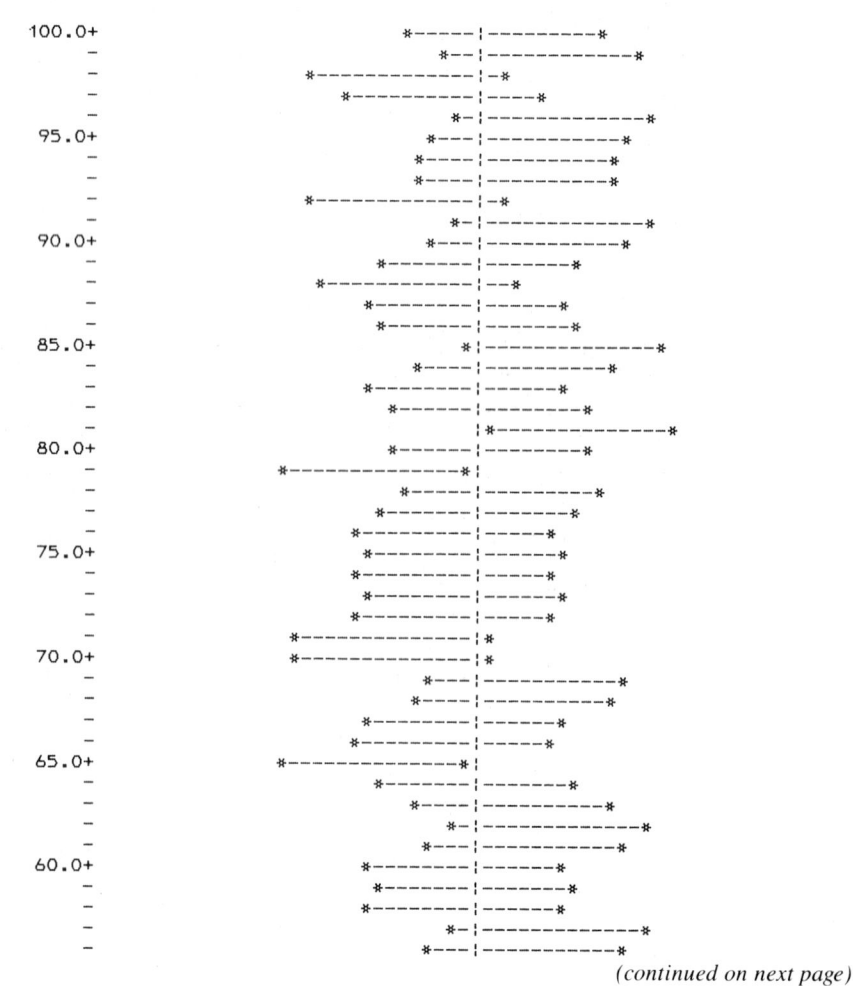

*(continued on next page)*

## SECTION 6-4 The Confidence Interval

**FIGURE 6-10** (*Continued*)

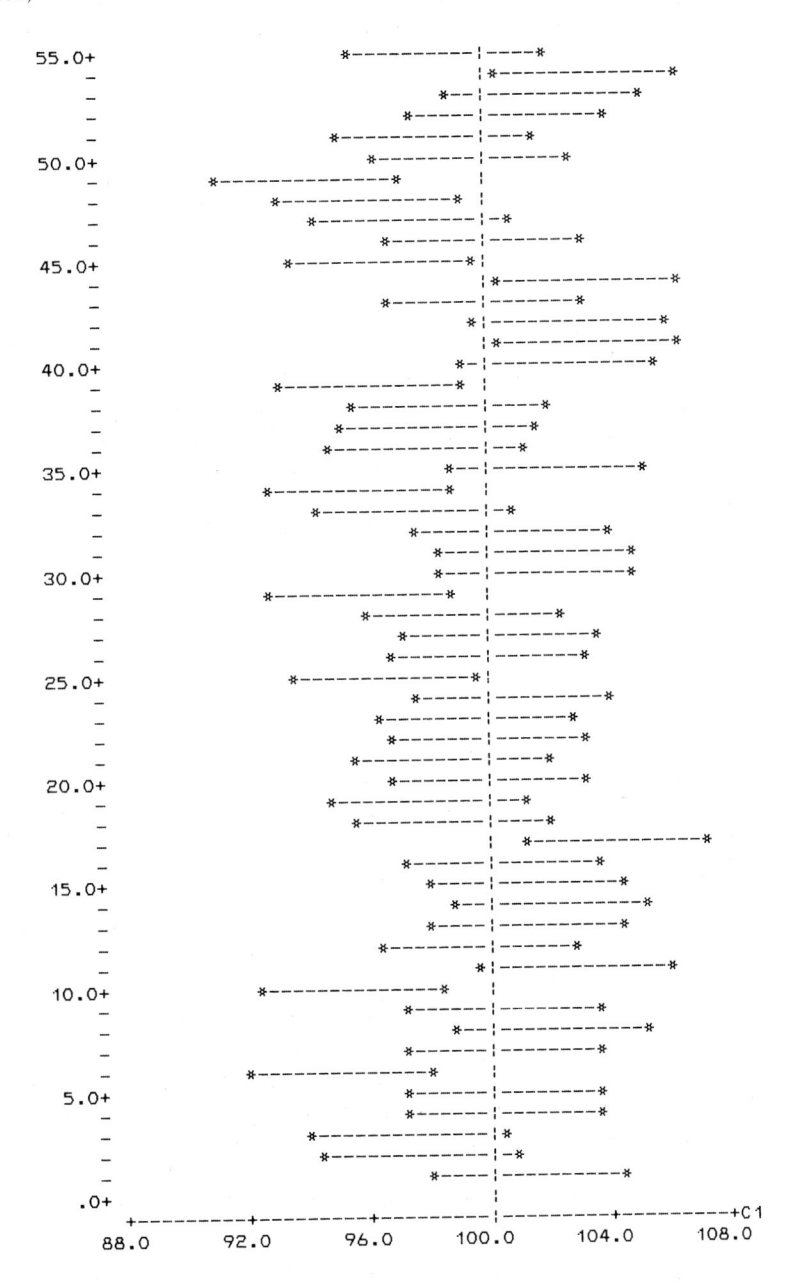

## Hand Calculations

Confidence intervals derive from the concept of the *distribution of sample means*. A distribution of sample means has its own standard deviation. This standard deviation, called the *standard error of the mean*, is needed in order to determine confidence intervals.

The sample that we draw is one possible sample of a population of sample means, and so it has a calculated mean and standard deviation. If we drew all possible samples of a certain size from a population, we would have a large number of samples. Each of these samples would have a mean value. If we then took all of the means that we had calculated and graphed them as a histogram, they would constitute a distribution of sample means.

This distribution of sample means has a measure of variance in the same way that each of our individual samples has a variance measure. For the individual samples, that measure of spread is called the standard deviation. For the distribution of sample means, that measure of spread is called the *standard error*. The formula for the standard error is:

$$\sigma_{\bar{x}} = \frac{\sigma}{\sqrt{n}}$$

Let's assume that we have taken a sample of 5 from a population with a mean of 3 and a population standard deviation of 1.58. The calculation of the standard error is then

$$\sigma_{\bar{x}} = \frac{1.58}{\sqrt{5}}$$
$$= 0.71$$

We now have all the data necessary to calculate a confidence interval by hand, including a sample mean, a standard deviation (which will be used to calculate the standard error), and a confidence level. The actual formula for an estimate of a confidence interval is

$$P[\bar{x} - Z_{\alpha/2}(\sigma_{\bar{x}}) \leq \mu \leq \bar{x} + Z_{\alpha/2}(\sigma_{\bar{x}})] = 1 - \alpha$$

This important formula should be memorized.

Now let's substitute our values into this formula. To calculate a 90% confidence interval, assuming a two-tail test and a Z-value of 1.65, we have

$$P[3 - 1.65(0.71) \leq \mu \leq 3 + 1.65(0.71)] = .90$$
$$P(1.83 \leq \mu \leq 4.17) = .90$$

Thus, around a mean value of 3, a 90% confidence interval would stretch from 1.83 to 4.17. We can verify this result with a Minitab command. If we loaded the 5 values into column 1, we would then command

```
MTB >ZINT 90 1.58 C1
```

The output of this command is

```
THE ASSUMED SIGMA = 1.58

         N      MEAN     STDEV    SE MEAN     90.0 PERCENT C.I.
C1       5      3.00     1.58     0.71        ( 1.84,     4.16)
```

Notice that we had to assume that the population standard deviation was 1.58 in order to run our test. The results indicate that the standard error of the mean is 0.71 and the confidence interval is 1.84 to 4.16. Thus, both Minitab and the hand calculations are in agreement.

## SUMMARY

In this chapter we have laid some very important groundwork for future chapters. The critical issue is the fact that we do not normally know the population standard deviation. Consequently, we must use a slightly different statistic known as a $t$ statistic, which we will do in Chapter 8.

# APPENDIX

Earlier in this chapter, we told you to assume that the means of 100 samples of size 30 were stored in column 1. At that time, we deferred the necessary calculations until after our discussion of confidence intervals. It is now time to show how to use Minitab to do the complete analysis.

We have already discussed a general command to draw a random sample: the **RANDom** command with the **NORMal** subcommand. That command can be expanded to use more than one column. For example, we could draw 15 samples of size 30 and put them in the first 15 columns:

```
MTB >RANDom 30, C1-C15;
SUBC> NORMal 100, 10.
```

We can use this same format to create multiple samples. Remember that we are dealing with randomly drawn samples. Thus if we draw a single random number and put it in row 1 of column 1, we then have a single random value. We could then draw another random sample and put it in row 1 of column 2, put another in row 1 of column 3, and so on. In other words, since the values drawn from the population are random, we can think of a sample as being in either a row or a column. Consequently, there are two approaches to this problem. Study the following two sets of commands:

```
MTB >RANDom 30, C1-C100;
SUBC> NORMal 100, 10.

MTB >RANDom 100, C1-C30;
SUBC> NORMal 100, 10.
```

In the first set of commands we have created 100 columns and placed 30 randomly drawn samples from our population into each column. Each column is a sample. In the second set of commands, we have created 30 columns, each with 100 rows. Here each row across all columns comprises a sample of size 30. The 100th row of each column is our 100th sample.

Using the second set of commands is simpler for two reasons. First, many computer installations of Minitab will not create 100 columns, and so the first command group will not work. Even if yours does allow the command, there will be no easy way to obtain the means of the values in each column and store those means in another column. The **DESCribe** command could be used in the form **DESC C1-C100** to calculate the value of each mean, but that would not store the values in a new column. Since other manipulations are needed, such storing is important.

The second set of commands avoids both of these problems. First, all computer installations of Minitab will generate at least a couple of hundred rows. Second, Minitab has a set of row commands, which are summarized below:

```
            RN C,...,C, put in C
            RNMIs C,...,C, put in C
```

```
                    RMEAn C,...,C, put in C
                    RMEDian C,...,C, put in C
                    RSTDev C,...,C, put in C
                    RMAX C,...,C, put in C
                    RMIN C,...,C, put in C
                    RSUM C,...,C, put in C
                    RSSQ C,...,C, put in C
                    RCOUnt C,...,C, put in C
```

Most of these commands are self-explanatory, although the first two may not be apparent. **RN** stands for the number of items in the row, and **RNMIs** the number of rows in which items are missing. Thus, **RN** and **RNMIs** commands together would equal the value determined by **RCOUnt**.

For our purposes, the command of interest is **RMEAn**. With this command we can find the mean of the 30 random observations in row 1 of columns 1 through 30 and put that value in column 31. The complete set of commands is

```
MTB >RAND 100 C1-C30;
SUBC> NORM 100, 10.
MTB >RMEA C1-C30, C31
```

Column 31 now contains the means of our 100 samples of size 30 from a population with mean 100 and standard deviation 10. In Figure 6-1, we simply moved column 31 to column 1, performed a **SORT** command on column 1, and put the resulting values in column 2.

## DISCUSSION QUESTIONS

1. Is the point estimate of the mean a good estimate of the mean for a population? Why or why not?
2. In order to use the Z score table when estimating population means, what must be true of the sample?
3. As the sample size increases, what happens to the interval estimate? Why?

## PROBLEMS

**6-1** A zoo curator is estimating the mean age of her tortoises. She knows that the mean age of all tortoises is 96.7 years old, that the standard deviation is 3, and that sample sizes of 60 were used to compute this mean. The zoo houses 60 tortoises.

**a)** Using Table 1 of Appendix B, compute the lower and upper limits for a Z interval at a 90% confidence level for a two-tail estimate.

**b)** Would you expect the mean age for the zoo's tortoises to be exactly 96.7?

**c)** What does the computed Z interval tell you about the age of the zoo's tortoises?

**6-2** A sample of the age of customers that frequent the Old Time Pizza Parlor appears in the following table:

| 21 | 19 | 20 | 23 | 18 | 23 | 20 |
| 21 | 22 | 18 | 21 | 22 | 19 | 8 |

## CHAPTER 6 Statistical Estimation

| | | | | | | |
|---|---|---|---|---|---|---|
| 20 | 20 | 21 | 20 | 23 | 34 | 21 |
| 22 | 22 | 20 | 22 | 21 | 5 | 19 |
| 21 | 19 | 21 | 21 | 35 | 18 | 19 |

**a)** Use Minitab to find the sample mean.
**b)** Find the Z interval using Minitab at the 95% confidence level with a population standard deviation of 5.
**c)** Define the population that the Old Time Pizza Parlor is sampling.
**d)** What does the Z interval tell you?

**6-3** Using Minitab's normal random data simulator, generate 30 observations with a population mean of 200 and a population standard deviation of 12.
**a)** Repeat the process of generating samples and collect 100 sample means. Store all 100 sample means in a single column.
**b)** Sort the sample means.
**c)** Create a histogram and dot plot of the sample means.
**d)** Using a 90% confidence level, what is the Z interval?
**e)** Using a 95% confidence level, what is the Z interval?
**f)** Which Z interval is wider, the 90% confidence level or the 95% confidence level? Why?
**g)** Examine the histogram or dot plot. Are 90% of the sample means included within the 90% Z interval?

**6-4** Use Minitab's normal random data simulator. Generate 45 observations with a population mean of 200 and a population standard deviation of 12.
**a)** Repeat the process of generating samples and collect 100 sample means. Store all 100 sample means in a single column.
**b)** Sort the sample means.
**c)** Create a histogram and dot plot of the sample means.
**d)** Compute the Z interval using a 90% confidence level.
**e)** Compare the Z interval just computed to the Z interval in Problem 6-3(d). What are your observations?
**f)** Compare the histogram and dot plot with those in Problem 6-3(c).

**6-5** Ann, a computer science major, is looking forward to graduating in May. Upon graduation, she hopes to find employment. To get an idea of the pay scale in the job market, she has taken a random sample of annual starting salaries for other students in her major.

| | | | | |
|---|---|---|---|---|
| $16,700 | $14,000 | $15,400 | $18,500 | $12,000 |
| 17,000 | 16,700 | 15,000 | 13,500 | 16,700 |
| 18,500 | 19,000 | 14,000 | 17,000 | 15,000 |
| 16,700 | 18,900 | 16,700 | 15,400 | 17,000 |
| 15,400 | 15,400 | 18,500 | 16,700 | 15,400 |
| 18,500 | 16,700 | 17,000 | 17,000 | 15,000 |

**a)** What average starting salary can Ann expect?
**b)** Create a histogram or dot plot of the data.
**c)** Examine the histogram or dot plot. What type of distribution does the data represent?
**d)** What can one infer about the population from the distribution of the sample data?
**e)** Using Minitab, find the Z interval with a population standard deviation of 1,700 and a confidence interval of 90%.
**f)** Based upon your analysis in (e), should Ann expect a starting salary of $19,000?

## PROBLEMS

g) Using Minitab, find the Z interval with a population standard deviation of 10 and a confidence interval of 90%. Discuss the consequences of entering this command with this particular set of data.

**6-6** Orchid Pleasures, a florist that specializes in orchid arrangements, feels that there will be a trend toward sending orchids as gifts instead of roses. Orchids tend to last longer and are less expensive than the traditional dozen roses. The following is randomly sampled data on the number of sprays of orchids purchased per order for 50 orders:

| No. of Sprays | 1 | 2 | 3 | 4 | 5 |
|---|---|---|---|---|---|
| Frequency | 1 | 11 | 28 | 9 | 1 |

a) What is the point estimate for the average number of sprays ordered?
b) What is the interval estimate for the average number of sprays ordered? Use a 95% confidence level and a population standard deviation of 0.75.
c) Which estimate should Orchid Pleasures rely on to determine the average number of sprays needed per day?
d) If Orchid Pleasures has 20 orders per day, what is the average number of sprays needed?

**6-7** Because of the farm industry's troubles, the concept of a farmers' market has been reconsidered. Many farmers have narrowed their market by growing specialty crops, such as the tomatillo used by Mexican-Americans. Following is a random sample of monthly tomatillo needs (in thousands of pounds) for the West Coast:

| | | | | | |
|---|---|---|---|---|---|
| 140 | 135 | 162 | 140 | 135 | 155 |
| 112 | 140 | 155 | 140 | 140 | 140 |
| 162 | 155 | 124 | 135 | 140 | 135 |
| 135 | 140 | 140 | 155 | 140 | 155 |
| 170 | 135 | 140 | 135 | 155 | 162 |
| 155 | 140 | 167 | 135 | 140 | 140 |

a) Using Table 1 of Appendix B, compute the Z interval for a one-tail test at a 95% confidence level. Assume a population mean of 140 and a standard deviation of 12.
b) Compute the Z interval in (a) with Minitab.
c) If storage costs were relatively low, how many tomatillos (in pounds) should farmers consider producing every month?

**6-8** There are two types of customers in the self-storage business: the apartment or condominium dweller and the business owner. For the first type of customer, the population mean for square feet of storage used per month is 100 and the standard deviation is 12. For the second type, the population mean is 150 square feet and the standard deviation is 8.
a) Examine the standard deviation for each type of customer. What does the standard deviation tell you about the mean for each group?
b) Generate a sample size of 100 for each type of customer and compute the Z interval for each group using a 90% confidence level.
c) Who would you recommend renting to, the apartment dweller or the business owner? Why?

**6-9** Brand name clothing manufacturers are displeased with how department stores are displaying their goods. Instead of creating a look and mood, department stores try to display as many clothes as possible in one area. Clothing manufacturers, on the other hand, believe they can sell more clothes by projecting an appropriate image. Randomly sampled data on the revenue per square foot from sales of brand name clothing at a department store and at a manufacturer's outlet follows:

**Department Store**

| $125 | $ 97 | $ 98 | $103 | $128 | $127 |
|---|---|---|---|---|---|
| 96 | 98 | 104 | 103 | 92 | 90 |
| 97 | 125 | 103 | 97 | 98 | 127 |
| 98 | 96 | 104 | 98 | 97 | 92 |
| 103 | 98 | 104 | 98 | 96 | 98 |
| 97 | 104 | 98 | 103 | 98 | 89 |

population standard deviation = 10

**Manufacturer's Outlet**

| $112 | $114 | $109 | $114 | $124 | $116 |
|---|---|---|---|---|---|
| 121 | 116 | 114 | 112 | 114 | 116 |
| 114 | 112 | 112 | 116 | 114 | 121 |
| 114 | 116 | 114 | 121 | 116 | 112 |
| 109 | 114 | 112 | 114 | 114 | 116 |
| 114 | 112 | 116 | 114 | 112 | 114 |

population standard deviation = 3

**a)** Compute the mean revenue per square foot for the department store and for the manufacturer's outlet.
**b)** Using Minitab, calculate the Z interval for both types of store using a 98% confidence level.
**c)** Create a histogram or dot plot for each data set using the **SAME** subcommand.
**d)** Which mean is more representative of its data set? (Use the histogram and the standard deviation.)
**e)** Are the clothing manufacturers right in thinking that they can sell more clothes?

**6-10** Sue, an elementary school teacher, asked each of 50 randomly sampled students to bring a pumpkin to carve for Halloween. The pumpkins would all be displayed on a table for parents to see on open house night. Sue wondered if the table could support all the weight. Sue knows that the population mean weight for pumpkins is 11 pounds and that the standard deviation is 2.

**a)** Using Table 1 of Appendix B, calculate the lower and upper limits for a Z interval at a 95% confidence level for a two-tail estimate.
**b)** If the table can support 500 pounds, should Sue be concerned? Why or why not?

**6-11** Chad's soccer team has a 7-0-1 record this season. Tomorrow his team faces its toughest opponent. A random sample of both team's scores for the last three years follows:

**Chad's Team**

| 6 | 3 | 5 | 5 | 4 | 6 |
|---|---|---|---|---|---|
| 4 | 5 | 6 | 7 | 5 | 4 |
| 5 | 4 | 4 | 5 | 6 | 5 |
| 5 | 6 | 5 | 6 | 4 | 5 |
| 4 | 5 | 6 | 6 | 5 | 4 |

population standard deviation = 1.0

**Opponent**

| 5 | 7 | 3 | 3 | 4 | 4 |
|---|---|---|---|---|---|
| 4 | 3 | 2 | 7 | 3 | 5 |
| 4 | 5 | 4 | 4 | 3 | 2 |
| 3 | 4 | 7 | 3 | 4 | 3 |
| 5 | 4 | 3 | 4 | 5 | 4 |

population standard deviation = 1.5

# PROBLEMS

    **a)** What is the average score for each team?
    **b)** Compute the Z interval at the 98% confidence level for a one-tail test.
    **c)** Should Chad worry about tomorrow's game?

**6-12** Randomly sampled data on the number of house calls made per day by two salespeople appears in the following:

| Worker 1 | | | | | Worker 2 | | | | |
|---|---|---|---|---|---|---|---|---|---|
| 8 | 14 | 10 | 8 | 8 | 10 | 7 | 8 | 10 | 12 |
| 10 | 12 | 14 | 10 | 10 | 7 | 10 | 16 | 8 | 10 |
| 12 | 10 | 12 | 12 | 8 | 10 | 8 | 10 | 8 | 10 |
| 10 | 7 | 8 | 10 | 10 | 8 | 10 | 8 | 10 | 8 |
| 8 | 8 | 10 | 12 | 10 | 10 | 12 | 8 | 10 | 10 |

    **a)** Given that the population standard deviation for both workers is 2 and that the confidence level is 98%, compute both workers' Z intervals using Minitab.
    **b)** Create a histogram and dot plot of the number of house calls made per day for each worker. Be sure to use the **SAME** subcommand to ensure a common basis for comparisons.
    **c)** If you needed to fire one of the workers, who would it be?

**6-13** Mitchell's Fishing Adventure takes interested people deep-sea fishing. Jerry wants to enter a marlin fishing contest and decides to examine the fishing records of two captains. Randomly sampled data on weight per marlin catch follows:

| Captain 1 | | | | | |
|---|---|---|---|---|---|
| 110 | 130 | 110 | 130 | 114 | 131 |
| 110 | 131 | 130 | 131 | 130 | 132 |
| 114 | 114 | 131 | 132 | 114 | 131 |
| 130 | 110 | 130 | 131 | 114 | 130 |
| 130 | 131 | 130 | 114 | 130 | 131 |

| Captain 2 | | | | | |
|---|---|---|---|---|---|
| 120 | 125 | 111 | 112 | 125 | 112 |
| 112 | 126 | 128 | 128 | 126 | 120 |
| 125 | 120 | 125 | 111 | 120 | 120 |
| 111 | 112 | 120 | 112 | 112 | 126 |
| 126 | 120 | 112 | 128 | 111 | 125 |

    **a)** What is the mean marlin catch for each captain?
    **b)** Based on the point estimate of the mean, who would Jerry want to fish with?
    **c)** Use Minitab to compute the Z interval at the 95% confidence level for each captain. The population standard deviation is 8.
    **d)** Based on the computed Z interval, who would Jerry hire?

**6-14** Two horses participating in today's race appeal to you. The mean finishing time for the first horse has a Z interval of 55 to 61 seconds; for the second horse, a Z interval of 54 to 59 seconds.
    **a)** What does the Z interval represent?
    **b)** Based on the Z interval, which horse would you bet on?

**6-15** A competitor would like to race against you on the quarter-mile drag strip. A random sample of his finishing times (in seconds) on the quarter mile is as follows:

| 12 | 11 | 14 | 12 | 14 | 12 |
| 10 | 12 | 13 | 12 | 12 | 13 |
| 12 | 13 | 14 | 11 | 12 | 13 |
| 13 | 14 | 11 | 12 | 13 | 12 |
| 12 | 13 | 12 | 11 | 11 | 13 |

**a)** Compute his Z interval using Minitab with a 99% confidence level and a population standard deviation of 1.

**b)** If your mean time in a quarter mile was 12.4 seconds with a standard deviation of 2.0, do you think you could win? Explain.

# INTRODUCTION TO HYPOTHESIS TESTING

*The following Minitab command will be introduced in this chapter:*

```
ZTEST mu = K, sigma = K, data in C;
  ALTernative hypothesis = K.
```
A command to test the null hypothesis against the alternative hypothesis; calculates the test statistic Z in either a one-sided or two-sided test

In the previous chapters we developed the techniques and concepts needed to perform statistical analysis. In this chapter we will introduce one of the major forms of analysis—testing statistical hypotheses. By the end of this chapter you should be able to perform simple tests on a group of data. In the next chapter we will compare two or more groups of data.

## 7-1 DEVELOPMENT OF A HYPOTHESIS

Suppose that you had an instructor who gave incredibly difficult true–false tests. In fact, they were so difficult that it was generally only by chance that a student got questions right or wrong. However, the instructor tended to make her tests either 80% true or 30% true. Before taking her 500-question final exam, you hear (from a reputable source) that the exam for the other section of the same course was 30% true. Thus, if you took the same exam and answered all questions false, you would get a score of 70%.

The day of the examination arrives. On the way to the test, you wonder, "What if she uses another exam?" You decide to look for questions that you can answer and then make up your mind. Remember, the questions tend to be so difficult that answering any question correctly seems to be a random process. Intuitively you decide that, since the first exam was a 30% true test, you should be at least 90% certain that the test is *not* 30% true before changing your strategy. Otherwise, you will keep your present strategy and answer all questions false.

After receiving your examination, you screen it for questions you can answer. There are 5, and all of them are true. With this information, let's try to figure out what you should do. With a knowledge of statistics, you can test what is called a hypothesis. A *hypothesis* is simply a statement that can be statistically tested. Normally we will have a *null hypothesis,* which is the statement that we think is true, and an *alternative hypothesis,* which is the statement that we will accept if we decide that our null hypothesis is not true. We will designate the null hypothesis as $H^*$ and the alternative hypothesis as $H^{**}$.

As you sit in the middle of this true–false exam, we can state what we know as follows:

**(1)** *Null Hypothesis* ($H^*$): The test is 30% true.
*Alternative Hypothesis* ($H^{**}$): The test is *not* 30% true.

**(2)** *Decision Rule*: If there is less than a 10% chance that $H^*$ is true, you will reject it and answer all the questions true. If you fail to reject $H^*$, then you will answer all the questions false.

**(3)** *Test Statistic*: If our $H^*$ is true, that is, if only 30% of the questions are true, then what is the probability of finding 5 questions at random that are all true? The probability can be calculated by referring to our probability tree in Chapter 4. Since these questions are independent events, our calculation is

$$P(H^*) = (.3)(.3)(.3)(.3)(.3)$$
$$= .00243$$
$$= 0.243\%$$

**(4)** *Decision*: The probability is very small of randomly finding 5 consecutive questions with true answers if the exam is only 30% true. Since our calculation shows the probability of $H^*$ being true is less than 10%, we will reject our null hypothesis and accept the alternative hypothesis. In this case, we will answer all questions true since we feel that the examination contains 80% true answers.

It is important to understand the procedure used in this illustration. This four-step system will be used throughout our discussion of statistical inference. You must first state your null hypothesis. Typically the null hypothesis is a statement of what you actually think will be the case. Normally it is stated positively. Thus, we would not usually state the null hypothesis for our illustration as "The test is not 30% true." The null hypothesis must also be phrased in such a way that it can be statistically tested. Thus, you cannot say as a hypothesis that you want to try to find out what percentage of the questions are true. (This is *estimation,* which was discussed in Chapter 6.)

SECTION 7-2 Errors in Testing Hypotheses

Our alternative hypothesis must also be stated in a certain way. Normally the alternative must include all other outcomes not specified by the null hypothesis. In the true–false test we stated the alternative as the test not being 30% true. Thus the two hypotheses cover all possible outcomes. In this very simple example, the null hypothesis *could* have been "The test is 80% true," because the problem had only two possible outcomes. However, always be careful that your null and alternative hypotheses cover all possible outcomes.

To illustrate this problem, assume for a minute that you were measuring the diameter of ball bearings manufactured by a machine. If all ball bearings are to have a diameter of 0.5 cm, there are three possible outcomes: The ball bearings are smaller than 0.5 cm, greater than 0.5 cm, or equal to (within tolerance limits) 0.5 cm. Your hypotheses must cover all three possible outcomes. If your null hypothesis is that the ball bearings have a mean diameter of 0.5 cm, then your alternative hypothesis must be that the diameters are *either* less than or greater than 0.5 cm.

Notice that your alternative hypothesis could not be that the ball bearings have a diameter of mean greater than 0.5 cm. What if the measurements show the bearings to be less than 0.5 cm?

However, your null hypothesis could be that the ball bearings are less than 0.5 cm in diameter; then the alternative hypothesis would have to be that the diameters of the ball bearings are equal to or greater than 0.5 cm. The important thing is that you state your null and alternative hypotheses to cover all possible outcomes. In most situations, there will be either two options (as in true or false) or three (as in less than, equal to, or greater than).

## 7-2 ERRORS IN TESTING HYPOTHESES

In our true–false examination, we could be very certain that the test was 80% true. However, there is still a chance that we were wrong. What is that chance? In this case, it is 0.243%. To determine a general rule, consider the following: Before you undertake the test of any hypothesis, one of two conditions exists: The null hypothesis is either true or false. That condition is a simple fact. However, we do not *know* if the hypothesis is true or false, and we draw a statistical inference from a sample. We then conclude whether we think that the null hypothesis is true or false. Our inference about the hypothesis does not change reality. The hypothesis is in fact either true or false.

Consequently, there are four possible outcomes:

(1) The null hypothesis is true and we conclude that it is true.
(2) The null hypothesis is true, but we conclude that it is false. Such a conclusion would be an error.
(3) The null hypothesis is false, and we conclude that it is false.
(4) The null hypothesis is false, but we conclude that it is true. Again we have made an error.

Table 7-1 illustrates these possible outcomes.

If $H^*$ is true, we can either correctly "fail to reject" $H^*$ (some might say "accept" $H^*$, but this is technically not correct) or incorrectly "reject" $H^*$. To control the possibility of making this error, called a *Type I error*, we establish a decision rule. In our example, we set the significance level, or alpha, at 10%; thus, if $H^*$ was correct, we would reject it as being incorrect only 10% of the time. Thus, when we set our significance level at 10%, 5%, or 1% and $H^*$ is true, we will be correct 90%, 95%, or 99% of the time, respectively. Conversely, we will also be incorrect 10%, 5%, or 1% of the time, respectively.

**TABLE 7-1**  Errors in Hypothesis Testing

|  | Actual Condition |  |
| --- | --- | --- |
| Statistical Inference | $H^*$ True | $H^*$ False |
| $H^*$ True | No error | Type II error |
| $H^*$ False | Type I error | No error |

## 7-3 POWER OF A TEST

If $H^*$ is false, we can either correctly reject $H^*$ or incorrectly fail to reject it. The *Type II error* (with its theoretical probability of beta) can also be controlled, but not by selecting the confidence level. The Type II error is controlled by selecting a "powerful" test. The *power of a statistical test is its ability to correctly reject a false hypothesis*. We control this power in two ways.

First, we select the most sensitive test available, being certain that we can meet the test's assumptions. It will become apparent as we proceed that different tests work for different problems. However, some tests are better than others. Normally, as a test statistic has more restraints, its power increases. Unfortunately, its usefulness also decreases, since it cannot be used in as many circumstances. As an example, you will soon learn about the $t$ test. In this test, if the data is paired—that is, if you have two sets of observations about the same item—you can use a more powerful test than if the data is not paired. The requirement of paired data is more constraining than unpaired. Generally, in statistical analysis if the constraints are greater, the test is more powerful.

Second, we control the dispersion of our sampling distribution through sample size. Remember that the standard deviation of the sampling distribution is inversely proportional to the square root of the sample size. Thus, as our sample becomes larger, the probability of committing a Type II error decreases.

The relationship of variance to sample size can be seen in Figure 7-1, which illustrates the fact that for any distribution, the larger the sample size, the less the probability of committing a Type II error. A sampling distribution for $n = 100$ will cluster much more

**FIGURE 7-1**  Effect of Sample Size on Variance

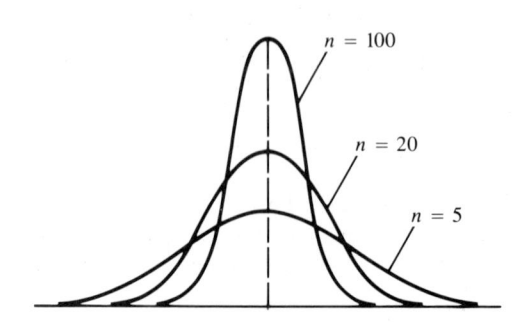

tightly around the mean than a sampling distribution for $n = 20$. When the variance is greater, there is a greater chance of drawing a sample from one of the tails of our distribution. When the variance is less, the probability of drawing a sample out of one of the tails is smaller.

You might automatically think that all problems are solved if you always use a 99% confidence level (that is, a 1% significance level), a very large sample size, and the most powerful test possible under the given circumstances. Although this statement is true in theory, in the real world numerous facts will prevent you from pursuing this course. The most obvious obstacle is cost. Larger samples always cost more than smaller samples. For example, when testing products, products must often be destroyed. As a result, you want to minimize sample size while maximizing the amount of information obtained. Also, you must be concerned about the cost of committing a Type I or Type II error. If the cost of an error is relatively small, then it is managerially imprudent to spend great sums of money to prevent such an error. In almost all business decisions, a compromise must be reached on the level of accuracy desired, the probability of error, the sample size, and the cost. Management experience is often essential in this decision process.

# 7-4 HYPOTHESIS TESTS

## The Binomial Test

We often have several discrete observations that form a binomial distribution. These situations are typically found when equipment or products either pass or do not pass an inspection. A supplier of rubber soles for Go-Fast's running shoes faces this situation. This supplier maintains that 95% of the soles will stretch 10 mm or more. This characteristic is essential for proper flexibility of the shoe.

What if we had to test the vendor's claim? We could set up the following hypotheses:

**(1)** $H^*$: The proportion of soles with at least 10 mm of stretch will be 95% or greater.
$H^{**}$: The proportion of soles with at least 10 mm of stretch will be less than 95%.

Notice that our hypotheses cover all possible outcomes. We have stated our $H^*$ as being equal to or greater than 10 mm, and $H^{**}$ as being less than 10 mm.

We now must formulate our decision rule:

**(2)** *Decision Rule*: If there is less than a 2% chance that $H^*$ is true, we will reject the shipment and find a new vendor. Otherwise, we will fail to reject $H^*$, indicating insufficient reason to change vendors, and we will accept the shipment. (The confidence level is thus 98%, and the significance level is 2%.)

At this point you might be asking, "If our hypothesis is that 95% of the soles are acceptable, why not use a 95% decision rule?" Remember that the hypothesis and the decision rule are two different things. In the hypothesis, we are concerned only with the soles. If 95% or more of the soles can stretch 10 mm, then the shipment of shoe soles is acceptable. However, if we decide to reject the shipment, we want to be 98% confident (2% error) in that decision.

Thus the situation could have been that 70% of the shoe soles were promised to stretch 10 mm, and we might have wanted to be 99% confident in our decision that 70% of the soles did indeed meet this criterion. Keeping these two concepts separate might be a bit confusing at first. Just remember, first we are making a statement about the population based on our sample (or, in the rubber sole example, we are given a claim) and then we are setting a confidence level to test that statement.

Let's now return to our problem. We test a rubber sole for stretch by putting it in a machine and seeing how far it will stretch before it rips. Thus, our testing process destroys the product. Because of the cost of the rubber soles, the cost of testing personnel, and the time involved, we have a sample of 20 soles for testing.

**(3)** *Test Statistic*: Out of 20 randomly selected soles from a very large shipment, 2 are faulty and do not pass the stretch test.

With these results, we fire up our computer terminal and call Minitab. The Bernoulli process effectively produces the binomial distribution. Remember, the hypothesis is that the soles are from a population that is 95% acceptable. Thus we wish to generate a distribution that has 95% acceptable items. We will then compare our results against such a distribution. With this in mind we execute the following command:

```
MTB >CDF;
SUBC> BINO 20, .95.
```

The results are shown in Figure 7-2. We are concerned with the column headed **P(X LESS OR = K)**, which gives the accumulated probability of various numbers of defective items. Reading up from the bottom of the table, we see that the probability of 19 or fewer soles being acceptable when the true population is 95% acceptable is .6415, or 64.15%. The probability of 18 good shoe soles or fewer is .2642, or 26.42%.

**FIGURE 7-2** Binomial Probabilities for a Sample of Size 20 with a 95% Probability of Acceptance

```
MTB >CDF;
SUBC> BINO 20 .95.

      BINOMIAL WITH N =   20   P = 0.950000
         K    P( X LESS OR = K)
        13         0.0000
        14         0.0003
        15         0.0026
        16         0.0159
        17         0.0755
        18         0.2642
        19         0.6415
```

**(4)** *Decision*: Fail to reject claim and accept the shipment.

Our initial decision rule was that if there was less than a 2% chance of 95% of the shipment being acceptable we would reject it. In this case, we can see that 2 defects in a sample of 20, with an assumed probability of 95% nondefective, fell well within the acceptable range. Figure 7-2 shows that if we had had 4 defects, we would have rejected the claim that the shipment was 95% good. In this case, 3 or fewer defects would have been acceptable.

### SECTION 7-4 Hypothesis Tests

The binomial test is a reasonably simple one but is very effective in many circumstances. Note, however, that there are many cases where it will not work effectively. At this point you might be asking yourself, "What have I gotten myself in for? I know that there are lots of different probability distributions. Will I have to learn a test for each of them?" The central limit theorem comes to the rescue here. If we take large enough samples (generally 30 or more), the central limit theorem assures us that the distribution of sample means will be normal. Thus, we really need a test for only one distribution—the normal distribution.

## The Z Test ▼▲▼▲▼▲▼▲▼▲▼▲▼▲▼▲▼▲▼▲▼▲▼▲▼▲▼▲▼▲▼▲▼▲▼▲▼▲▼▲▼▲▼▲

The first test that we will use for the normal distribution is a largely theoretical test, the Z test. You will understand why we say "theoretical" when we review the assumptions required for the Z test.

**(1)** We must have a random sample.
**(2)** We need to be sampling from a normal population.
**(3)** We must know the *population* standard deviation.
**(4)** We must collect data that is either interval or ratio.

The first assumption is easily met if we are careful in our sampling technique. The second is assured by the central limit theorem if we collect a large enough sample. The fourth assumption simply requires that the number be actual counts or measurements and not simply classes or rankings of items. The real problem lies in the third assumption. In order to know the population standard deviation, we must have conducted a census of the population. If we had conducted a census, that is, measured every item, we would not have to test a hypothesis about some population parameter—we would know the parameter with certainty. However, all we ever really know is the *sample* standard deviation, not the population standard deviation.

The Z test, however, is a most valuable test. First, it allows the testing of hypotheses about normal distributions. Second, with large samples, the law of large numbers forces the sample standard deviation to approximate the population standard deviation so closely that we can actually use the Z test anyway with very little loss of accuracy.

Suppose, for example, that Go-Fast has been accused of age discrimination in its hiring practices. Since the management of Go-Fast is conscientious, they have decided to test the hypothesis that they have not discriminated. If they have not discriminated, then the average age of an employee at Go-Fast should be about the same as the average age of the general population. Also, since the average age of the population tends to behave as a random variable, we should expect the average age of employees at Go-Fast to behave like a random variable also.

After reviewing the figures from the Bureau of Labor Statistics, we determine that the average age of the work force is 31. Here is the format of our hypothesis test:

**(1)** $H^*$: The average age of employees at Go-Fast is 31.
$H^{**}$: The average age of employees at Go-Fast is not 31.

What we are really testing here is whether the average age of the work force at Go-Fast could reasonably occur in a random sampling of the entire work force.

This is a good time to illustrate possible questions concerning the assumption of randomness and representativeness. First, are the local population and work force similar to the national population and work force? For example, the average age of the population in Florida is generally much higher than the average age of the population in many other states because of the high percentage of retired individuals there.

Second, people in the local work force and in the national work force may be noticeably different. Even if the average age distributions of the local and national populations are similar, the age distributions of people seeking work may be different in the two populations. For example, in a college town many individuals between 18 and 22 years of age do not seek work because they are full-time students. In a noncollege town, many individuals in this age group are seeking work. In our example here, we will assume that the national figures are representative of the local figures, but in general such an assumption should not be casually made.

Another point to notice is that this particular hypothesis is an "equal to" hypothesis. If our results are either too low or too high, we will reject the hypothesis. This is different than our previous examples, in which we were interested in only one extreme of the distribution. In terms of the curve of the normal distribution, this is a two-tail test, since too many observations at either end of the distribution will cause us to reject the hypothesis.

Let's assume that we want to be at least 90% confident that our work force is different from the national work force (that is, we are setting our alpha at 10%) before we can reject the null hypothesis that our employee ages match those of the national work force population. If the Bureau of Labor Statistics figures indicate a national standard deviation of 4 years, then we could calculate our Z test statistic as follows:

$$Z = \frac{\text{(observed sample mean)} - \text{(hypothesized population mean)}}{\text{(population standard deviation)}/\sqrt{n}}$$

From our Z-values in Appendix B, Table 1, we can determine the appropriate critical value. If there is less than a 10% chance of our sample coming from a population with a mean of 31 and a standard deviation of 4, our Z statistic will be equal to or greater than 1.65. Now we can establish our decision rule.

**(2)** *Decision Rule*: We will reject $H^*$ if it has less than a 10% chance of being true ($Z \geq 1.65$).

Now we want to compute our test statistic. Luckily, if we have the sample data (and we do), Minitab will do the work for us. We use the general command

```
ZTESt mu = K, sigma = K, data in C;
  ALTernative hypothesis = K.
```

This command is not as complicated as it looks. Let's look at each item of data individually. First, we need to tell Minitab the required value for the hypothesized mean. It is against this value that your data will be tested. In our case, we will be testing our data against a hypothesized mean of 31.

The second value required is the population standard deviation. This, of course, is the problem with Z test statistics in the first place. The `ZTESt` command does not calculate a sample standard deviation from your sample data (other tests discussed in the next chapter will do this) but simply uses the population standard deviation figure you supply. In this case, we know the population standard deviation to be 4 from the data from the Bureau of Labor Statistics.

### SECTION 7-4 Hypothesis Tests

The next input required for the `ZTESt` command is the number of the column in which your data is stored. We have read the data on ages into column 1. (You will need to load the Go-Fast data into your worksheet, or your instructor might have already set up a central file. We have read the data for ages into column 1 and named that column `'AGES'`.)

The `ZTESt` (and all other Minitab tests of hypotheses) follow the same command format used in creating distributions. In hypothesis testing, there is a command line that specifies the mean, standard deviation, and location of the data. In the subcommand line we tell the computer which type of hypothesis we are testing. There are three possible types: "less than" tests, "equal to" tests, and "greater than" tests. The Minitab codes for these three types of alternative hypotheses are as follows:

| Hypothesis Type | Code |
|---|---|
| Less than | −1 |
| Equal to | 0 |
| Greater than | +1 |

You should memorize these code values, because the same codes are used in the commands for several different tests in addition to Z test statistics.

With all the information supplied, our command looks as follows:

```
MTB >ZTESt mu = 31, sigma = 4, data in C1;
SUBC> ALTernative hypothesis = 0.
```

This command could be shortened in several ways. The usual form is

```
MTB >ZTES 31, 4, C1;
SUBC> ALT = 0.
```

(If we had only been interested in the ages of our employees being less than the national average, then instead of specifying a `0` for the alternative hypothesis, we would have specified a `−1`. If we had wanted a greater-than test, we would have entered a `1`.)

Returning to our problem, the output from our command is shown in Figure 7-3. Notice that the printout includes the following:

**(1)** a description of the data column,
**(2)** a statement of the test hypothesis,
**(3)** the Z statistic, and
**(4)** the probability of $H^*$ being true.

For our data, the Z test statistic of 1.800 and the significance level of .0722 (normally indicated as $p$) are important. Minitab always assumes a 95% confidence level (alpha = .05) for all calculations. The printout indicates that we cannot reject alpha at .05.

**FIGURE 7-3** Minitab Output of Z Test for Ages of Go-Fast Employees

```
MTB >ZTES 31, 7, C1-C5;
SUBC> ALT 0.
    AGE      N = 100     MEAN =     31.720    ST.DEV. =    3.80

    TEST OF MU =    31.0000 VS. MU N.E.    31.0000
    THE ASSUMED SIGMA =       4.0000
    Z =  1.800
    THE TEST IS SIGNIFICANT AT   .0722
    CANNOT REJECT AT ALPHA = 0.05
```

**CHAPTER 7** Introduction to Hypothesis Testing

The *p*-value, or the level at which the statistic becomes significant, is of primary interest to us. We had specified our own decision rule of alpha = .10, or a 10% level of significance. In this case, the Minitab printout indicates that we cannot reject the null hypothesis at .05, but we can at .10. Thus we complete steps 3 and 4:

**(3)** *Test Statistic*: Z = 1.800.

**(4)** *Decision*: Reject the null hypothesis at a .10 level of significance because the test probability is .0722.

An important issue of ethics arises here. From the printout for this problem, we see that the *p*-value is .0722, which means that for any alpha value above 7% we should reject H*, and for any alpha value below 7% we should not. You cannot change your decision rule *after* you have seen the results of your calculations. You must determine your decision rules before starting your analysis. If you have a good, logical reason for specifying an alpha of 10% to start with, then stay with it. Never go back and change your alpha level so that the results of your analysis will be reversed. Such manipulation is one way in which unethical practitioners manipulate statistical analysis. Whenever you see some statistical tests for a problem at one level of significance and other statistical tests for the same problem at other levels of significance, it is a sign of an unethical analysis.

One final note about the general command form for **ZTESt**: You can test multiple columns in a single command. For example, if we had similar data in columns 1 through 5 (with a common population mean and standard deviation), then we could enter the command

```
MTB >ZTES 31, 7, C1-C5;
SUBC> ALT = 0.
```

## 7-5 FORMULAS AND HAND CALCULATIONS

The computation of a Z statistic for any normal curve is a simple variation of the computation for the standard normal curve. The standard normal curve formula is

$$Z = \frac{x - \mu}{\sigma}$$

This formula converts a distance along the base of any normal curve into the distance along the base of a standard normal curve. Thus, we could easily look up the area from Appendix B, Table 1.

Now we are using a sample mean instead of a single observation as our test statistic. Consequently, we must standardize the distance with a standard deviation of the sample means (the standard error of the mean). The hand calculation of the standard error of the mean was developed in Chapter 6. At that time, our calculation was

$$Z = \frac{\bar{x} - \mu}{\sigma_{\bar{x}}}$$

Using the Go-Fast data on the ages of employees, we can reconstruct Minitab's computation of the observed Z. Substituting into this formula, we obtain

$$Z = \frac{31.72 - 31}{4/\sqrt{100}} = \frac{0.72}{0.4} = 1.8$$

# PROBLEMS

As you can see, using this computation, we can use the Z test anywhere. This will be handy in many circumstances, such as when you have only one simple problem to do and do not want to use a computer. To make such hand computations easier, you should memorize the most commonly used Z-values so that you have a sense of the value of Z that you have calculated. For example, you know that the value of 1.65 is a 90% two-tail or a 95% one-tail Z-value. Also, a Z of 1.96 represents 2.5%; thus if you are doing a 95% two-tail test, then a value in the range of 1.96 would be desirable. As you do a variety of problems, two or three of the more common Z-values will become familiar to you.

# SUMMARY

In this chapter, we have discussed a general framework for statistical inference. You should be able to test hypotheses on binomial and normal distributions.

From the central limit theorem we can use the Z statistic when we have samples of size greater than 30 to 50. The major problem with the Z statistic is that the population standard deviation, which is seldom known, must be used. In subsequent chapters we will relax this requirement. However, the basic concepts behind the Z statistic apply to measures used in later chapters.

# DISCUSSION QUESTIONS

1. How must the null and alternative hypotheses be stated?
2. What is the appropriate alpha level for a Z test?
3. What is a Type I error, and how can the probability of committing one be reduced?
4. What is a Type II error, and how can the probability of committing one be reduced?
5. What is the major problem with using the Z test?
6. How can you overcome the Z test's major problem?

# PROBLEMS

**7-1** Unbearably Cute is a company that manufactures stuffed animals. It has just started to produce a new line of baby bears designed to appeal to expectant mothers. Unbearably Cute believes that 90% or more of the bears produced will weigh at least three pounds. A three-pound bear is necessary to ensure that the animal will sit upright.

**a)** Set up a binomial test. If there is less than a 5% chance that the null hypothesis is true, you must reject $H^*$.

**b)** You have taken a sample of 30 baby bears in which 4 weighed less than three pounds. Use the `BINOmial` subcommand to generate data with a sample of 30 and a confidence level of .90.

**c)** Compare the output to your decision rule. What is your conclusion?

**7-2** All of 250,000 copies of *Sun & Fitness* were distributed to bookstores across the country. The publisher expects 98% or more of the copies to be sold within a month. If this happens, the publisher will print 100,000 additional copies.

**a)** Set up a binomial test. If there is less than a 2% chance that $H^*$ is true, the publisher will not print the additional 100,000 copies.

**b)** Out of a sample of 100 copies previously earmarked for sampling, 5 had not been sold. Use the `BINOmial` subcommand to generate data with a sample size of 100 and a confidence level of 98%. (Most computers will not be able to compute the usual command. The command must be altered a bit.)

**c)** Compare the output to your decision rule. Will the publisher print the additional copies?

**7-3** The monthly industry average for the number of personal computers sold per salesperson at medium-sized retail stores is 19. The population standard deviation is 6. Computer Warehouse is the largest computer store in the Midwest. The philosophy behind its business is volume selling at low prices. The managers at Computer Warehouse believe that their average monthly sales of personal computers per salesperson is 19 or more. The following are the monthly PC sales for 30 of their salespeople:

| | | | | | |
|---|---|---|---|---|---|
| 19 | 25 | 21 | 16 | 19 | 21 |
| 17 | 20 | 21 | 26 | 37 | 36 |
| 19 | 24 | 23 | 30 | 18 | 22 |
| 22 | 29 | 25 | 21 | 32 | 14 |
| 20 | 22 | 24 | 35 | 26 | 20 |

**a)** Set up a Z test. If there is less than a 5% chance that $H^*$ is true, the managers must reject $H^*$.

**b)** Apply the Z test command to the data.

**c)** What are your conclusions?

**d)** Since the mean of the sample is 23.47, should you have tested for significance? Relate your answer to the *p*-value in this problem.

**7-4** For the utilities industry, the average yearly expenditure for capital expansion is $150,000 with a population standard deviation of $16,000. West Public Power feels that it may not be expanding rapidly enough to keep pace with the rest of the industry. Thus, its average expenditures may be $150,000 or less. West Public Power's yearly expenditures (in thousands of dollars adjusted to current dollars) appear in the following table:

| | | | | | |
|---|---|---|---|---|---|
| 145 | 130 | 140 | 159 | 180 | 150 |
| 175 | 160 | 157 | 170 | 154 | 152 |
| 160 | 155 | 162 | 153 | 143 | 173 |
| 150 | 145 | 151 | 148 | 157 | 140 |
| 202 | 150 | 130 | 149 | 160 | 122 |

**a)** Set up the null hypothesis and the alternative hypothesis. If there is less than a 10% chance that $H^*$ is true, West Public Power does not want to increase its expenditure for capital expansion.

**b)** Compute the test statistic using Minitab.

**c)** Should West Public Power spend more on capital expansion?

**7-5** John D. Wealthy III established a generous scholarship fund for outstanding sophomores at State College. However, he stipulated that an average of 8% or more of State College's

# PROBLEMS 179

sophomores must have grade-point averages of 3.5 or better before he would award the first scholarship. State College believed it could meet that requirement. Assume that in the United States, college sophomores with a 3.5 or better represent 8% of their class on average. The population standard deviation is 2.3. The following is a sample from State College of the proportion of sophomores from various years ending the year with a 3.5 or better.

| 10% | 6% | 7% | 4%  | 7%  | 11% | 7%  |
|-----|----|----|-----|-----|-----|-----|
| 9   | 4  | 7  | 9   | 8   | 9   | 8   |
| 6   | 7  | 9  | 12  | 5   | 10  | 7   |
| 3   | 7  | 8  | 6   | 11  | 8   | 14  |
| 6   | 8  | 9  | 5   | 7   | 9   | 8   |

**a)** Set up a Z test. If there is less than a 10% chance that $H^*$ is true, the scholarship money cannot be awarded.
**b)** Compute the test statistic using Minitab.
**c)** Will awards from the scholarship fund begin?
**d)** Assume that after the first scholarship was awarded, a statistician discovered that there was less than a 10% chance that an average of 8% or more of the sophomores had a 3.5 or better. What type of error was made?

**7-6** A nursery that supplies indoor plants to offices is expecting a large order for miniature palm trees. The nursery feels it can fill the order. However, its plants are of varying heights. The office ordering the plants has agreed to accept the order if the average height of the palm trees equals the average height of the plants the office already has. The population average is 26 inches, and the population standard deviation is 3. Following is a random sample of the height (in inches) of the palm trees:

| 32 | 30 | 29 | 28 | 25 | 24 |
|----|----|----|----|----|----|
| 26 | 30 | 23 | 25 | 24 | 32 |
| 25 | 27 | 14 | 30 | 26 | 27 |
| 24 | 25 | 37 | 26 | 27 | 28 |
| 24 | 30 | 31 | 29 | 27 | 26 |

**a)** Establish the null hypothesis and the alternative hypothesis. If there is less than a 5% chance that $H^*$ is true, then $H^*$ must be rejected.
**b)** Apply the Z test command to the data.
**c)** Will the nursery be able to fill the order?

**7-7** A bank loan officer is not very pleased with the 1985 interest coverage ratio of Slow Start, Inc. The loan officer felt that if Slow Start's average interest coverage ratio was equal to or greater than the industry average of 4.5, then he would not be concerned. The population standard deviation is 1.2. The managers at Slow Start are confident that they can ease the loan officer's mind. A sample of Slow Start's interest coverage ratios follows:

| 3.0 | 4.7 | 3.2 | 4.5 | 3.3 | 5.0 |
|-----|-----|-----|-----|-----|-----|
| 5.5 | 6.0 | 4.0 | 1.2 | 2.0 | 3.3 |
| 4.7 | 4.0 | 5.0 | 5.0 | 4.9 | 1.4 |
| 4.7 | 4.6 | 3.9 | 4.6 | 3.4 | 4.2 |
| 4.7 | 4.9 | 4.9 | 3.2 | 4.7 | 4.0 |
| 2.9 | 3.0 | 4.1 | 3.3 | 2.1 | 4.5 |
| 4.5 | 2.2 | 1.0 | 3.9 | 2.2 | 2.7 |

a) Set up the test hypotheses. If there is less than a 2% chance that the null hypothesis is true, $H^*$ must be rejected.
b) Do a Z test.
c) Should the loan officer be concerned?

**7-8** The quality board is determining the bacteria level in Daisy Dairy's milk. Daisy Dairy believes it can pass the quality check by having an average of 0.20% or less bacteria in its milk. The population standard deviation is 0.04%, and the population average is 0.20%. The following is a sample of the bacteria levels in Daisy Dairy's milk:

| 0.19% | 0.16% | 0.24% | 0.24% | 0.22% | 0.23% |
|---|---|---|---|---|---|
| 0.22 | 0.25 | 0.19 | 0.17 | 0.25 | 0.20 |
| 0.18 | 0.23 | 0.23 | 0.24 | 0.26 | 0.24 |
| 0.22 | 0.16 | 0.27 | 0.09 | 0.19 | 0.25 |
| 0.25 | 0.20 | 0.19 | 0.21 | 0.18 | 0.21 |

a) Set up a Z test. If there is less than a 2% chance that $H^*$ is true, it must be rejected.
b) Using Minitab, compute the Z statistic.
c) Did the milk pass the quality check?
d) Why would the confidence level be so stringent?

**7-9** In one West Coast state, garbage is generally collected by three-worker teams operating one truck. Other states use one worker to operate one truck. The people in the West Coast state believe that their system is more cost-effective. If so, the average cost per home for garbage collection in the West Coast state would be less than in the other states. The average cost for the other states is $10 per house, and the population standard deviation is $2. The following is the cost per home for the West Coast state in several areas:

| $12 | $10 | $14 | $10 | $12 | $ 9 |
|---|---|---|---|---|---|
| 11 | 8 | 7 | 9 | 8 | 9 |
| 10 | 11 | 12 | 14 | 12 | 8 |
| 11 | 12 | 12 | 14 | 12 | 10 |
| 11 | 11 | 9 | 8 | 10 | 12 |

a) Set up a Z test. Use an alpha of .10.
b) Use Minitab to determine the Z statistic.
c) Should the West Coast state switch methods?

**7-10** Style, Inc., is a national franchise chain of beauty salons. One year after a franchise shop opens, Style expects the average number of customers worked on by each stylist every week to be equal to or greater than the weekly average of the other franchises. The average for Style, Inc., is 124 per week, and the population standard deviation is 10.3. Following is randomly sampled data on the number of weekly customers worked on per stylist for one month at Style's newest franchise:

| 120 | 110 | 116 | 124 | 130 | 140 |
|---|---|---|---|---|---|
| 111 | 115 | 105 | 126 | 128 | 114 |
| 116 | 112 | 120 | 124 | 132 | 144 |
| 117 | 110 | 109 | 130 | 133 | 116 |
| 120 | 130 | 140 | 135 | 131 | 120 |

## PROBLEMS

a) Set up a Z test. Use a 90% confidence level.
b) Compute the Z statistic using Minitab.
c) Does the new franchise need help with its operations?

**7-11** A tomato farmer developed a new system for lengthening his harvest period from mid-October to mid-November. By covering the tomato plants with a plastic sheet in November, the farmer felt that the average low temperature under the cover equaled the average low temperature for the month of October. The average low temperature in October is 42° F, and the population standard deviation is 2.2° F. Following is a sample of temperature readings (in degrees Fahrenheit) under the cover:

| 45 | 37 | 40 | 38 | 42 | 43 |
| 44 | 39 | 41 | 45 | 43 | 40 |
| 42 | 44 | 43 | 40 | 42 | 39 |
| 43 | 41 | 42 | 44 | 44 | 45 |
| 40 | 42 | 39 | 41 | 38 | 42 |

a) Set up a Z test. Use an alpha of .05.
b) Use Minitab to compute the Z statistic.
c) Is the tomato farmer right?

**7-12** First City Bank wants its monthly proportion of average marketable securities to be equal to or less than the average proportion held by all other banks. The population average is 10% and the population standard deviation is 2%. The following data are the percentages of First City Bank's marketable securities for past months:

| 10 | 14 | 10 | 9  | 10 | 12 |
| 8  | 9  | 10 | 8  | 9  | 8  |
| 11 | 9  | 14 | 14 | 9  | 9  |
| 12 | 12 | 11 | 10 | 14 | 10 |
| 12 | 8  | 12 | 14 | 11 | 8  |

a) Set up a Z test. Use a confidence level of 90%.
b) Compute the Z statistic using Minitab.
c) Is the bank happy with the outcome?

**7-13** Refer to Problem 7-12. After conducting its study, First City Bank found it had made a Type I error.
a) How do you decrease the chance of the occurrence of a Type I error?
b) Redo Problem 7-12 using 95% and 98% confidence levels.
c) Discuss the results that you obtained. Why did these changes occur?

**7-14** A new method has been developed to measure the amount of lint in the air. A clothing manufacturing company believes that the amount of lint in the air in its sewing department is low enough to not require the wearing of masks. The acceptable level of lint is an average amount equal to or less than the industry average of 4%. The population standard deviation is 1.5%. The following data is a random sample of levels of lint for the sewing department at different times:

| 4% | 4% | 6% | 4% | 3% | 4% |
| 2  | 1  | 2  | 5  | 4  | 5  |
| 6  | 5  | 4  | 2  | 5  | 2  |
| 4  | 5  | 4  | 3  | 3  | 3  |
| 5  | 4  | 5  | 7  | 5  | 6  |

    **a)** Establish the null and alternative hypotheses. The alpha to be used is .02.
    **b)** Compute the Z statistic using Minitab.
    **c)** Are masks necessary?
    **d)** What are the possible reasons for using a small alpha?

**7-15** A Type II error was made in Problem 7-14.
    **a)** What is a Type II error?
    **b)** How can it be minimized?
    **c)** Do a Z test similar to the one done in Problem 7-14. Following is a larger random sample of the level of lint in the sewing department:

| 3% | 3% | 5% | 5% | 4% | 3% | 4% | 5% | 6% | 7% |
|----|----|----|----|----|----|----|----|----|----|
| 4  | 5  | 6  | 4  | 2  | 1  | 2  | 4  | 3  | 6  |
| 5  | 3  | 4  | 3  | 4  | 5  | 5  | 7  | 7  | 6  |
| 1  | 2  | 4  | 6  | 5  | 2  | 5  | 7  | 5  | 8  |
| 6  | 3  | 5  | 4  | 3  | 4  | 5  | 6  | 4  | 6  |
| 3  | 4  | 6  | 6  | 4  | 5  | 3  | 5  | 6  | 6  |

    **d)** Are masks required?

# STATISTICAL INFERENCE: ONE- AND TWO-SAMPLE CASES

*The following Minitab commands will be introduced in this chapter:*

```
TTESt mu = K, data in C;
  ALTernative hypothesis = K.
```
A command to calculate Student's *t* test statistic for either a one-tail or two-tail test

```
TINTerval with conf level = K, data in C
```
A command to calculate the two-sided *t* confidence interval for the population mean

```
COPY C,...,C into C,...,C;
  USE rows K,...,K;
  USE C = values K,...,K;
  OMIT rows K,...,K;
  OMIT C = values K,...,K.
```
A command to copy certain value(s) or a range of values from a given column to a new column; corresponding value(s) in additional columns can also be copied to new columns; rows can also be specified, and rows or values can be omitted

```
TWOSample t, K percent confidence, data in C and C;
  ALTernative hypothesis = K;
  POOLed.
```
A command to compute the *t* test statistic to compare two samples for either a one-tail or two-tail test; variances of the two samples can be equal (pooled) or unequal

# CHAPTER 8 Statistical Inference: One- and Two-Sample Cases

In Chapter 7, we introduced the concept of statistical inference through use of the Z test. The Z test is conceptual because its rigorous assumptions are seldom met. However, the Z test does provide necessary insight into the process of hypothesis testing.

Often in statistical analysis we have data from a single sample and must compare the mean of that population with some sample value. This is a case of single-sample data. Just as often we have data from two samples and want to compare the mean values of the two samples and draw conclusions about the population or populations from which they were drawn. This is the case of a two-sample test.

The single-sample and two-sample cases are studied by using a *t* test. Cases of more than two samples can be analyzed by using an *F* test. We will discuss the tests used for one and two samples in this chapter and the tests for more than two cases in Chapter 9. The tests discussed in Chapter 9 can be used for two-sample testing, but usually are not. Chapter 9 will also discuss testing hypotheses when the data is not interval.

## 8-1 ASSUMPTIONS OF THE *t* TEST

The *t* test is a powerful and robust test. By "robust" we mean that it can reject false hypotheses effectively and be used under a wide variety of circumstances. After studying the constraints of the Z test, it is nice to look at a practical test with assumptions that are easily met.

In order to use the *t* test, we will need to satisfy several assumptions:

**(1)** *The population should be normally distributed.* We need some empirical evidence of normality if we are using a small sample, or we will call upon the central limit theorem if we have a large sample.

**(2)** *The sample collected must be random.* Notice how often this assumption enters into our analyses. If you have not taken your sample in a way that eliminates systematic bias in your data, then the sample is worthless. This is consistently our most important assumption in statistical analysis and is typically given too little attention.

**(3)** *The data must be interval or ratio.* Thus, if you have true–false answers or rankings from a questionnaire, you cannot use a *t* test for your analysis. In metric data, differences between each value in the numeric set represent measurable amounts. For example, the difference between the ages of 36 and 37 is a meaningful amount and is equal to the difference between the ages of 44 and 45.

**(4)** *We must have an unbiased estimate of the population standard deviation.* This was our main problem with the Z test. The Z test requires the actual value of the population standard deviation. For the *t* test, all we need is an unbiased *estimate* of the population standard deviation, not the actual value. This unbiased estimate is not as difficult to obtain as you might think.

In Chapter 3, we determined that to obtain the population standard deviation, you added up the squared mean difference values, divided them by the sample size, and then took the square root of that calculation. However, an interesting change is made when you are taking samples *from* a population instead of using the entire population to determine the population standard deviation.

Suppose, for example, that we had a population of 10,000 values and knew the mean and standard deviation of that population. If we took 500 samples of size 50 and found their mean and standard deviation and then estimated the population standard deviation based upon these samples, our estimate would consistently be slightly too low. However, if for each time we took a sample of 50, we divided by 1 less than 50 (i.e., 49), our estimate of the population standard deviation would be very accurate. In fact, we would have an unbiased estimate.

Thus, in single-sample statistical analysis, we obtain an unbiased estimate of the population standard deviation by dividing our sample standard deviation by the sample size minus 1 ($n - 1$). Don't be overly confused by this issue: Minitab always assumes we are working with samples and automatically divides by the proper value to provide an unbiased estimate. In other statistical procedures, you will have to correct your estimates by dividing by the sample size minus 1 or more. Minitab will always compute the proper divisor automatically, as do most other statistical packages. (In many statistical discussions, this issue comes under the heading of "degrees of freedom." Whenever you run across this term, it refers to a way of correcting an estimate to make it unbiased. Different techniques require different correction factors.)

## 8-2 STUDENT'S *t* TEST

We thought that we would use the proper name for the *t* test when we introduced it—Student's *t* test. It is not called that because it is for students. The originator of the test, W. S. Gossett, was prohibited by his company, Guinness Breweries, from publishing his work under his own name. He chose the pen name Student—thus the name of the test.

As has been mentioned, the *t* test can be used for a wide variety of statistical testing. We can use it to test the mean of a single sample against a predetermined value or for testing two means against each other. When you are testing for a change in the mean values between two samples, you can make one of three different assumptions. You can be working with (1) paired data, (2) unpaired data with equal variances, and (3) unpaired data with unequal variances. Each of these possibilities will be discussed separately. Let's start by looking at a test for a single mean.

In Chapter 7, we did a *Z* test to determine whether the mean age of Go-Fast employees was different from a proposed population mean of 31. In that case, we assumed that we knew the population standard deviation to be 4. With the *t* test we can relax the assumption of knowing the population standard deviation and instead use the standard deviation of the sample as an unbiased estimate of the population standard deviation.

The *t* test command in Minitab has the following general form:

```
TTESt of mu = K, data in C;
ALTernative hypothesis = K.
```

This general form of the **TTESt** command is very similar to that of the **ZTESt** command in Chapter 7; the only difference is that we do not specify the standard deviation. Instead, Minitab will use the sample data to calculate an unbiased estimate of the population standard deviation from the sample standard deviation. We have loaded the data on the ages of the Go-Fast employees into column 1. With that data, our **TTESt** command would be

```
MTB >TTESt of mu = 31, data in C1;
SUBC> ALTernative hypothesis = 0.
```

This command can be shortened to

```
MTB >TTES 31, C1;
SUBC> ALT 0.
```

The alternative hypothesis codes remain the same as for the `ZTESt` command (−1 for "less than," 0 for "equal to," and 1 for "greater than" null hypotheses).

Figure 8-1 shows the output from the command above; it is very similar to Figure 7-3. Despite the similarities, the differences are important. In Figure 7-3, the Z-value is 1.800; in Figure 8-1, the *t*-value is 1.90. This is an important point. *The t test is an approximation of the Z test.* We are not testing anything new—we are only using a test with more realistic assumptions. When the sample size is small, the critical *t*-value for significance is much larger than the critical Z-value for the same test. However, as the sample size becomes larger and larger, the *t*-value will closely approximate the Z-value.

Appendix B, Table 2, is a general Student's *t* table. Minitab will always calculate the exact *t*-value and tell you if it is high enough to be significant for your sample size. However, you might need a *t* table to look up a significant value on other occasions. In Appendix B, Table 2, the far left column marked "df" stands for "degrees of freedom"; in *t* tests, as stated above, the degrees of freedom is the sample size minus 1. In this problem, our sample size is 100, thus our degrees of freedom is 99. If you look down the "df" column, you will notice that the table only goes to 29 and then jumps to Z as the final value. This illustrates the point we just made: As the sample size becomes larger, the Z-value and the *t*-value become identical. Other statistical tables provide a *t*-value for 60, 100, or 200 degrees of freedom, but in all cases the values become closer and closer to the equivalent Z-value.

The values in the final row, labeled "Z," are the same as those found in Table 6-1 for Z-values. If we are doing a *t* test with a 95% confidence level on the Go-Fast age data, we will have a total alpha of .05. Since this is a two-tail test, we will place 2.5% of our possible error in each tail. Looking down the "0.025" column to the final value, you see that our critical value must be 1.96 or greater. Since our calculated *t*-value is 1.90, we just barely miss; nevertheless, we still cannot reject the null hypothesis. The Minitab printout in Figure 8-1 indicates that our test is significant at the .061 level, which is just barely greater than .05. If we had chosen to run this test at the 90% level with a corresponding total alpha of .10 (.05 alpha in each tail), our critical *t*-value would have been 1.65; in that case, we would have rejected the null hypothesis. In the Minitab subcommand line, we indicate if our test is a one-tail or two-tail test by our specification of the alternative hypothesis. Minitab automatically determines the proper level of statistical significance. Since Minitab always

**FIGURE 8-1** Minitab *t* Test of Null Hypothesis That Go-Fast Employee Age Is Equal to 31

```
MTB >TTES 31 C1;
SUBC> ALT 0.

TEST OF MU = 31.000 VS MU N.E. 31.000

            N      MEAN     STDEV    SE MEAN        T    P VALUE
AGE       100    31.720     3.798      0.380     1.90     0.061
```

**SECTION 8-3** Interval Estimates ▶ **187**

calculates whether a *t*-value is significant, you will little need to use the *t* table in this course. However, you may have to read the table in other courses or after you graduate.

## 8-3 INTERVAL ESTIMATES

The similarity between *t* and *Z* exists in the area of interval estimates as well. In Chapter 6, we estimated an interval that was 95% likely to contain the average age of the population with the `ZINTerval` command. Now we will use a `TINTerval` command that has the same format. The command, used for the Student *t* and the paired *t* (described later), is

```
TINTerval with conf level = K, data in C
```

As you can see, the only difference between this command and the `ZINTerval` command of Chapter 6 is that we do not specify the population standard deviation; instead Minitab calculates that value based upon our sample. In our Go-Fast age problem, the command for a 90% confidence interval estimate is

```
MTB >TINTerval with conf level = 90, data in C1
```

This command produced the output shown in Figure 8-2. We can calculate virtually any level of confidence interval with this command by simply specifying the appropriate value in the command. Figure 8-2 can be compared with Figure 6-8. Although the interval values in the two tables are very close, the values in Figure 8-2, based on the *t* statistic, are slightly smaller. Why? The reason is the standard deviation value. In Figure 6-8 we were assuming a standard deviation value of 4. In Figure 8-2, the standard deviation value is 3.798. If the standard deviation is smaller, then we have less spread or variation; thus we have a smaller interval at the same level of confidence. We are still 90% likely to have created a confidence interval that contains the true population mean. Figure 8-3, which shows the comparable values for 95% and 99% confidence intervals, can be compared with

**FIGURE 8-2** Minitab 90% Confidence Interval Estimate of Age of Go-Fast Employees

```
MTB >TINT 90 C1

             N      MEAN    STDEV   SE MEAN    90.0 PERCENT C.I.
AGE         100    31.720   3.798    0.380    ( 31.089,  32.351)
```

**FIGURE 8-3** Minitab 95% and 99% Confidence Interval Estimates of Age of Go-Fast Employees

**95% Interval Estimate**

```
MTB >TINT 95 C1

             N      MEAN    STDEV   SE MEAN    95.0 PERCENT C.I.
AGE         100    31.720   3.798    0.380    ( 30.966,  32.474)
```

**99% Interval Estimate**

```
MTB >TINT 99 C1

             N      MEAN    STDEV   SE MEAN    99.0 PERCENT C.I.
AGE         100    31.720   3.798    0.380    ( 30.722,  32.718)
```

Figure 6-9. In both cases, the intervals created by the *t* test are slightly smaller because the standard deviation value used is slightly lower.

## 8-4 PAIRED *t* TEST

Suppose that a training consultant contends that she can increase the productivity of Go-Fast workers through a one-day seminar. As an experiment, we randomly select 10 employees and send them to her seminar. The productivity levels of these workers, both before and after the seminar, are shown in Table 8-1.

**TABLE 8-1** Productivity of Random Sample of Go-Fast Workers

| Worker | Productivity Before Seminar | Productivity After Seminar |
|---|---|---|
| 1 | 180 | 185 |
| 2 | 210 | 210 |
| 3 | 200 | 205 |
| 4 | 215 | 220 |
| 5 | 170 | 200 |
| 6 | 185 | 210 |
| 7 | 175 | 170 |
| 8 | 210 | 220 |
| 9 | 195 | 210 |
| 10 | 180 | 185 |

Notice that these values are two sets of observations on the *same* individuals. This is known as *paired data,* and we can perform a very powerful *t* test on such data. Notice that if we simply compared the productivity levels of one group of employees with those of another group of employees, it would not be a paired comparison.

In Table 8-1, we are interested in determining if the productivity of our employees has increased. Our null hypothesis is thus going to be a one-tail test. If the productivity of the employees has remained the same or decreased, then the seminar is not worthwhile. Our hypothesis test proceeds like this:

(1) $H^*$: There is no positive change. (This covers both a negative change and no change.)
$H^{**}$: There is a positive change.

(2) *Decision Rule*: We want to be at least 95% certain that the null hypothesis is false before we recommend that all Go-Fast employees attend the seminar. Because the seminar is expensive, we set the alpha level at .05.

(3) *Test Statistic*: We will use a paired *t* test.

If we read the productivity data before the seminar into column 1 and after the seminar into column 2, we can compute the differences between the two columns and place them in column 3. Column 3 thus represents the net gains (or losses) in productivity due to the seminar. We can then perform a *t* test on column 3 *against* a mean of 0. Since column 3 contains only differences, a test against 0 accurately assesses the significance of the change

## SECTION 8-4 Paired *t* Tests

in productivity due to the seminar. We wish to perform a one-tail test for a "greater than" alternative hypothesis. The commands are:

```
MTB >LET C3 = C2 - C1
MTB >TTESt mu = 0, data in C3;
SUBC> ALTernative hypothesis = 1.
```

This produces the output in Figure 8-4. Notice that the level of significance is greater than .05. In fact, our *t*-value of 2.75 provides a level of significance at .011, which is well above our requirement of .05.

**FIGURE 8-4** Minitab Paired One-Tail *t* Test on Go-Fast Employee Productivity before and after Seminar

```
MTB >TTES 0 C3;
SUBC> ALT 1.

TEST OF MU = 0.000 VS MU G.T. 0.000

          N       MEAN     STDEV    SE MEAN      T     P VALUE
C3       10      9.500    10.916     3.452     2.75     0.011
```

**(4)** *Decision*: Since we are more than 95% certain that the null hypothesis is false, we reject it and accept the alternative hypothesis.

This does not necessarily mean that we will recommend that the Go-Fast employees attend the seminar. We would still need to compare the cost of the seminar to the increase in productivity. All we know is that there was a significant increase; our analysis did not consider any economic factors. The entire cost of the seminar would include

- One-day production loss for each employee
- The seminar fee
- Transportation to and from the seminar
- Meals and other expenses

We may need more than a simple increase in productivity. For instance, we may decide that the average increase must be at least 7 units before the increase in productivity justifies the costs involved. How would you test the problem that the average increase must be at least 7?

In our original paired *t* test, we tested against a mean of 0. We can simply rerun the test against a hypothesized mean increase of 7. The command would be

```
MTB >TTESt mu = 7, data in C3;
SUBC> ALTernative hypothesis = 1.
```

This command provides the output shown in Figure 8-5.

**FIGURE 8-5** Minitab Paired *t* Test of Increase of Go-Fast Employee Productivity Being Greater Than 7

```
MTB >TTES 7 C3;
SUBC> ALT 1.

TEST OF MU = 7.000 VS MU G.T. 7.000

          N       MEAN     STDEV    SE MEAN      T     P VALUE
C3       10      9.500    10.916     3.452     0.72     0.24
```

As you can see, the *t* statistic is only 0.72, and the significance level a meager .24. Here we are not confident that the average gain is statistically greater than 7 units per employee. Notice that the mean gain was 9.500. Why would our test against 7 show up so weakly? The answer lies in the standard deviation. Notice that the standard deviation is 10.916, which is very high. If you look back at the raw data, you can see that two employees (with increases of 30 and 25) account for a good bit of the total increase. Thus our conclusion seems consistent with the likelihood that the average gain is not statistically greater than 7.

The paired *t* test is still a one-sample test. In all cases, you first subtract one set of observations from another and then compare the differences against a predetermined value. Each subject tested must be chosen randomly. In our example above, if we had asked for volunteers, we might have gotten a sample of employees who were easily motivated. Clearly this strategy would not be effective and would lead to a biased test. Keep in mind that the *t* test cannot detect such a bias. As far as the computer is concerned, good data and bad data look the same.

## 8-5 TWO-SAMPLE *t* TESTS

Suppose that a Go-Fast floor manager complains that the older workers are not doing their fair share on the production line. As a consultant, it seems strange to you that age should make a difference in productivity, but you decide to test the hypothesis. Let's divide the work force into two groups: (1) workers with ages up to and including the mean age and (2) workers who are older than the mean age. Now we will conduct a hypothesis test.

(1) *H\**: There is no difference between the average productivity of younger and older workers (mean of population 1 = mean of population 2).
*H\*\**: There is a difference between the productivity levels of younger and older workers.

Initially we will simply do a two-tail test where "less than" and "greater than" are included in the alternative hypothesis. We could do a one-tail test of just "less than" for the older workers if we only wanted to test the floor manager's assumption, but at first we will simply determine if there is any difference between the two groups.

(2) *Decision Rule*: If there is less than a 1% chance that the null hypothesis is true, then we will reject it and accept the alternative hypothesis.

(3) *Test Statistic*: We will test for a difference in the mean values of the productivity levels of younger and older workers.

For this problem, we will read employee ages into column 1 and productivity into column 3. Based upon age, we want to group productivity data into two columns. Notice that the columns will probably not be of equal length and we certainly do not have paired data; thus we will need a new type of *t* test. However, before we discuss the type of test to use, we must break up column 3 based on the values in column 1.

We can do this maneuver with an extension of the `COPY` command in Minitab. With the `COPY` command we can select a single value or range of values from a column and place those values in another area of your worksheet. At the same time you can carry along the corresponding values from yet other columns. The general form of the command is

```
COPY C,...,C into C,...C;
```

### SECTION 8-5 Two-Sample *t* Tests

with the subcommands

```
USE rows K,...,K;
USE C = values K,...,K;
OMIT rows K,...,K;
OMIT C = values K,...,K.
```

Although the command format looks complicated, it is really quite simple. The subcommands are not necessary, and you use only those subcommands that you desire. In our illustration, we know the mean age to be 31.720. We want to copy all values between 0 and 31.720—the ages of the younger workers. Examine the following command line:

```
MTB >COPY C1 and C3 into C11 and C12;
SUBC> USE C1 = 0:31.720.
```

These command lines select all the younger ages from `C1` and the corresponding values for productivity in `C3`. These values from `C1` are placed in `C11` and the corresponding values from `C3` are placed in `C12`. In the subcommand line, we have specified that values between 0 and 31.720 are to be chosen from `C1`. Remember, when using the `COPY` command you must specify new columns for your primary column and any corresponding columns. Column 11 now contains only younger employees and column 12 contains their productivity levels.

For the older employees, we will copy their ages and carry along the corresponding productivity values. The command line is

```
MTB >COPY C1 and C3 into C13 and C14;
SUBC> USE C1 = 31.721:99.99.
```

We have selected all the older workers from `C1` and placed their ages in `C13`. Their respective productivity values from `C3` have been carried over and placed in `C14`. To summarize, `C11` contains the ages of younger employees; `C12`, the productivity levels of younger employees; `C13`, the ages of older employees; and `C14`, the productivity levels of older employees.

To keep things straight, we have named columns 12 and 14 `'YOUNGER'` and `'OLDER'`, respectively. The Minitab command is

```
MTB >NAME C12 'YOUNGER', C14 'OLDER'
```

Minitab provides a *t* test command for two samples that is similar in format to previous commands we have studied. There are a general command line and optional subcommands. In reality, two different *t* tests are involved. One is the standard test and the other is called a *pooled t* test. The general form of the command is

```
TWOSample t, K percent confidence, data in C and C;
    ALTernative hypothesis = K;
    POOLed.
```

The pooled subcommand assumes that the standard deviation of the two groups is the same. The standard `TWOSample` command without the pooled subcommand does not make this assumption and so is more conservative. However, if you can assume that the standard deviations are essentially the same, the pooled subcommand provides a more powerful test. When sample sizes are greater than 30, the difference between the two commands becomes minimal. However, with small samples the different commands provide noticeably different results, and care must be taken to use the proper command. If in doubt, the `TWOSample` command *without* the pooled subcommand is safer; however, it will underestimate slightly.

For our present analysis, we will use the `TWOSample` test. As mentioned previously, whenever a Minitab command has more than one possible subcommand, the final subcommand ends with a period; all previous subcommands end with a semicolon. The semicolon tells the system that more subcommands follow. The period informs the system not to look for more subcommands and to execute the specified operation.

For the *t* tests, you must specify the confidence level you desire. In our present example, the command line would be

```
MTB >TWOSample, 99 percent confidence, data in C12 and C14;
SUBC> ALTernative hypothesis = 0.
```

This command indicates a 99% confidence level, a two-tail test (since our alternative hypothesis code is 0), and that the data is in columns 12 and 14. On a two-tail test ("equal to"), it does not matter in which order you specify the columns. We would not use the pooled subcommand here because we would not assume that the variances of the two groups are equal.

Figure 8-6 shows the results from the above command. You should notice that the average productivity for the older workers is actually higher than that for the younger employees. The *t*-value is $-7.59$ and is quite significant. The difference between the standard deviations of the two groups is great enough to justify our use of the `TWOSample` command without the pooled assumption. We can now return to the last item in our decision process:

**(4)** *Decision*: There is less than a 1% chance that the null hypothesis is true, so we reject it. It appears that the older (and probably more experienced) employees are more productive.

Let's review the assumptions of the `TWOSample` *t* test commands.

**(1)** The samples are randomly selected.
**(2)** The data is interval type.
**(3)** The two samples are independent.
**(4)** The parent distribution is normally distributed.

(In this case, the fourth assumption is taken care of by the central limit theorem.)

A good exercise here is to rerun the data with different confidence levels and with pooling. Figure 8-7 has the results of specifying 95% and 90% confidence levels with the

**FIGURE 8-6** 99% Two-Sample *t* Test on Productivity of Older and Younger Go-Fast Employees

```
MTB >NAME C12 'YOUNGER' C14 'OLDER'
MTB >TWOS 99 C12 C14;
SUBC> ALT 0.

TWOSAMPLE T FOR YOUNGER VS OLDER

              N      MEAN    STDEV   SE MEAN
YOUNGER      43     185.2     25.9     4.0
OLDER        57     220.8     19.0     2.5

99 PCT CI FOR MU YOUNGER - MU OLDER: (-48.0, -23.2)
TTEST MU YOUNGER = MU OLDER (VS NE): T=-7.59 P=0.0000 DF=73.8
```

**SECTION 8-5** Two-Sample *t* Tests

**FIGURE 8-7** 95% and 90% Two-Sample *t* Tests on Productivity of Older and Younger Go-Fast Employees

**95% Confidence Interval**

```
MTB >TWOS 95 C12 C14;
SUBC> ALT 0.

TWOSAMPLE T FOR YOUNGER VS OLDER

             N      MEAN    STDEV    SE MEAN
YOUNGER     43     185.2     25.9      4.0
OLDER       57     220.8     19.0      2.5

95 PCT CI FOR MU YOUNGER - MU OLDER: (-44.9, -26.2)
TTEST MU YOUNGER = MU OLDER (VS NE): T=-7.59 P=0.0000 DF=73.8
```

**90% Confidence Interval**

```
MTB >TWOS 90 C12 C14;
SUBC> ALT 0.

TWOSAMPLE T FOR YOUNGER VS OLDER

             N      MEAN    STDEV    SE MEAN
YOUNGER     43     185.2     25.9      4.0
OLDER       57     220.8     19.0      2.5

90 PCT CI FOR MU YOUNGER - MU OLDER: (-43.4, -27.7)
TTEST MU YOUNGER = MU OLDER (VS NE): T=-7.59 P=0.0000 DF=73.8
```

**TWOSample** command. Figure 8-8 shows the results of 99%, 95%, and 90% confidence intervals with the pooled *t* test subcommand. Notice that Minitab will run pooled or unpooled tests; it is up to you to decide if the proper procedure has been used—it is *not* up to the computer.

If we go back to the example of the productivity seminar, we might notice a very interesting fact. This problem could have been analyzed using any of the *t* test commands. In fact, Figure 8-9 contains the results of running the analysis on the data using a paired *t*, then an unpooled two-sample *t*, and finally a two-sample *t* with the pooled assumption.

There are two important things to note from Figure 8-9. First, Minitab ran all three tests—it cannot tell you which test is proper and which is not. That is very much a human decision. Keep in mind that if you have entered data properly into columns and have executed a command line correctly, then in all likelihood Minitab will execute the calculations *whether they make managerial sense or not!*

Second, note the results in Figure 8-9. The paired *t* allows you to reject at an alpha of .05, but the other two tests do not. In this case, all three commands are technically correct. It might be possible to use a **TWOSample** or **POOLed** subcommand and to run a *t* test on this data set. (This would violate the assumption of independent samples.) Which test should you use? The general rule is this: Always use the strongest test for which you meet the constraints. The paired *t* test has the most constraints, but it also provides the most powerful and robust analysis. The pooled *t* test has fewer constraints but provides less power. Finally, the two-sample *t* test has the fewest constraints but also the least power.

These two points are very important. A Minitab printout does not mean that you have run the proper analysis. Remember the old computer saying "Garbage in, garbage out." Also, there are many occasions in statistics when a particular statistic can be properly calculated in more than one way. The proper choice depends on your assumptions and the

**FIGURE 8-8** 99%, 95%, and 90% Pooled Two-Sample $t$ Tests on Productivity of Older and Younger Go-Fast Employees

**99% Confidence Interval**

```
MTB >TWOS 99 C12 C14;
SUBC> ALT 0;
SUBC> POOL.

TWOSAMPLE T FOR YOUNGER VS OLDER
           N      MEAN     STDEV    SE MEAN
YOUNGER   43     185.2     25.9      4.0
OLDER     57     220.8     19.0      2.5

99 PCT CI FOR MU YOUNGER - MU OLDER: (-47.4, -23.8)
TTEST MU YOUNGER = MU OLDER (VS NE): T=-7.92 P=0.0000 DF=98.0
```

**95% Confidence Interval**

```
MTB >TWOS 95 C12 C14;
SUBC> ALT 0;
SUBC> POOL.

TWOSAMPLE T FOR YOUNGER VS OLDER
           N      MEAN     STDEV    SE MEAN
YOUNGER   43     185.2     25.9      4.0
OLDER     57     220.8     19.0      2.5

95 PCT CI FOR MU YOUNGER - MU OLDER: (-44.5, -26.6)
TTEST MU YOUNGER = MU OLDER (VS NE): T=-7.92 P=0.0000 DF=98.0
```

**90% Confidence Interval**

```
MTB >TWOS 90 C12 C14;
SUBC> ALT 0;
SUBC> POOL.

TWOSAMPLE T FOR YOUNGER VS OLDER
           N      MEAN     STDEV    SE MEAN
YOUNGER   43     185.2     25.9      4.0
OLDER     57     220.8     19.0      2.5

90 PCT CI FOR MU YOUNGER - MU OLDER: (-43.0, -28.1)
TTEST MU YOUNGER = MU OLDER (VS NE): T=-7.92 P=0.0000 DF=98.0
```

constraints you can effectively meet. Generally, the greater the constraints, the greater the power of the statistical test. Unfortunately, the greater the constraints, the narrower the range of applicability.

## 8-6 FORMULAS AND HAND CALCULATIONS

In this chapter, we have discussed four different $t$ tests. For each test, we will now give the computation of the standard deviation and standard error, the degrees of freedom, the formulation of the test, and, where appropriate, the confidence interval estimate.

In the last few chapters we have developed the ideas of the standard deviation of the population and the standard error of the sample mean. However, we are now dealing with samples and not the entire population. There is a slight difference in the formulas for calculating the standard deviation for a population and the standard deviation for a sample. The latter statistic is needed in this chapter, and we will give that formula here. We will

**SECTION 8-6** Formulas and Hand Calculations

**FIGURE 8-9** Different Minitab *t* Tests on the Productivity Seminar Data

**Paired *t* Results**

```
MTB >TTES 0 C3;
SUBC> ALT 1.

TEST OF MU = 0.000 VS MU G.T. 0.000

              N       MEAN     STDEV    SE MEAN        T     P VALUE
    C3       10      9.500    10.916      3.452     2.75       0.011
```

**Two-Sample *t* Results**

```
MTB >TWOS 95 C1 C2;
SUBC> ALT 1.

TWOSAMPLE T FOR C1 VS C2
         N      MEAN     STDEV    SE MEAN
   C1   10     192.0      16.2        5.1
   C2   10     201.5      16.5        5.2

95 PCT CI FOR MU C1 - MU C2: (-24.9, 5.9)
TTEST MU C1 = MU C2 (VS GT): T=-1.30 P=0.89 DF=18.0
```

**Pooled *t* Results**

```
MTB >TWOS 95 C1 C2;
SUBC> ALT 1;
SUBC> POOL.

TWOSAMPLE T FOR C1 VS C2
         N      MEAN     STDEV    SE MEAN
   C1   10     192.0      16.2        5.1
   C2   10     201.5      16.5        5.2

95 PCT CI FOR MU C1 - MU C2: (-24.9, 5.9)
TTEST MU C1 = MU C2 (VS GT): T=-1.30 P=0.89 DF=18.0
```

then calculate the standard error of the mean. These two statistics will be used in the other formulas that follow.

## Standard Deviation for a Sample

The formula given previously for the standard deviation of a population is

$$\sigma = \sqrt{\frac{\sum_{i=1}^{N}(x_i - \bar{x})^2}{N}}$$

Theorists have shown, however, that when we calculate the standard deviation of a subset of a population, that is, a sample, the value obtained by the above formula is on the average too small. This bias can be corrected by slightly changing the formula. The technical description is called "correcting for the loss of degrees of freedom." It is not necessary for us to go into detail, but basically when you calculate the standard deviation of a sample, you must correct for the loss of 1 degree of freedom. In short, instead of dividing by *n* (the sample size), we divide by *n* − 1. (In contrast to the previous formula, we use *n* for sample size instead of *N*, which represents the population size.) The unbiased formula thus becomes

$$s = \sqrt{\frac{\sum_{i=1}^{n}(x_i - \bar{x})^2}{n-1}}$$

Minitab always uses this formula to compute the standard deviation.

To illustrate, we can compute the standard deviation corrected for the loss of 1 degree of freedom for the following sample data. If we have the numbers

1    2    3    4    5

our calculations would be as follows (note that the mean equals 3):

| $x$ | $\bar{x}$ | $x - \bar{x}$ | $(x - \bar{x})^2$ |
|---|---|---|---|
| 1 | 3 | −2 | 4 |
| 2 | 3 | −1 | 1 |
| 3 | 3 | 0 | 0 |
| 4 | 3 | 1 | 1 |
| 5 | 3 | 2 | 4 |

Substituting into the formula, we obtain

$$s = \sqrt{\frac{4 + 1 + 0 + 1 + 4}{5 - 1}} = \sqrt{\frac{10}{4}}$$
$$= 1.58$$

Remember, whenever you calculate a standard deviation of a sample, simply divide by the number in the sample minus 1.

## Standard Error of the Mean

When we construct an interval estimate, we are really working with the value of one particular sample mean and trying to determine how close that sample mean is to the population mean. The particular sample mean we use is one among many possible sample means. If we drew 1,000 samples of size $n$ from a given population, we would get 1,000 different sample means. These 1,000 values (means) themselves form a distribution—a distribution of sample means. Estimates of confidence intervals derive from this concept. We could use the 1,000 samples and find a mean of these means (typically called "$x$ double bar"). We could then find the standard deviation of this distribution of sample means, which is called the standard error of the mean. However, since we have an *unbiased* estimator of the population standard deviation from our single sample, we need not go to the trouble of taking an additional large number of samples. We can calculate the standard error of the mean by

$$s_{\bar{x}} = \frac{s}{\sqrt{n}}$$

### SECTION 8-6 Formulas and Hand Calculations

For the example given, our standard error estimate is

$$s_{\bar{x}} = \frac{1.58}{\sqrt{5}} = 0.71$$

This value is used in estimates of confidence intervals. Remember, confidence interval estimates are based on the distribution of sample means. The issue in confidence intervals is "How likely is our sample to represent the true population?" If we drew another sample, the same question would still apply. When dealing with distributions of sample means, we use a standard error based on mean values, not on the individual values from a single sample.

## t Tests

We have discussed four different types of $t$ tests: (1) Student's $t$ test, (2) the paired $t$ test, (3) the pooled $t$ test, and (4) the two-sample $t$ test. For each test we need a formula for calculating the appropriate $t$ statistic and a formula for calculating the associated confidence interval.

▶▶▶▶▶ **Student's t Test** Student's $t$ test, like the Z test, takes the form

$$t_{n-1} = \frac{\bar{x} - \mu_x}{s_{\bar{x}}}$$

The noticeable difference between the $t$ and the Z is that we must use an unbiased estimate of the true population standard deviation in calculating the $t$. As was illustrated, we divide by $n - 1$ instead of $n$ to obtain our unbiased estimate. Thus we use the formula

$$s = \sqrt{\frac{\sum_{i=1}^{n}(x_i - \bar{x})^2}{n-1}}$$

With this value calculated, we can then calculate the standard error of the mean using the formula

$$s_{\bar{x}} = \frac{s}{\sqrt{n}}$$

We now have all of the necessary parts to calculate a $t$ test statistic. For the sample data 1, 2, 3, 4, and 5, we calculated a standard deviation of 1.58 and a standard error of 0.71. If we want to test the mean of this sample data against a hypothesized population mean of 2, our calculation would look as follows:

$$t_{n-1} = \frac{\bar{x} - \mu}{s_{\bar{x}}}$$

$$t_4 = \frac{3 - 2}{0.71} = 1.41$$

The confidence interval for Student's $t$ test follows the general form of the confidence interval for the Z test discussed in Chapter 6. However, the standard error of the mean and the $t$-value used are based on the proper number of degrees of freedom. The general formula is

$$P[\bar{x} - t_{n-1}(s_{\bar{x}}) \leq \mu \leq \bar{x} + t_{n-1}(s_{\bar{x}})] = 1 - \alpha$$

We can now compute a 90% confidence interval estimate of the population mean of 3 in this problem using the appropriate $t$-value. Since we are calculating a 90% interval, we will put 5% in each tail of the confidence interval. A $t$-value for 4 degrees of freedom and 5% area is 2.132.

$$P[3 - 2.132(.071) \leq \mu \leq 3 + 2.132(.071)] = .90$$
$$P(1.48628 \leq \mu \leq 4.51372) = .90$$

▶▶▶▶▶▶ **Paired $t$ Test** There are no new concepts in the formulas for the paired $t$. In the paired $t$ test, we subtract each set of observations to obtain a single column of values. This single column of values is then tested against a hypothesized mean of 0 (or some other appropriate hypothesized value) in exactly the same way as the Student $t$. Basically, once we have subtracted the two sets of values and obtained a single set of observations, then the Student $t$ procedures are used.

▶▶▶▶▶▶ **Two-Sample Method with Equal Variances (POOLed Subcommand)** When the population variances for two samples are about equal (that is, when we would use the pooled $t$ subcommand in Minitab), we can pool the variances of the two samples when computing the standard error of the mean. We pool the variances by the rather ornery-looking formula

$$s_p^2 = \frac{n_1 \left[ \sum_{i=1}^{n}(x_i - x_1) \right]^2 + n_2 \left[ \sum_{i=1}^{n}(x_i - x_2) \right]^2}{n_1 + n_2 - 2}$$

and compute the standard deviation by

$$s_p = \sqrt{s_p^2}$$

This delivers an unbiased estimate of the variance of the pooled population with degrees of freedom

$$df = n_1 + n_2 - 2$$

With all the appropriate parts in place, we can now compute the $t$-value by

$$t_{n_1 + n_2 - 2} = \frac{\bar{x}_1 - \bar{x}_2}{s_p \sqrt{(1/n_1) + (1/n_2)}}$$

### SECTION 8·6 Formulas and Hand Calculations

The confidence interval estimate for this method is

$$P[(\bar{x}_1 - \bar{x}_2) - t_{n_1+n_2-1}(s_p\sqrt{(1/n_1) + (1/n_2)}) \le \mu_1 - \mu_2 \le (\bar{x}_1 - \bar{x}_2) + t_{n_1+n_2-1}(s_p\sqrt{(1/n_1) + (1/n_2)})] = 1 - \alpha$$

Let's use the preceding formulas on the following data sets:

| Sample 1 | Sample 2 |
|----------|----------|
| 6 | 8 |
| 7 | 9 |
| 8 | 10 |
| 9 | 11 |
| 10 | 12 |

$$s_p^2 = \frac{5(1.41) + 5(1.41)}{5 + 5 - 2}$$

$$= \frac{14.1}{8}$$

$$= 1.768$$

$$s_p = 1.330$$

$$df = 5 + 5 - 2 = 8$$

$$t = \frac{8 - 10}{1.330\sqrt{\frac{1}{5} + \frac{1}{5}}}$$

$$= -2.38$$

A 90% confidence interval estimate would use a *t*-value (obtained from the tables) of 1.860 based on 8 degrees of freedom. The computations would be

$$P[-2 - 1.860(1.330\sqrt{\tfrac{1}{5} + \tfrac{1}{5}}) \le \mu_1 - \mu_2 \le -2 + 1.860(1.330\sqrt{\tfrac{1}{5} + \tfrac{1}{5}})] = .90$$
$$P(-3.564 \le \mu_1 - \mu_2 \le -0.435) = .90$$

▶▶▶▶▶**Two-Sample Method with Unequal Variances** When it is appropriate to use the two-sample *t* test without pooling, we must use another method to compute the standard error. In this case, the formula is

$$s_{\bar{x}_1 - \bar{x}_2} = \sqrt{\frac{s_1^2}{n_1} + \frac{s_2^2}{n_2}}$$

Once we have the standard error, we still need to determine the correct degrees of freedom. Unfortunately, the calculation of the degrees of freedom for unequal-variance models is very complex. If you look at one of the Minitab printouts for unequal variances, you will notice that the degrees of freedom value is either indicated as approximate or is expressed as a decimal value (i.e., 7.6 degrees of freedom) depending on the version of Minitab you are using. Although Minitab will use the correct formula to determine the proper number of degrees of freedom, we can use a decision rule that is much less complex. If you have two small samples, use $n - 1$ for the smaller of the two samples. For example, if sample 1 has 8 observations and sample 2 has 11, then use $8 - 1$, or 7, degrees of freedom for your value.

The use of $n - 1$ for the smaller of the two samples will produce a more conservative estimate in all cases; thus, you will not err by underestimating. As a statistical rule of thumb, it is always appropriate to use a more conservative estimate if there is a chance of error.

When the combined size of the two samples is greater than 30, we can use the same formula for calculating the degrees of freedom as we used for equal variances. Thus, the formula is

$$df = n_1 + n_2 - 2$$

(Remember, Minitab calculates the degrees of freedom by a different method, and its values will be slightly different. Trust us, you don't want to see the real formula! The approximation method is close enough.)

Now we can compute the $t$-value by

$$t_{df} = \frac{\bar{x}_1 - \bar{x}_2}{s_{\bar{x}_1 - \bar{x}_2}}$$

and the confidence interval estimate by

$$P[(\bar{x}_1 - \bar{x}_2) - t_{df}\, s_{\bar{x}_1 - \bar{x}_2} \leq \mu_1 - \mu_2 \leq (\bar{x}_1 - \bar{x}_2) + t_{df}\, s_{\bar{x}_1 - \bar{x}_2}] = 1 - \alpha$$

Let's now use the following data to calculate the $t$-value and confidence interval estimate, assuming unequal variances.

| Sample 1 | Sample 2 |
|---|---|
| 7 | 7 |
| 8 | 9 |
| 9 | 11 |
| 10 | 13 |
| 11 | 15 |

$$s_{\bar{x}_1 - \bar{x}_2} = \sqrt{\frac{10/4}{5} + \frac{40/4}{5}}$$

$$= 1.58$$

$$t = \frac{9 - 11}{1.58}$$

$$= -1.27$$

$$df \cong 5 - 1$$

$$\cong 4$$

The 90% confidence interval estimate is

$$P[-2 - 2.132(1.58) \leq \mu_1 - \mu_2 \leq -2 + 2.132(1.58)] = .90$$
$$P[-5.369 \leq \mu_1 - \mu_2 \leq 1.369] = .90$$

# SUMMARY

As shown in Figure 8-9, we can calculate any of the possible $t$ tests. Thus, it is important that we use the proper test. Although a more powerful test is always preferred over a less powerful test, we must first determine if the assumptions behind a particular test can be met. Then and only then can the test be properly used.

When you are analyzing someone else's work, this is an area of critical importance. All of the statistical analysis can look great, and the interpretation might follow from the statistics presented; however, if the wrong test was used, then the entire exercise is a waste of time. In general, few errors are made in calculating statistical values. The errors are almost always in the selection of the test and the assumptions needed for that test.

# DISCUSSION QUESTIONS

1. What assumptions must be satisfied in order to conduct a $t$ test?
2. How do you obtain an unbiased estimate of the population standard deviation?
3. What is the difference between a two-sample and a pooled $t$ test?
4. What rule should one follow when choosing the proper test to use?

# PROBLEMS

**8-1** Tone-up America, a store selling fitness equipment, has a limited amount of floor space. To overcome this problem, Tone-up America has installed several monitors, in which customers can view its large equipment. Because of this efficient use of floor space, the company believes that its mean revenue per square foot is greater than the industry average of $112 per week. Following is a random sample of Tone-up America's weekly revenue per square foot:

| 110 | 82  | 94  | 119 | 115 | 118 |
| 95  | 112 | 117 | 101 | 112 | 116 |
| 82  | 120 | 109 | 98  | 101 | 120 |
| 114 | 85  | 102 | 120 | 109 | 116 |
| 97  | 90  | 108 | 116 | 116 | 104 |

a) Set up the null and alternative hypotheses and perform a $t$ test. Use a confidence level of 90%. Use Minitab to perform the $t$ test.
b) What are your conclusions?
c) Was this a one-tail or a two-tail test?
d) Is it necessary to compute a confidence interval? Why or why not?

**8-2** A local dentist, Dr. Ed Task, advertises in a neighborhood newspaper. Most dentists in his area report that they see on average 2 new patients a month. Dr. Task hopes to have 2 or more new patients a month. The following table is a random sample of the number of new patients seen by Dr. Task per month after he began advertising:

| 3 | 2 | 0 | 0 | 1 | 2 | 4 |
|---|---|---|---|---|---|---|
| 0 | 1 | 1 | 4 | 1 | 1 | 3 |
| 2 | 1 | 3 | 0 | 1 | 0 | 0 |
| 1 | 2 | 4 | 0 | 0 | 3 | 1 |
| 1 | 4 | 2 | 1 | 1 | 0 | 1 |

**a)** Perform a *t* test. Use an alpha of .01.
**b)** Do you think Dr. Task should continue to advertise in the neighborhood newspaper? Why or why not?

**8-3** A luggage manufacturer has received several letters from mothers complaining that the special children's line of suitcases does not hold up well during travel. The mothers felt that the suitcases do not support the weight of other objects piled on top of them. To test this theory, the manufacturer randomly sampled 30 children's suitcases. Weights were placed on the suitcases to measure the maximum amount of weight each could support before breaking. The following is the maximum amount of weight (in pounds) supported by each of the 30 suitcases:

| 425 | 350 | 520 | 440 | 375 | 425 |
|-----|-----|-----|-----|-----|-----|
| 450 | 449 | 395 | 460 | 435 | 440 |
| 456 | 458 | 490 | 349 | 445 | 430 |
| 500 | 457 | 452 | 415 | 460 | 444 |
| 495 | 454 | 382 | 425 | 490 | 435 |

**a)** The manufacturer believes that the children's line holds up just as well or better than its other small suitcases. Test the mean maximum amount of weight supported by the children's suitcases against a mean of 445 pounds for the manufacturer's other line of small suitcases. Use a confidence level of 90%.
**b)** Is the children's line as durable as the manufacturer's other small lines?

**8-4** A newly formed pharmaceutical company is manufacturing a generic drug that has recently lost its patent protection. The company wants the capsule to have a standard length of half an inch. The following is a random sample of the lengths (in inches) of the first batch of pills:

| 0.501 | 0.500 | 0.503 | 0.502 | 0.500 | 0.502 |
|-------|-------|-------|-------|-------|-------|
| 0.502 | 0.498 | 0.500 | 0.504 | 0.500 | 0.499 |
| 0.497 | 0.501 | 0.500 | 0.503 | 0.506 | 0.502 |
| 0.496 | 0.498 | 0.502 | 0.502 | 0.500 | 0.500 |
| 0.498 | 0.503 | 0.505 | 0.500 | 0.501 | 0.499 |
| 0.501 | 0.502 | 0.501 | 0.504 | 0.505 | 0.502 |

**a)** Set up the null and alternative hypotheses. Use a confidence level of 90%. Do a *t* test.
**b)** Is the pharmaceutical company meeting its standards?
**c)** Calculate the **TINTerval** using Minitab.

# PROBLEMS

d) What does the **TINTerval** tell you?

e) How can the interval help you with further testing?

**8-5** Since metal expands slightly when subjected to heat, Weights and Measures, Inc., must test its metal tape measures for accuracy. At a temperature of 105° F, the company is 90% confident that its 7-foot tape measures will, on average, expand to 7 feet, 0.005 inch. Following is the amount of expansion per tape measure (in inches):

| | | | | | |
|---|---|---|---|---|---|
| 0.001 | 0.003 | 0.010 | 0.015 | 0.004 | 0.002 |
| 0.011 | 0.001 | 0.002 | 0.005 | 0.012 | 0.001 |
| 0.009 | 0.008 | 0.005 | 0.001 | 0.004 | 0.014 |
| 0.016 | 0.007 | 0.001 | 0.004 | 0.003 | 0.012 |
| 0.012 | 0.002 | 0.004 | 0.001 | 0.002 | 0.010 |

a) Set up and perform a $t$ test using Minitab. Use a 90% confidence level.

b) Calculate the **TINTerval** using Minitab.

c) What are your conclusions?

d) How can the **TINTerval** help with future testing?

**8-6** A garment manufacturer employs many single mothers. An increasing amount of absenteeism has spurred the company to open its own day care center in hopes of changing that trend. Following is data on randomly sampled female employees of the number of days absent in a three-month period before and after the day care center opened:

| Employee | Before Opening | After Opening |
|---|---|---|
| 1 | 4 | 5 |
| 2 | 3 | 1 |
| 3 | 5 | 2 |
| 4 | 3 | 3 |
| 5 | 3 | 4 |
| 6 | 4 | 1 |
| 7 | 2 | 2 |
| 8 | 2 | 1 |
| 9 | 2 | 0 |
| 10 | 3 | 2 |
| 11 | 2 | 3 |
| 12 | 4 | 2 |
| 13 | 5 | 2 |
| 14 | 4 | 1 |
| 15 | 2 | 0 |
| 16 | 3 | 2 |
| 17 | 4 | 5 |
| 18 | 2 | 1 |
| 19 | 6 | 3 |
| 20 | 4 | 1 |
| 21 | 0 | 5 |
| 22 | 2 | 1 |
| 23 | 4 | 4 |
| 24 | 5 | 6 |

*(continued on next page)*

*(Continued)*

| Employee | Before Opening | After Opening |
|---|---|---|
| 25 | 3 | 1 |
| 26 | 2 | 2 |
| 27 | 3 | 0 |
| 28 | 0 | 4 |
| 29 | 2 | 1 |
| 30 | 4 | 0 |

**a)** If absenteeism had increased, the day care center would have had to be closed. Set up the *t* test using a 95% confidence level. Perform the *t* test.
**b)** What are your conclusions?
**c)** Calculate the `TINTerval`.
**d)** If another sample was taken on the number of days absent for these same employees, when would you need to perform another *t* test? (Use the *t* interval.)

**8-7** Refer to Problem 8-6. Assume that the manufacturer requires a mean decrease in absenteeism of at least 1 day.
**a)** Set up and perform the *t* test using a confidence level of 95%.
**b)** Should the manufacturer keep the day care center open?

**8-8** Your printing machines have been breaking down quite often. The manufacturer suggests that they be checked or tuned up every month. The following chart presents data on the number of printed pages produced before breakdown before and after the start of regular tune-ups:

| Machine | Before Tune-ups | After Tune-ups |
|---|---|---|
| 1 | 1,200 | 1,440 |
| 2 | 1,100 | 920 |
| 3 | 1,000 | 1,500 |
| 4 | 1,400 | 2,020 |
| 5 | 2,000 | 1,700 |
| 6 | 2,100 | 2,500 |
| 7 | 1,010 | 1,440 |
| 8 | 1,600 | 1,800 |
| 9 | 1,250 | 1,400 |
| 10 | 1,400 | 1,700 |
| 11 | 1,220 | 1,400 |
| 12 | 1,550 | 1,800 |
| 13 | 1,600 | 1,440 |
| 14 | 1,700 | 1,800 |
| 15 | 1,460 | 1,500 |
| 16 | 1,240 | 1,600 |
| 17 | 1,400 | 1,100 |
| 18 | 1,780 | 1,480 |
| 19 | 2,010 | 1,712 |
| 20 | 1,900 | 2,020 |
| 21 | 1,980 | 2,100 |
| 22 | 2,300 | 2,200 |

*(continued on next page)*

(*Continued*)

| Machine | Before Tune-ups | After Tune-ups |
|---|---|---|
| 23 | 1,600 | 1,800 |
| 24 | 1,400 | 1,600 |
| 25 | 1,100 | 1,400 |
| 26 | 1,500 | 1,400 |
| 27 | 1,700 | 1,920 |
| 28 | 1,460 | 1,670 |
| 29 | 1,590 | 2,020 |
| 30 | 1,610 | 1,900 |

**a)** If the number of printed pages remained the same or decreased, you would discontinue the tune-ups. Set up and perform the *t* test using an alpha of .05.

**b)** Should you continue the tune-ups?

**8-9** Refer to Problem 8-8. To continue the practice of regular tune-ups, you require a mean increase in printed pages of 160 or more.

**a)** Set up and perform a *t* test. Use an alpha of .05.

**b)** Should you continue the practice of regular tune-ups?

**c)** What other factors might you need to consider concerning the practice of regular tune-ups?

**8-10** Because of a government regulation, you are required to install an air-conditioning system for your employees. However, part of your business is to house tropical plants that must be kept in warm air. To ascertain the effect that the new air-conditioning system has on the plants, you compare the mean number of plants in a box of 10 that look acceptable for sale before and after the installation of the air-conditioner.

| Box | Before Installation | After Installation |
|---|---|---|
| 1 | 10 | 9 |
| 2 | 9 | 8 |
| 3 | 9 | 10 |
| 4 | 8 | 9 |
| 5 | 10 | 8 |
| 6 | 10 | 7 |
| 7 | 8 | 7 |
| 8 | 10 | 10 |
| 9 | 9 | 8 |
| 10 | 10 | 8 |
| 11 | 8 | 7 |
| 12 | 10 | 9 |
| 13 | 10 | 10 |
| 14 | 9 | 9 |
| 15 | 10 | 8 |
| 16 | 9 | 7 |
| 17 | 8 | 8 |
| 18 | 10 | 10 |
| 19 | 10 | 7 |
| 20 | 9 | 10 |

(*continued on next page*)

*(Continued)*

| Box | Before Installation | After Installation |
|---|---|---|
| 21 | 9 | 9 |
| 22 | 10 | 10 |
| 23 | 10 | 10 |
| 24 | 8 | 9 |
| 25 | 10 | 8 |
| 26 | 10 | 10 |
| 27 | 8 | 6 |
| 28 | 8 | 7 |
| 29 | 9 | 9 |
| 30 | 10 | 9 |

**a)** You hope that the air-conditioning system has no effect on the number of plants acceptable for sale. Set up and perform a paired $t$ test using a 99% confidence level.
**b)** Calculate the $t$ interval using a 90% confidence level.
**c)** Should you house the plants elsewhere?
**d)** If you were skeptical of the results and took another sample, over what range would you know that the mean difference did not equal 0?

**8-11** The local telephone company has been accused of charging big businesses more than small businesses for phone calls. Big businesses contend that there are a lot of extra charges that make their phone calls more expensive. Following is a random sample of cost per call (in cents) for both big businesses and small businesses (costs include both direct and indirect charges):

Big Business

| | | | | | |
|---|---|---|---|---|---|
| 5 | 3 | 2 | 7 | 4 | 3 |
| 5 | 4 | 3 | 2 | 4 | 4 |
| 3 | 3 | 4 | 2 | 2 | 3 |
| 4 | 4 | 3 | 2 | 3 | 3 |
| 4 | 3 | 4 | 2 | | |

Small Business

| | | | | | |
|---|---|---|---|---|---|
| 4 | 5 | 2 | 2 | 4 | 5 |
| 3 | 3 | 4 | 3 | 6 | 6 |
| 4 | 5 | 2 | 5 | 4 | 3 |
| 3 | 3 | 3 | 2 | 4 | 5 |
| 4 | 2 | 3 | 3 | 2 | 3 |

**a)** Test to see if the costs per call are equal. Use a confidence level of 90%. What are your conclusions?
**b)** Calculate the $t$ interval for both businesses.
**c)** Why does the $t$ interval for big business have a slightly smaller range?

**8-12** You and a friend have been hired to move cartons from a loading deck to a nearby truck. Your friend demands that you carry heavier cartons since he believes that the bulk of his cartons so far have weighed more. The following chart gives data on random samples of the weight of the boxes carried by you and your friend:

You

| | | | | | |
|---|---|---|---|---|---|
| 22 | 12 | 14 | 20 | 21 | 29 |
| 19 | 21 | 25 | 20 | 11 | 30 |
| 25 | 24 | 24 | 26 | 23 | 23 |
| 29 | 24 | 23 | 25 | 24 | 26 |
| 25 | 25 | 23 | 22 | 20 | 15 |

Friend

| | | | | | |
|---|---|---|---|---|---|
| 20 | 22 | 21 | 16 | 19 | 27 |
| 26 | 20 | 21 | 25 | 19 | 20 |
| 21 | 24 | 23 | 20 | 21 | 18 |
| 20 | 24 | 20 | 19 | 23 | 23 |
| 26 | 27 | 25 | 20 | 19 | 18 |

**a)** Perform a two-sample $t$ test. Use a confidence level of 90%.
**b)** Was your friend right?
**c)** Why is a two-sample $t$ test the best test to use here?

**8-13** A household goods manufacturer produces two brands of car wax, Glow Coat and Super Wax. The manufacturer wants to test both products side by side to see whether they last equally long. Data on the number of days that water beads when placed onto a car hood waxed with each type of wax follows:

| Glow Coat |    |    |    |    |
|---|---|---|---|---|
| 45 | 60 | 50 | 55 | 60 |
| 62 | 58 | 61 | 55 | 56 |
| 62 | 53 | 56 | 54 | 50 |
| 46 | 59 | 60 | 64 | 63 |
| 60 | 60 | 52 | 54 | 55 |
| 57 | 58 | 59 | 60 | 60 |

| Super Wax |    |    |    |    |
|---|---|---|---|---|
| 62 | 50 | 49 | 55 | 57 |
| 56 | 54 | 55 | 54 | 60 |
| 44 | 40 | 42 | 55 | 58 |
| 61 | 50 | 49 | 54 | 53 |
| 61 | 55 | 57 | 54 | 53 |
| 60 | 60 | 47 | 49 | 60 |

**a)** Set up and perform a two-sample $t$ test. Use a 95% confidence level.
**b)** Do both waxes last equally long?

**8-14** Refer to Problem 8-13. The managers of the company wish to phase out one of the car waxes. Most of the managers believe Glow Coat is not as long lasting as Super Wax.
**a)** Test the managers' hypothesis using an alpha of .05.
**b)** Which product should be phased out?

**8-15** A very exclusive school admits young teenagers with IQs of 130 and above. A study is being done to test whether the mean IQs are about even at both campuses of the school. The following are samples of the student IQs at both campuses (the standard deviations for both campuses are the same):

| Campus 1 |    |    |    |    |    |
|---|---|---|---|---|---|
| 135 | 146 | 130 | 144 | 130 | 150 |
| 130 | 132 | 145 | 146 | 150 | 137 |
| 132 | 145 | 142 | 136 | 148 | 130 |
| 136 | 143 | 141 | 132 | 146 | 148 |
| 142 | 130 | 135 | 141 | 136 | 131 |

| Campus 2 |    |    |    |    |    |
|---|---|---|---|---|---|
| 146 | 140 | 144 | 152 | 139 | 136 |
| 153 | 150 | 138 | 146 | 139 | 147 |
| 148 | 152 | 139 | 138 | 152 | 134 |
| 149 | 145 | 138 | 135 | 156 | 136 |
| 137 | 144 | 138 | 137 | 137 | 139 |

**a)** Set up and perform a pooled $t$ test on whether the mean IQs are equal. Use a confidence level of 95%.
**b)** What are your conclusions?
**c)** Campus 1 is supposed to have the higher IQ. Perform a pooled $t$ test using a 95% confidence level.

**8-16** Minority workers in your company claim that they are not being paid as much as their white co-workers. To test this claim, you divide your work force and their respective weekly pay (in dollars) into two groups as follows:

| Minority Workers | | | | | White Workers | | | | |
|---|---|---|---|---|---|---|---|---|---|
| 425 | 400 | 450 | 440 | 482 | 430 | 445 | 458 | 490 | 410 |
| 450 | 445 | 490 | 410 | 456 | 435 | 444 | 412 | 416 | 455 |
| 470 | 453 | 425 | 450 | 421 | 470 | 425 | 434 | 452 | 402 |
| 482 | 446 | 452 | 430 | 458 | 431 | 420 | 402 | 399 | 438 |
| 445 | 412 | 457 | 420 | 475 | 482 | 456 | 501 | 480 | 412 |
| 455 | 465 | 445 | 475 | 456 | 449 | 448 | 445 | 438 | 437 |
| | | | | | 462 | 433 | 428 | 450 | 422 |
| | | | | | 414 | 445 | 460 | 422 | 435 |
| | | | | | 412 | 435 | 468 | 414 | 455 |
| | | | | | 420 | 431 | 426 | 434 | 465 |

**a)** You believe that the variances of the two groups are the same and that there is no difference between the average pay of a minority worker and the average pay of a white worker. Use a confidence level of 90%. What are your conclusions?
**b)** Compute the $t$ intervals for each group.
**c)** In the future, if other samples are taken and the mean for each group falls outside its respective $t$ interval, what would that indicate?

**8-17** Refer to Problem 8-16 on minority workers. The standard deviations are known to be equal.
  **a)** Test the minority workers' claim that they earn less money than their white co-workers. Set up and perform a one-tail pooled $t$ test using a confidence level of 90%.
  **b)** What are your conclusions?

**8-18** A new, faster-rising yeast has been developed by one of your competitors. You have already done some preliminary studies and believe the standard deviations to be equal. The following are samples of the rising times (in minutes) for your product and your competitor's product:

| Your Yeast | | | | | | Competitor's Yeast | | | | | |
|---|---|---|---|---|---|---|---|---|---|---|---|
| 12 | 11 | 11 | 12 | 10 | 12 | 11 | 10 | 14 | 10 | 11 | 12 |
| 14 | 12 | 10 | 11 | 12 | 14 | 11 | 12 | 12 | 11 | 9 | 11 |
| 10 | 11 | 10 | 12 | 14 | 12 | 12 | 10 | 10 | 10 | 12 | 14 |
| 12 | 12 | 11 | 10 | 12 | 10 | 11 | 12 | 14 | 10 | 10 | 11 |
| 11 | 11 | 12 | 11 | 10 | 12 | 11 | 12 | 12 | 11 | 11 | 10 |

**a)** You believe that your yeast rises faster than your competitor's does. Do a pooled $t$ test using an alpha of .02.
**b)** Were you right?
**c)** Would it have been better for you to use a two-sample $t$ test?

# STATISTICAL INFERENCE: MULTIPLE-SAMPLE CASES

*The following Minitab commands will be introduced in this chapter:*

```
AOVOneway on data in C,...,C
```
A command to perform a one-way (one-factor) analysis of variance

```
CODE (K,...,K) to K for C,...,C, put in C,...,C
```
A command to recode a value(s) in a given column into any other value specified and to place the recoded values in a new column

```
TABLe data in C,...,C;
  CHISquare [K].
```
A command to form a table of observed and expected values for each cell and to calculate a chi-square test on the data

```
CHISquare on table in C,...,C
```
A command to calculate a chi-square test on summary data found in specified columns

In Chapter 7, we introduced the concept of statistical inference using the Z test, a theoretical test of hypotheses. The Z test is theoretical because its rigorous assumptions are seldom met. However, the Z test does provide the necessary insight into the process of testing hypotheses.

In Chapter 8 we used a *t* test based on single samples and two samples to evaluate hypotheses. Cases of more than two samples can be effectively analyzed with an *F* test, which we will study in this chapter. Another issue that we will address in this chapter is which statistical test to use when our data is not interval or ratio. Such tests are called nonparametric tests.

## 9-1 ANALYSIS OF VARIANCE

In Chapter 8, we developed four *t* tests, which assumed normal populations and random data. We must still observe those assumptions when working with more than two samples. The only additional assumption is that the various populations have approximately equal variances. Thus, when we have two or more samples and can meet the assumptions of the pooled *t* test, we can use the analysis of variance (or *F* test).

### Minitab and the Analysis of Variance

The *F* test can be used with just two sample groups, in which case it will produce the same results as a *t* test. Of course, when we have three or more samples, we must use the *F* test, because the *t* test will handle only two groups. However, for the moment, we will conduct a two-sample *F* test so that you can see how the *F* test works and compare its results with those obtained from the *t* test.

The general form of the analysis of variance command in Minitab is quite simple:

```
AOVOneway on data in C,...,C
```

This command performs what is called a *one-way analysis of variance*. Any number of columns two or greater can be specified. To use this command, we must observe the following assumptions:

(1) A random sample has been obtained from each population.
(2) All populations have approximately the same variance.
(3) Each population has a normal distribution.

The *F* test is used primarily to answer the question "Do all the populations sampled have an equal mean value?" The *F* test will test *only* for equality and cannot be used to test "greater than" or "less than" hypotheses.

We can repeat our previous analysis comparing the productivity of older and younger employees. In that analysis we had the productivity rates of younger employees in column 12 of our worksheet and the productivity rates of older employees in column 14. Thus the command is

```
MTB >AOVOneway on data in C12, C14
```

This command produces the printout shown in Figure 9-1.

## SECTION 9-1 Analysis of Variance

**FIGURE 9-1** Analysis of Variance for Productivity Levels of Go-Fast Employees Based upon Age

```
MTB >AOVO C12 C14

ANALYSIS OF VARIANCE
SOURCE      DF         SS         MS         F
FACTOR       1      30988      30988      62.70
ERROR       98      48437        494
TOTAL       99      79425
                                      INDIVIDUAL 95 PCT CI'S FOR MEAN
                                      BASED ON POOLED STDEV
LEVEL        N       MEAN      STDEV   -+---------+---------+---------+-----

YOUNGER     43     185.23      25.93   (---*----)
OLDER       57     220.79      18.99                              (---*---)
                                       -+---------+---------+---------+-----

POOLED STDEV =     22.23               180       195       210       225
```

There are several interesting features about Figure 9-1. The first item of importance is the *F ratio*, found at the top right of the printout. The *F* ratio is always the square of the *t*-value found in the pooled *t* test. However, we will need a new table to interpret the *F*-value. This is one of the few cases where Minitab does not print the level of significance; for each **AOVO** command you execute, you must consult an *F* table to find the appropriate level of significance.

Appendix B, Table 3, is a set of three *F* tables, one each for a .05, .025, and .01 level of significance. In each table the *F*-value changes dramatically as the degrees of freedom changes. For the *F*, there is a degrees of freedom calculation for both the numerator and the denominator. At the top of Figure 9-1, the **DF** column represents the degrees of freedom. The number of degrees of freedom for the heading **FACTOR** is the numerator and the number of degrees of freedom for the heading **ERROR** is the denominator. In this case the degrees of freedom are 1 and 98.

We can now determine the *F*-value needed to determine if the calculated value of 62.70 is significant. If we use the *F* table for an alpha of .05, we look at the first column, because it represents 1 degree of freedom in the numerator. We then go to a row near the bottom of the chart to get the degrees of freedom for the denominator. We cannot get a value for 98 degrees of freedom exactly, so we will use the row for 120, which is the closest value. The critical *F*-value is determined to be 3.92 or greater. Thus our *F* ratio of 62.70 is quite significant.

Another interesting feature of the output of the Minitab command for the analysis of variance is the plot of a 95% confidence interval for the means of the samples. This plot is based on an average of the standard deviations of the samples based on the size of each sample. This averaging process provides an estimate of the pooled standard deviation, and a 95% confidence interval for each mean is calculated using the specific value for each mean and the *common* pooled standard deviation.

In this example, the difference between the interval estimates for the two groups is quite dramatic and shows up clearly in the plot. The analysis of variance simultaneously tests for differences in several samples but, as noted earlier, does not test for directionality ("greater than" or "less than" hypotheses). Frequently, however, a visual analysis can show which group has a greater mean. When analyzing more than two groups, it is difficult to determine which sample is sufficiently out of line to make the *F* ratio significant.

Let's now take a look at a three-sample analysis of variance. The same individual who thought that the older employees were not doing their fair share has decided to try one more time. This time he feels that workers with a high school education are more productive than workers with less than a high school education or workers with more than a high school education. You again decide to test his contention statistically by breaking down the productivity records into three groups, namely, productivity levels for workers with less than 12 years of schooling, productivity levels for workers with exactly 12 years of schooling, and productivity levels for workers with more than 12 years of schooling.

We can use the **COPY** command for this problem. We can copy data for employees with less than 12 years of schooling and their productivity to other columns, and similarly for the other two groups. We have read the data on schooling into column 2 and on productivity into column 3. Thus, the command lines are as follows:

```
MTB >COPY C2 C3 into C11 C12;
SUBC> USE C2 = 1:11.
MTB >COPY C2 C3 into C13 C14;
SUBC> USE C2 = 12.
MTB >COPY C2 C3 into C15 C16;
SUBC> USE C2 = 13:20.
```

We now have the lower-educational-level employees in **C11** and their productivity in **C12**, the employees with exactly 12 years of schooling in **C13** and their productivity in **C14**, and the employees with more than 12 years of schooling in **C15** and their productivity levels in **C16**. To make these groups easy to identify, we will name **C12**, **C14**, and **C16** as follows:

```
MTB >NAME C12 'UNDER 12', C14 'EQUAL 12', C16 'OVER 12'
```

With our columns now created and named, we can execute our **AOVOneway** command and produce the output shown in Figure 9-2. The command is

```
MTB >AOVOneway C12, C14, C16
```

In this case we have a very small $F$ ratio with 2 and 97 degrees of freedom. Turning to the $F$ tables in Appendix B, Table 3, we see that 2 and 97 degrees of freedom require an $F$ ratio value of 3.07 (using 2 and 120 degrees of freeedom) to be significant at the .05 level. In this case, there does not appear to be a significant difference between the mean values of

**FIGURE 9-2** Analysis of Variance for Productivity Levels of Go-Fast Employees Based upon Years of Education Completed

```
MTB >AOVO C12 C14 C16

ANALYSIS OF VARIANCE
SOURCE       DF         SS         MS         F
FACTOR        2        315        157       0.19
ERROR        97      79110        816
TOTAL        99      79425
                                             INDIVIDUAL 95 PCT CI'S FOR MEAN
                                             BASED ON POOLED STDEV
LEVEL         N       MEAN      STDEV    ----+---------+---------+---------+-
UNDER12      30     207.67      23.26              (---------------*---------------)
EQUAL12      36     203.33      32.16         (--------------*--------------)
OVER12       34     205.88      28.72           (--------------*--------------)
                                             ----+---------+---------+---------+-
POOLED STDEV =       28.56                      196.0     203.0     210.0     217.0
```

**SECTION 9-1** Analysis of Variance

our three samples. Thus, we cannot reject the null hypothesis that our three samples have statistically different means. The visual plotting of the 95% pooled estimates of mean values also supports this conclusion; you can easily see the large overlapping among the mean values.

This analysis of variance provides strong evidence that the productivity of Go-Fast employees does not differ by the level of schooling. The same fellow who thought older workers were not as productive as younger has struck out again!

## Hand Calculations for Analysis of Variance

Perhaps the easiest way to visualize the analysis of variance computations is through the construction of the ANOVA (analysis of variance) table. A typical ANOVA table is shown in Table 9-1.

**TABLE 9-1** A Typical ANOVA Table

| Source | df | SS | MS (SS/df) | F |
|---|---|---|---|---|
| Among | | | | |
| Within | | | | |
| Total | | | | |

This table is similar in layout to Figures 9-1 and 9-2 of this chapter. However, there is one minor difference. Minitab uses the terms **FACTOR** and **ERROR** to indicate the two portions of the variance shown in our ANOVA table as "Among" and "Within." We have chosen to use *among* and *within* simply to illustrate the point that different statisticians use different terms for the same concepts. As you go from one statistics book to another (or from one journal article to another), the use of different terminology will present a slight but manageable problem.

Let's start by calculating the total SS (total sum of squares) value. This value is found by the formula

$$\text{total SS} = \sum x^2 - \frac{\sum x^2}{n_{\text{total}}}$$

You might recognize that this computation is the numerator of the fraction used to determine the sample variance except that you do not have to calculate $\bar{x}$ (the grand mean of all the samples). When we made our previous calculation, it was called the sum of squares and appeared as

$$\text{sum of squares} = \sum (x - \bar{x})^2$$

The total SS is nothing more than the numerator of the pooled variance of all the samples.

Next, we need to calculate the among SS value as follows:

$$\text{among SS} = \left[ \frac{(CT)_1^2}{n_1} + \frac{(CT)_2^2}{n_2} + \cdots + \frac{(CT)_k^2}{n_k} \right] - \frac{(\sum x)^2}{n_{\text{total}}}$$

where $(CT)_1$ is the column total for column 1 and $n_1$ is the number in that column.

Finally, we must calculate within SS. There is a simple way and a complicated way to figure the within SS value. We should start with the more formal (complicated) formula:

$$\text{within SS} = \sum x^2 - \left[ \frac{(CT)_1^2}{n_1} + \frac{(CT)_2^2}{n_2} + \cdots + \frac{(CT)_k^2}{n_k} \right]$$

Although this formula can be used to calculate the value for within SS directly, note that

$$\text{among SS} + \text{within SS} = \text{total SS}$$

Thus, if we have any two of these three values, we can easily obtain the third.

To fill in the "df" (degrees of freedom) column, we must compute the appropriate degrees of freedom for each portion in the table. The following formulas can be used:

$$\text{total df} = n_{\text{total}} - 1$$
$$\text{among df} = k - 1$$
$$\text{within df} = n_{\text{total}} - k$$

where $n_{\text{total}}$ is the number in all the samples combined and $k$ is the number of samples.

We can easily compute the "MS" (mean square) column by dividing the "SS" column by the respective values in the "df" column. The $F$-value is simply

$$F = \frac{\text{MS among}}{\text{MS within}}$$

Let's work an example to solidify our understanding of this process. Let's perform an analysis of variance on the following data:

| Sample 1 | Sample 2 | Sample 3 |
|---|---|---|
| 3 | 1 | 10 |
| 4 | 2 | 11 |
| 5 | 3 | 12 |
| 6 | 4 | 13 |
| 7 | 5 | 14 |

We must first find the squared value for each observation and then obtain the totals of the individual values and of the squared values. Our example now looks as follows (the squared values are in parentheses):

|  | Sample 1 | Sample 2 | Sample 3 | Total |
|---|---|---|---|---|
|  | 3 (9) | 1 (1) | 10 (100) |  |
|  | 4 (16) | 2 (4) | 11 (121) |  |
|  | 5 (25) | 3 (9) | 12 (144) |  |
|  | 6 (36) | 4 (16) | 13 (169) |  |
|  | 7 (49) | 5 (25) | 14 (196) |  |
| $x$ total | 25 | 15 | 60 | 100 |
| $x^2$ totals | (135) | (55) | (730) | (920) |

# SECTION 9-2 Chi-Square Test

$$\text{total SS} = 920 - \frac{100^2}{15} = 920 - 666.6667 = 253.3333$$

$$\text{among SS} = \frac{25^2}{5} + \frac{15^2}{5} + \frac{60^2}{5} - 666.6667$$
$$= 125 + 45 + 720 - 666.6667$$
$$= 890 - 666.6667$$
$$= 223.333$$

$$\text{within SS} = \text{total SS} - \text{among SS}$$
$$= 253.3333 - 223.3333$$
$$= 30$$

(We can also calculate within SS directly from the formula.)

The degree of freedom values are

$$\text{total df} = 15 - 1 = 14$$
$$\text{among df} = 3 - 1 = 2$$
$$\text{within df} = 15 - 3 = 12$$

We can now construct the complete ANOVA table, shown in Table 9-2.

**TABLE 9-2** ANOVA Table for Example

| Source | df | SS | MS (SS/df) | F |
|---|---|---|---|---|
| Among | 2 | 223.3333 | 111.6667 | 44.6667 |
| Within | 12 | 30.0 | 2.5 | |
| Total | 14 | 253.3333 | | |

## 9-2 CHI-SQUARE TEST

In the statistics described in this chapter and Chapter 8, the samples had to be either from normal populations or of sufficient size so that the central limit theorem applied. It is important to keep reminding yourself of this assumption. If we use a sample of less than 30, we must be fairly certain that it is drawn from a normally distributed population. This fact can be ascertained rationally or by inspection of the data. For example, you might use the `HISTogram` or `DOTPlot` command to plot the data in the sample. The other fundamental assumption for our statistics has been that we have metric data rather than nominal or ordinal.

What if you are unable to satisfy these requirements? There are occasions when you need a test when such assumptions do not apply. Such tests are usually referred to as *distribution free* or *nonparametric* tests. Although most of these tests are beyond the scope of this text, one popular test (and the one requiring the fewest assumptions) will be covered. That test is the chi-square test of independence.

### Minitab and Chi-Square

To better understand what the chi-square test measures, let's return to Go-Fast. The following example is slightly artificial but should serve our purpose. Assume for the moment that we do not have actual data on age and productivity but have kept data on whether someone's age and productivity are above or below the average. In other words, we have only two columns of numbers, and all values are either zeros or ones. We will code a below-average score as 0 and an above-average score as 1. Notice that this data is nominal rather than metric.

In our test of independence, we will now be answering the question "Are age and productivity related?" If these two variables are not related, then we expect the percentage of young workers with high productivity to be equal to the percentage of old workers with high productivity. That is, the variables should be independent. However, if the variables are related, then the percentage of young workers with high productivity should be different from the percentage of old workers with high productivity.

*It is important to realize that this is a test for independence but not direction.* If the percentage of young workers with high productivity is about the same as the percentage of old workers with high productivity, then one variable does not depend on the other. We will simultaneously be testing the relationship between young employees and low productivity. All employees are either high or low producers; thus we will really build two percentage values for the young employees (percent high producers and percent low producers). We will then take those two percentage values and compare them with the percentage value of high and low producers among old employees.

However, before we can complete our analysis, we must recode some of our data for the hypothetical case. This can be done using the **CODE** command in Minitab, which allows you to change any value or group of values. The general command form is

```
CODE (K,...,K) to K for C,...,C, put in C,...,C
```

Again, this command is not as complicated as it looks. Assume we have entered the data for ages into column 1 and for productivity into column 3. With this information and the fact that the mean age is 31.72 years, we will recode all values between 0 and 31.72 as 0. Then we will recode all values above 31.72 to 1. For recoding ages, our two command lines will look as follows if we want to put the new values into column 10:

```
MTB >CODE (0:31.72) to 0 for C1, put in C10
MTB >CODE (31.73:99.99) to 1 for C10, put in C10
```

Take a minute to study these commands. Notice that the second command is executed on column 10, *not* column 1. That is because our first command created a complete column of numbers, but only half of those numbers are zeros. The other half are still their original values. With this completed, we can create column 11 with our productivity data.

```
MTB >CODE (0:205.5) to 0 for C3, put in C11
MTB >CODE (205.6:999.99) to 1 for C11, put in C11
```

To keep things clear, we will name our columns:

```
MTB >NAME C10 'AGE 1', C11 'PRODUCT 1'
```

Figure 9-3 lists the raw data in columns 10 and 11 as created by the above commands. As you can see, the columns are nothing but zeros and ones. If we wanted, we could now go through this raw data and count how many of our young employees are low producers (rows with a 0 in column 10 and a 0 in column 11), and how many are high producers (rows with a 0 in column 10 and a 1 in column 11), and similarly for the old employees.

## SECTION 9-2 Chi-Square Test

**FIGURE 9-3** Printout of Coded Values for Employee Ages and Productivity

```
MTB >NAME C10 'AGE1' C11 'PRODUCT1'
MTB >PRIN C10 C11
 ROW   AGE1   PRODUCT1              ROW   AGE1   PRODUCT1
   1     1       1                   51     1       1
   2     1       1                   52     0       0
   3     1       1                   53     1       1
   4     1       1                   54     1       1
   5     0       0                   55     0       0
   6     0       0                   56     0       0
   7     0       0                   57     0       0
   8     1       0                   58     0       0
   9     1       1                   59     1       1
  10     1       1                   60     1       0
  11     1       0                   61     0       1
  12     1       1                   62     1       1
  13     0       1                   63     1       1
  14     0       1                   64     0       0
  15     1       1                   65     1       0
  16     1       1                   66     0       0
  17     1       0                   67     1       0
  18     0       1                   68     1       1
  19     0       0                   69     1       1
  20     1       0                   70     1       1
  21     1       1                   71     1       0
  22     0       1                   72     0       0
  23     0       0                   73     1       1
  24     1       0                   74     1       1
  25     1       1                   75     1       0
  26     1       1                   76     1       1
  27     1       1                   77     0       0
  28     0       0                   78     1       0
  29     1       0                   79     0       0
  30     1       1                   80     1       0
  31     1       0                   81     0       0
  32     0       0                   82     1       1
  33     0       0                   83     1       1
  34     1       0                   84     0       0
  35     1       1                   85     0       0
  36     0       0                   86     1       1
  37     0       1                   87     1       1
  38     1       1                   88     1       1
  39     1       1                   89     0       0
  40     0       0                   90     0       0
  41     0       0                   91     1       0
  42     1       1                   92     1       1
  43     0       0                   93     1       1
  44     1       1                   94     0       1
  45     0       0                   95     0       0
  46     1       1                   96     1       1
  47     0       0                   97     0       0
  48     0       0                   98     1       1
  49     0       0                   99     0       0
  50     0       0                  100     0       0
```

Such a count of the data reveals 36 young–low producers, 7 young–high producers, 16 old–low producers, and 41 old–high producers. With Minitab we can calculate a *chi-square test* from either our raw data (our zeros and ones) or our summary data (36, 7, 16, and 41). Clearly, it would be inconvenient if we had to count up a couple of thousand individual data points before we could run our tests, so let's look at running a chi-square test on the raw data first. Minitab has a **TABLe** command that will take zeros and ones, create a table, and then run a chi-square on the table.

This entire process and the corresponding statistical analysis can be carried out using the **TABLe** command and **CHISquare** subcommand. This command set builds a contingency table of your data and calculates a chi-square statistic. It will be easier to explain the table after we have built one. The Minitab command is

TABLE data in C,...,C;
CHISquare K.

The **TABLe** command simply calculates the cross tabulations for the data given. The **CHISquare** subcommand can be stated at two different levels. If **K** = 2, then both observed and expected counts are printed in each cell. If **K** is omitted or set equal to 1, then only the observed counts are printed. We will always be setting **K** equal to 2 in our illustrations.

Returning to our problem, the command lines would be:

MTB >TABLe data in C10, C11;
SUBC> CHISquare 2.

This command creates a contingency table and calculates the chi-square statistic as shown in Figure 9-4. Let's first look at the contingency table. As indicated at the top of Figure 9-4, the rows are derived from our age data (column 10). The columns are our productivity data. The numbers at the top of each cell in the table are the actual values obtained from the data (also called the *observed* values). For example, the first cell contains the value 36. This means that there were 36 employees who were both young (coded as 0) and low producers (coded as 0). This is the same value we found by our physical count of the data points when we were looking at the raw data. The value of 7 for the first row and second column is also what we obtained in our count of the data.

In the second row of cells, the first cell, which is the intersection of the row marked **1** and the column marked **0**, indicates the number of old employees who were low producers. In this case, there were 16 such employees. There were 41 employees who were old and high producers (the second cell in the second row). The column and row total values found

**FIGURE 9-4** Chi-Square Test for Independence between Age and Productivity Level of Go-Fast Employees

```
           MTB >TABL C10 C11;
           SUBC> CHIS 2.


               ROWS: AGE1     COLUMNS: PRODUCT1

                          0         1        ALL

                 0       36         7         43
                      22.36     20.64      43.00

                 1       16        41         57
                      29.64     27.36      57.00

               ALL       52        48        100
                      52.00     48.00     100.00

              CHI-SQUARE =     30.412    WITH D.F. =      1

                 CELL CONTENTS --
                              COUNT
                              EXP FREQ
```

## SECTION 9-2 Chi-Square Test

at the far right column and bottom row are the sums of the rows and columns, respectively. Thus, if you add 36 and 7 (across the top of the first row), the total is 43, which is the number of employees identified as young.

Under the observed value in each cell is a second value. This value is the expected value. In other words, if there was no relationship between the two variables, then the second set of values would be expected. For example, in the first cell, we would have expected approximately 22 employees (about half of 43) to be both younger and lower producers. In fact, we found 36 employees in this cell. The question then becomes "Is that difference statistically significant?" In the second cell, we would have expected approximately 21 employees to be younger and higher producers, but in fact we found 7. Again, we want to know if this difference is significant.

To determine significance, we proceed as follows: If you look at the totals for the columns, you will note that 52 employees are classified under the 0 column (low producers), and 48 classified under the 1 column (high producers). Since these values represent a 52%–48% split between high and low producers, we should expect about the same split of employees falling into each cell based on age. The first row total of 43 employees is then split between the two cells according to this ratio; that is, 52% of the 43 employees go into the 0 cell, and 48% of the 43 employees go into the 1 cell. These are our expected values.

The chi-square calculation is then a process of evaluating the differences between the actual values and the expected values for each cell. The total chi-square is 30.41, and the degrees of freedom for this statistic is equal to 1. We must then turn to Appendix B, Table 4, and determine the significance of 30.41. Values are found in Table 4 in the same general way as in Table 2 of Appendix B. The desired alpha levels are listed across the top and the degrees of freedom are listed down the left margin. For an alpha of .05, the chi-square value must be 3.84 or greater. Thus our chi-square value of 30.41 is certainly significant. The hypothesis that these two variables are independent can be rejected, and the alternative hypothesis that they are related accepted.

If the chi-square test has fewer assumptions, why not use it all the time instead of $t$ or $F$? The reason is simple. Since there are fewer assumptions, there is also less power. Remember, power is the ability of a test to identify a false null hypothesis. The significance level of the chi-square test is always lower than that of the $t$ test or $F$ test. Rather than attempt a long technical explanation of why that is the case, just consider the following logical explanation. In interval data, we use all relationships found between the numbers. However, when we use nominal data, we use only the one relationship of above or below a specified value; we do not use the actual values within the data set. In a very real sense, we are using less of the information contained within the sample; thus we have a less powerful test.

The chi-square test is not without assumptions. The primary assumption pertains to the expected number of observations per cell. You must have at least five expected observations for each cell. If you do not, then the chi-square statistic tends to become inflated and thus invalid. If you have such data, Minitab will complete the analysis but it will not print a total chi-square value. It will also print a message informing you that one or more of your cells does not have enough observations for the test to be valid.

Before we conclude this discussion of the chi-square, we will show how to use summary data instead of raw data with Minitab to compute chi-square. To continue our present illustration, let's assume that we are given the data in Table 9-3.

**TABLE 9-3** Summary Data on Productivity of Go-Fast Workers by Age

|  | Productivity | |
|---|---|---|
|  | Young Workers | Old Workers |
| Young Workers | 36 | 7 |
| Old Workers | 16 | 41 |

For summary data, Minitab provides a **CHISquare** command, which has the form

**CHISquare on table in C,...,C**

We will read the summary data into columns 20 and 21 and then perform the **CHISquare** command.

```
MTB >READ C20 C21
DATA>36 7
DATA>16 41
DATA>END
MTB >CHISquare on table in C20, C21
```

This group of commands produces the output shown in Figure 9-5. As you can see, the format is only slightly different from that in Figure 9-4. The most noticeable difference is that the **CHISquare** command shows you the contribution made by each cell to the difference between the observed and expected values; however, the total chi-square value is the same. Since you will frequently have summary data, this command will be useful in performing a chi-square test quickly and easily.

**FIGURE 9-5** Chi-Square Test Using Summary Data for Independence between Age and Productivity Level of Go-Fast Employees

```
MTB >CHIS C20 C21

Expected counts are printed below observed counts

            C20      C21    Total
    1        36        7      43
            22.4     20.6

    2        16       41      57
            29.6     27.4

Total        52       48     100

ChiSq =   8.32 +   9.01 +
          6.28 +   6.80 = 30.41
df = 1
```

## Hand Calculations and Chi-Square ▼▲▼▲▼▲▼▲▼▲▼▲▼▲▼▲▼▲▼▲▼▲▼▲

The chi-square statistic is computed by

$$\chi^2 = \frac{\sum(\text{observed value} - \text{expected value})^2}{\text{expected value}}$$

with

$$df = (\text{number of rows} - 1)(\text{number of columns} - 1)$$

or, for a goodness-of-fit test,

$$df = \text{number of cells} - 1$$

### SECTION 9-2 Chi-Square Test

Let's compute the chi-square statistic for the following data set (observed values):

|   | X | Y | Z | Total |
|---|---|---|---|---|
| A | 100 | 120 | 80 | 300 |
| B | 50 | 60 | 20 | 130 |
| C | 150 | 120 | 100 | 370 |
| Total | 300 | 300 | 200 | 800 |

The expected frequencies of cell AX are computed as follows:

$$\text{expected value of cell } AX = \frac{\text{row total of } A \times \text{column total of } X}{\text{grand total}}$$

$$= \frac{300 \times 300}{800}$$

$$= 112.5$$

Using this same formula, we can calculate the other expected frequencies. The results of all these calculations are as follows:

|   | X | Y | Z | Total |
|---|---|---|---|---|
| A | 100 (112.5) | 120 (112.5) | 80 (75.0) | 300 |
| B | 50 (48.8) | 60 (48.8) | 20 (32.5) | 130 |
| C | 150 (138.8) | 120 (138.8) | 100 (92.5) | 370 |
| Total | 300 | 300 | 200 | 800 |

In each cell, the observed frequency is listed first and the expected frequency is listed second in parentheses. With these calculations made, the portion of the total chi-square contributed by each cell can be calculated as follows:

$$\chi^2_{AX} = \frac{(\text{observed value} - \text{expected value})^2}{\text{expected value}}$$

$$= \frac{(100 - 112.5)^2}{112.5}$$

$$= 1.39$$

In this same way, we can calculate the portion of the chi-square that each cell contributes to the total. Our final calculations will look as follows:

$$\chi^2 = 1.39 + 0.50 + 0.33 + 0.03 + 2.60 + 4.81 + 0.91 + 2.53 + 0.61$$
$$= 13.71$$
$$df = (3 - 1)(3 - 1) = 4$$

This result can be verified by using Minitab. If you read your data into columns 1 through 3 and execute a `CHISquare` command, your results will be as shown in Figure 9-6.

## 9-3 NONPARAMETRIC TEST CHART

In this chapter we have briefly discussed only one of the many nonparametric statistical tests. Further discussion of this topic is beyond the scope of this text. However, we would like to suggest the range of nonparametric tests that are available in Table 9-4, which is

**FIGURE 9-6** Minitab Solution to Hand Calculation Example of Chi-Square

```
MTB >CHIS C1-C3

Expected counts are printed below observed counts

            C1         C2         C3      Total
   1       100        120         80        300
         112.5      112.5       75.0

   2        50         60         20        130
          48.7       48.7       32.5

   3       150        120        100        370
         138.8      138.8       92.5

Total      300        300        200        800

ChiSq =   1.39 +    0.50 +    0.33 +
          0.03 +    2.60 +    4.81 +
          0.91 +    2.53 +    0.61 = 13.71

df = 4
```

adapted from the classic 1956 text *Nonparametric Statistics for the Behavioral Sciences* by Sidney Siegel (New York: McGraw-Hill). Minitab supports a few of these tests, but most are used only in advanced statistical problems. Many of these tests require different sets of assumptions.

# SUMMARY

The critical issue in statistical analysis is meeting the assumptions for the test you wish to perform. As shown in Chapter 8, we can calculate many statistical tests; our concern must always be that we choose the proper one. When we are forced to choose between a less powerful test (like chi-square) and a possibly invalid test (such as a $t$ test or $F$ test when its assumptions may not be satisfied), we must always use the test that we know is valid. Our goal is to use the most powerful test that is legitimate.

In Chapters 8 and 9 we have introduced the most common tests of hypotheses. There are hundreds more. If the tests presented here are not sufficient, refer to texts about experimental design or nonparametric statistics. However, in the business world, $t$, $F$, and chi-square will enable you to do many types of evaluations.

We have emphasized that each test requires certain assumptions. If these assumptions are not met, the adage "Garbage in, garbage out" will apply. Always consider how the sample was collected, the probable normality of the population, and the test statistic used when you evaluate a statistical analysis.

# DISCUSSION QUESTIONS

1. What assumptions are required in a one-way analysis of variance?
2. What is an analysis of variance and what does it test?
3. What is a chi-square test?
4. What do the analysis of variance and chi-square tests have in common?

**TABLE 9-4** Nonparametric Statistical Tests

| Type of Measurement | One-Sample Case | Two-Sample Case – Related Samples | Two-Sample Case – Independent Samples | k-Sample Case – Related Samples | k-Sample Case – Independent Samples | Nonparametric Measure of Correlation |
|---|---|---|---|---|---|---|
| Nominal | Binomial test<br>Chi-square | McNemar test for significance of changes | Fisher exact probability test<br>Chi-square for two independent samples | Cochran $Q$ test | Chi-square test of $k$ independent samples | Contingency coefficient |
| Ordinal | Kolmogorov–Smirnov one-sample test<br>One-sample runs test | Sign test<br>Wilcoxon matched-pairs signed-ranks test | Median test<br>Mann–Whitney $U$ test<br>Kolmogorov–Smirnov two-sample test<br>Wald–Wolfowitz runs test<br>Moses test of extreme reactions | Friedman two-way analysis of variance | Extension of median test<br>Kruskal–Wallis one-way analysis of variance | Spearman rank-order correlation coefficient<br>Kendall rank-order correlation coefficient<br>Kendall partial rank-order correlation coefficient<br>Kendall coefficient of concordance |
| Interval or Ratio (Metric) | | Walsh test<br>Randomization test for matched pairs | Randomization test for two independent samples | | | |

Adapted from *Nonparametric Statistics for the Behavioral Sciences* by Sidney Siegel, McGraw-Hill, 1956, inside front cover.

# PROBLEMS

**9-1** The past five years have been extremely profitable for ABC Records, and it seemed fitting that the company's three record promoters be evaluated. The president of ABC Records commented that the company showed a substantial increase in profits over this time but that none of the promoters deserved a promotion because the mean sales of records for all the promoters were about the same. Following are randomly sampled monthly sales (in millions of dollars) for each promoter over the last five years:

| Promoter 1 | | | Promoter 2 | | | Promoter 3 | | |
|---|---|---|---|---|---|---|---|---|
| 1.20 | 0.62 | 0.40 | 0.67 | 0.23 | 0.66 | 0.62 | 0.43 | 0.44 |
| 0.44 | 0.12 | 0.58 | 0.67 | 0.59 | 0.90 | 0.23 | 0.47 | 0.48 |
| 0.54 | 0.58 | 0.62 | 0.49 | 0.57 | 0.67 | 0.57 | 1.40 | 0.46 |
| 0.46 | 0.57 | 0.63 | 0.48 | 0.09 | 0.52 | 0.49 | 0.68 | 0.42 |
| 0.58 | 0.59 | 0.57 | 0.69 | 0.71 | 0.44 | 0.55 | 0.46 | 0.50 |
| 0.55 | 0.45 | 0.66 | 0.59 | 0.48 | 0.65 | 0.89 | 0.52 | 0.47 |
| 0.52 | 0.57 | 0.41 | 0.50 | 0.62 | 0.48 | 0.39 | 0.55 | 0.49 |
| 0.22 | 0.60 | 0.48 | 0.52 | 0.65 | 0.33 | 0.37 | 0.45 | 0.50 |
| 0.43 | 0.55 | 0.40 | 0.53 | 0.57 | 0.63 | 0.62 | 0.49 | 0.58 |
| 0.30 | 0.62 | 0.49 | 0.77 | 0.59 | 0.65 | 0.51 | 0.39 | 0.43 |

**a)** Set up and perform an analysis of variance. Do not refer to the $F$ table just yet.
**b)** Examine the plot of the 95% confidence interval of the means. Do you believe that the three samples have statistically similar means or does at least one promoter have a statistically different mean?
**c)** Complete your analysis. Use a 95% confidence level. Should the president promote one of the record promoters?

**9-2** Arabian horses have increased in value in recent years. As a result, the market has attracted many investors who wish to breed these valuable animals. Arab Farms foresees an oversupply of these horses in the near future. Thus, Arab Farms wishes to divest itself of one of its three farms. Following is a random sample of total sales (in thousands of dollars) by each farm in various months:

| Farm 1 | | | Farm 2 | | | Farm 3 | | |
|---|---|---|---|---|---|---|---|---|
| 250 | 235 | 302 | 427 | 398 | 920 | 298 | 301 | 500 |
| 289 | 287 | 291 | 358 | 401 | 905 | 310 | 206 | 122 |
| 802 | 320 | 351 | 401 | 333 | 401 | 315 | 489 | 322 |
| 165 | 275 | 310 | 499 | 422 | 502 | 289 | 310 | 311 |
| 489 | 289 | 298 | 457 | 389 | 512 | 245 | 316 | 216 |
| 150 | 445 | 105 | 500 | 256 | 444 | 256 | 342 | 899 |
| 265 | 333 | 301 | 480 | 598 | 357 | 344 | 258 | 284 |
| 260 | 457 | 120 | 352 | 395 | 456 | 444 | 134 | 254 |
| 378 | 289 | 325 | 107 | 435 | 298 | 311 | 254 | 301 |
| 356 | 280 | 340 | 623 | 287 | 323 | 325 | 312 | 232 |

**a)** Test the hypothesis that all three farms have the same mean sales. Perform an analysis of variance using a 99% confidence level.
**b)** Examine the plot of the 95% confidence interval of the means. What do you notice?
**c)** Should Arab Farms sell one of its farms based on lowest average sales? If not, what could it do?

**9-3** A local tropical fish breeder has been in business since he was 15 years old. In recent years the business has not been as profitable as it was in its first years. The breeder believes that the problem stems from his carrying varieties of fish with small profit margins. Following is a random sample of the profits earned (in cents) per dozen sold of his least profitable fishes:

# PROBLEMS

| Black Sharks | | | | | |
|---|---|---|---|---|---|
| 12 | 19 | 20 | 22 | 14 | 24 |
| 28 | 12 | 27 | 22 | 25 | 15 |
| 14 | 32 | 19 | 21 | 20 | 17 |
| 20 | 14 | 17 | 18 | 28 | 22 |
| 20 | 14 | 16 | 27 | 26 | 22 |
| 24 | 28 | 14 | 12 | 22 | 18 |
| 27 | 24 | 22 | 12 | 11 | 22 |
| 26 | 12 | 27 | 25 | 16 | 15 |
| 18 | 20 | | | | |

| Angelfish | | | | | |
|---|---|---|---|---|---|
| 30 | 25 | 32 | 33 | 35 | 34 |
| 28 | 22 | 29 | 34 | 22 | 24 |
| 33 | 35 | 24 | 25 | 20 | 19 |
| 29 | 34 | 36 | 33 | 32 | 30 |
| 28 | 21 | 23 | 30 | 31 | 33 |

| Catfish | | | | | |
|---|---|---|---|---|---|
| 30 | 18 | 22 | 32 | 12 | 11 |
| 16 | 33 | 18 | 25 | 28 | 17 |
| 20 | 21 | 16 | 34 | 14 | 15 |
| 24 | 20 | 16 | 34 | 21 | 24 |
| 23 | 20 | 17 | 15 | 38 | 35 |
| 15 | 33 | 21 | 24 | 20 | 19 |
| 15 | 12 | 24 | 23 | 22 | 25 |

| Clown Loaches | | | | | |
|---|---|---|---|---|---|
| 12 | 15 | 18 | 24 | 20 | 11 |
| 24 | 21 | 28 | 19 | 10 | 12 |
| 18 | 19 | 22 | 20 | 27 | 10 |
| 20 | 18 | 17 | 23 | 15 | 22 |
| 20 | 24 | 22 | 21 | 25 | 24 |
| 11 | 27 | 23 | 26 | 12 | 28 |
| 20 | 21 | 16 | 24 | 19 | 18 |

**a)** Are the four types of fish equally profitable? Use a 95% confidence level.
**b)** Which fish should he stop selling?

**9-4** Every year travel agencies use bogus names to reserve airline seats in hopes of booking a customer later. This is done primarily for the Christmas season when the airlines are the busiest. Often, however, the agencies fail to book customers, thus decreasing profits for the airlines. Four travel agencies are under investigation. The number of cancellations made each day by each agency two weeks prior to a flight date for 40 days in December and January follows:

| Agency 1 | | | | |
|---|---|---|---|---|
| 10 | 12 | 8 | 8 | 0 |
| 15 | 2 | 6 | 11 | 14 |
| 0 | 14 | 12 | 9 | 9 |
| 10 | 12 | 2 | 11 | 14 |
| 6 | 18 | 12 | 14 | 15 |
| 12 | 15 | 15 | 16 | 11 |
| 7 | 7 | 12 | 14 | 8 |
| 11 | 7 | 12 | 11 | 9 |

| Agency 2 | | | | |
|---|---|---|---|---|
| 9 | 6 | 5 | 11 | 12 |
| 2 | 10 | 9 | 14 | 8 |
| 9 | 12 | 14 | 9 | 9 |
| 4 | 10 | 14 | 16 | 12 |
| 8 | 7 | 4 | 10 | 11 |
| 11 | 8 | 10 | 5 | 5 |
| 14 | 8 | 7 | 10 | 0 |
| 7 | 17 | 11 | 12 | 11 |

| Agency 3 | | | | |
|---|---|---|---|---|
| 11 | 12 | 14 | 8 | 7 |
| 14 | 9 | 8 | 4 | 5 |
| 7 | 0 | 7 | 11 | 8 |
| 17 | 0 | 10 | 9 | 6 |
| 4 | 5 | 5 | 12 | 12 |
| 6 | 9 | 7 | 7 | 11 |
| 12 | 14 | 5 | 4 | 6 |
| 9 | 9 | 7 | 11 | 3 |

| Agency 4 | | | | |
|---|---|---|---|---|
| 4 | 12 | 15 | 11 | 3 |
| 5 | 18 | 6 | 4 | 14 |
| 0 | 6 | 2 | 8 | 9 |
| 10 | 12 | 14 | 8 | 9 |
| 6 | 9 | 10 | 12 | 11 |
| 15 | 5 | 16 | 4 | 4 |
| 7 | 10 | 12 | 8 | 6 |
| 18 | 7 | 2 | 14 | 9 |

**a)** Test to see if the agencies' mean numbers of cancellations are equal. Use a 99% confidence level.

**b)** If the numbers of cancellations are the same, does that rule out bookings for bogus customers? Why or why not?

**9-5** You are evaluating the purchase of a solar heating unit to help keep your heating costs down. Three different models are within your price range. Following is a random sample of the monthly cost of heating (in dollars) for each unit:

| Unit 1 | | | | | Unit 2 | | | | | Unit 3 | | | |
|---|---|---|---|---|---|---|---|---|---|---|---|---|---|
| 90 | 87 | 88 | 97 | | 92 | 98 | 72 | 70 | | 89 | 85 | 79 | 67 |
| 80 | 81 | 78 | 80 | | 74 | 78 | 70 | 72 | | 80 | 85 | 84 | 79 |
| 72 | 91 | 89 | 71 | | 68 | 74 | 70 | 71 | | 89 | 85 | 90 | 86 |
| 88 | 72 | 87 | 90 | | 78 | 70 | 88 | 82 | | 85 | 89 | 92 | 78 |
| 79 | 86 | 91 | 94 | | 67 | 85 | 90 | 76 | | 79 | 87 | 89 | 88 |
| 86 | 70 | 75 | 80 | | 73 | 70 | 78 | 72 | | 88 | 86 | 77 | 69 |
| 70 | 92 | 55 | 82 | | 90 | 68 | 70 | 85 | | 84 | 91 | 86 | 87 |
| 85 | 80 | 89 | 76 | | 79 | 72 | 70 | 68 | | 93 | 90 | 80 | 85 |
| 108 | 90 | 95 | | | 69 | 72 | 82 | 71 | | 87 | 87 | 87 | 85 |
| | | | | | 97 | 89 | 70 | 87 | | 81 | | | |
| | | | | | 69 | 60 | 74 | 74 | | | | | |
| | | | | | 71 | 71 | 74 | | | | | | |

**a)** Evaluate the data using analysis of variance. Use an alpha of .025.

**b)** Examine the plot of the 95% confidence interval of the means. Which, if any, of the sample means is different from the others?

**c)** Which solar heating unit should you invest in?

**9-6** Three overweight couples decided to go to a weight loss clinic together. Each couple believed they could lose the most average weight per day. Following is a random sample of the pounds lost per day over a three-month period:

| Couple 1 | | | | Couple 2 | | | | Couple 3 | | |
|---|---|---|---|---|---|---|---|---|---|---|
| 8 | 12 | 8 | | 6 | 5 | 10 | | 4 | 9 | 3 |
| 7 | 10 | 11 | | 9 | 5 | 5 | | 10 | 2 | 3 |
| 5 | 6 | 9 | | 9 | 6 | 8 | | 4 | 12 | 2 |
| 2 | 6 | 7 | | 5 | 8 | 7 | | 3 | 3 | 6 |
| 2 | 6 | 8 | | 8 | 6 | 7 | | 2 | 5 | 1 |
| 10 | 15 | 9 | | 7 | 2 | 9 | | 4 | 5 | 2 |
| 7 | 8 | 2 | | 8 | 5 | 8 | | 1 | 5 | 5 |
| 5 | 2 | 1 | | 7 | 8 | 7 | | 4 | 2 | 1 |
| 4 | 7 | 12 | | 6 | 6 | 7 | | 2 | 4 | 2 |
| 3 | 6 | 9 | | 6 | 9 | 5 | | 1 | 2 | 6 |

**a)** Determine if each couple on average lost the same amount of weight. Use an alpha of .025.

**b)** If at least one sample has a statistically different mean, determine who lost the most weight. In case of a tie, perform a pooled $t$ test. Use a 95% confidence level.

## PROBLEMS

**9-7** Last month your younger sister graduated from high school. She has asked you to help her choose her major in college. She has narrowed the possibilities to accounting, finance, and management. You decide to evaluate the starting salaries offered to recent college graduates in each major. The following chart is a random sample of starting salaries (in thousands of dollars):

| Accounting | | | Finance | | | Management | | |
|---|---|---|---|---|---|---|---|---|
| 17.5 | 18.9 | 16.9 | 17.0 | 16.8 | 16.3 | 16.3 | 16.8 | 19.2 |
| 19.0 | 17.5 | 15.9 | 17.1 | 17.8 | 17.2 | 16.3 | 16.5 | 16.2 |
| 18.4 | 17.2 | 17.1 | 16.5 | 16.2 | 17.8 | 17.5 | 17.1 | 17.0 |
| 19.2 | 18.4 | 18.2 | 16.7 | 17.2 | 17.9 | 15.9 | 16.8 | 16.2 |
| 17.7 | 19.4 | 18.2 | 17.8 | 19.3 | 16.9 | 17.3 | 16.2 | 16.3 |
| 19.5 | 16.3 | 18.2 | 16.8 | 16.7 | 17.4 | 16.4 | 16.3 | 16.5 |
| 18.2 | 18.2 | 17.2 | 17.7 | 17.2 | 17.0 | 18.7 | 18.5 | 16.3 |
| 18.2 | 18.0 | 16.6 | 16.7 | 18.1 | 17.2 | 16.8 | 17.3 | 18.2 |
| 19.1 | 17.6 | 18.3 | 17.3 | 19.0 | 16.5 | 16.8 | 16.2 | 16.3 |
| 16.8 | 19.0 | 17.8 | 17.1 | 18.2 | 16.4 | 16.1 | 18.2 | 17.5 |
| 17.8 | 19.0 | 19.3 | 16.0 | 18.2 | 17.5 | 16.7 | 17.0 | 17.3 |
| 16.8 | 16.7 | | 16.8 | 17.0 | 17.1 | | | |
| | | | 17.3 | 17.4 | 17.9 | | | |

**a)** Is the average salary in each area the same? Use a 95% confidence level.
**b)** Which major should your sister choose to get the highest starting salary? Why?

**9-8** Your company intends to introduce a new line of single-serving vegetables. Test markets have been established on the West Coast, on the East Coast, and in the Midwest. Following is a random sample of sales per day in each test market:

| West Coast | | | East Coast | | | Midwest | | |
|---|---|---|---|---|---|---|---|---|
| 112 | 116 | 121 | 127 | 140 | 125 | 132 | 122 | 124 |
| 128 | 125 | 122 | 135 | 119 | 127 | 122 | 120 | 124 |
| 122 | 118 | 120 | 132 | 124 | 142 | 124 | 119 | 119 |
| 125 | 129 | 115 | 122 | 128 | 130 | 123 | 121 | 130 |
| 120 | 134 | 119 | 135 | 131 | 128 | 121 | 122 | 124 |
| 126 | 125 | 125 | 139 | 129 | 134 | 122 | 124 | 127 |
| 126 | 125 | 117 | 140 | 127 | 120 | 130 | 130 | 122 |
| 120 | 125 | 124 | 124 | 121 | 142 | 120 | 126 | 128 |
| 125 | 119 | 131 | 128 | 126 | 121 | 127 | 123 | 127 |
| 119 | 124 | 124 | 132 | 129 | 124 | 124 | 125 | 127 |
| 121 | 124 | 120 | 136 | 139 | 131 | 125 | | |
| 121 | 123 | 124 | 128 | 128 | 141 | | | |
| 125 | 127 | | 140 | 134 | 128 | | | |
| | | | 143 | 122 | 132 | | | |
| | | | 125 | 140 | 135 | | | |
| | | | 119 | 130 | 137 | | | |
| | | | 136 | 133 | | | | |

**a)** Do all three areas have the same mean sales? Use a 99% confidence level.
**b)** Where would your company want to market?

**9-9** The coach of your high school football team believes that out of three placekickers, Joe kicks the farthest. You are one of the three. You believe that each of you averages about the same distance. If this was true, the coach would have to base his decision on some other criterion. The following is a random sample of distances kicked by each contender (in yards):

| You | | | Larry | | | Joe | | |
|---|---|---|---|---|---|---|---|---|
| 46 | 44 | 45 | 52 | 50 | 44 | 41 | 52 | 54 |
| 50 | 52 | 45 | 44 | 40 | 55 | 45 | 52 | 49 |
| 45 | 52 | 50 | 54 | 41 | 40 | 44 | 54 | 49 |
| 45 | 45 | 48 | 42 | 40 | 41 | 46 | 53 | 44 |
| 48 | 46 | 46 | 43 | 40 | 44 | 55 | 52 | 49 |
| 44 | 48 | 49 | 50 | 52 | 55 | 51 | 50 | 54 |
| 51 | 56 | 55 | 52 | 50 | 51 | 44 | 55 | 46 |
| 44 | 40 | 48 | 45 | 44 | 47 | 55 | 52 | 46 |
| 44 | 46 | 49 | 46 | 49 | 48 | 45 | 47 | 50 |
| 50 | 48 | 47 | 46 | 45 | 52 | 44 | 48 | 52 |
| 45 | 40 | 41 | 48 | 47 | 45 | 48 | 47 | 50 |
| 45 | 40 | 42 | 48 | 48 | 42 | 39 | | |
| 52 | 50 | 42 | 45 | 42 | 47 | | | |
| | | | 44 | 43 | 45 | | | |
| | | | 48 | 46 | 42 | | | |

Must the coach make a decision based on some other criterion? Use an alpha of .05.

**9-10** You are a young executive paying too many taxes. As a result, you wish to evaluate investments for tax breaks. After some consideration, you decide to invest in real estate. Four pieces of land appeal to you, located in Arizona, New Mexico, Colorado, and Maine. Each seems equally attractive, but you can only afford to invest in one of them. The following is a random sample of housing starts in each of the areas:

| Arizona | | | | New Mexico | | | |
|---|---|---|---|---|---|---|---|
| 159 | 158 | 162 | 166 | 157 | 154 | 149 | 142 |
| 155 | 158 | 140 | 157 | 158 | 161 | 153 | 144 |
| 154 | 158 | 154 | 151 | 148 | 152 | 151 | 140 |
| 144 | 152 | 155 | 154 | 162 | 166 | 155 | 154 |
| 160 | 145 | 140 | 142 | 142 | 153 | 141 | 140 |
| 152 | 145 | 153 | 158 | 156 | 162 | 163 | 154 |
| 146 | 162 | 168 | 164 | 150 | 151 | 156 | 147 |
| 168 | 158 | 154 | 147 | 155 | 152 | 156 | 161 |
| 142 | 144 | 152 | 156 | 148 | 147 | 158 | 152 |
| | | | | 143 | 142 | | |

| Colorado | | | | Maine | | | |
|---|---|---|---|---|---|---|---|
| 133 | 162 | 160 | 132 | 135 | 133 | 145 | 152 |
| 142 | 125 | 116 | 150 | 133 | 152 | 120 | 124 |
| 135 | 124 | 122 | 120 | 120 | 118 | 154 | 160 |

*(continued on next page)*

# PROBLEMS

*(Continued)*

| Colorado | | | | Maine | | | |
|---|---|---|---|---|---|---|---|
| 145 | 120 | 132 | 131 | 128 | 130 | 127 | 126 |
| 150 | 126 | 122 | 119 | 132 | 118 | 140 | 142 |
| 130 | 122 | 124 | 138 | 139 | 135 | 130 | 150 |
| 144 | 150 | 151 | 133 | 160 | 132 | 124 | 133 |
| 124 | 135 | 152 | 150 | 134 | 133 | 138 | 133 |
| 160 | | | | 152 | 132 | | |

**a)** Assume that each piece of property costs the same. You believe that land appreciation is tied directly to the average number of housing starts in the area. Determine if the average housing starts in each of the areas is the same. Use a 99% confidence level.

**b)** Examine the plot of the 95% confidence interval of the means. If no single property has a higher average number of housing starts than any other, perform a pooled $t$ test to see if one property has a significantly greater mean than the other. Use a 99% confidence level.

**c)** Which piece of property would you invest in based on your findings?

**9-11** The following data on the age and speed (in seconds per furlong) of racehorses was given to you:

| Age | Speed | Age | Speed | Age | Speed | Age | Speed |
|---|---|---|---|---|---|---|---|
| 2 | 6 | 1 | 4 | 2 | 4 | 2 | 7 |
| 3 | 5 | 3 | 5 | 4 | 8 | 1 | 5 |
| 1 | 4 | 2 | 5 | 2 | 7 | 2 | 4 |
| 6 | 8 | 6 | 7 | 1 | 5 | 6 | 6 |
| 5 | 7 | 2 | 5 | 4 | 4 | 5 | 4 |
| 2 | 5 | 2 | 5 | 1 | 4 | 5 | 4 |
| 2 | 4 | 1 | 5 | 1 | 5 | 2 | 4 |
| 5 | 6 | 3 | 6 | 2 | 4 | 1 | 5 |
| 5 | 5 | 4 | 7 | 6 | 4 | 4 | 5 |
| 1 | 4 | 6 | 7 | 1 | 5 | 6 | 5 |
| 4 | 4 | 3 | 4 | 2 | 6 | 1 | 4 |
| 1 | 4 | 2 | 7 | 2 | 4 | 2 | 5 |
| 6 | 7 | 2 | 4 | 6 | 6 | 5 | 6 |
| 2 | 4 | 2 | 7 | 4 | 5 | 1 | 6 |
| 3 | 8 | 4 | 5 | 3 | 4 | 4 | 8 |
| 1 | 5 | 1 | 4 | 4 | 7 | 5 | 7 |
| 2 | 5 | 1 | 4 | 1 | 7 | 1 | 5 |
| 3 | 7 | 6 | 5 | 2 | 8 | 2 | 4 |
| 1 | 5 | 2 | 5 | 1 | 4 | 5 | 8 |
| 3 | 6 | 6 | 4 | 2 | 5 | 1 | 4 |
| 6 | 5 | 1 | 5 | 2 | 4 | 6 | 5 |
| 2 | 5 | 5 | 4 | 5 | 7 | 1 | 4 |
| 1 | 6 | 1 | 5 | 1 | 5 | 2 | 6 |
| 5 | 8 | 1 | 7 | 3 | 4 | 5 | 7 |
| | | | | | | 1 | 4 |

Test the dependence of the variables speed and age. Use an alpha of .05.

**9-12** A marketing research firm in your community is testing the relationship between level of education and starting salary. Student counselors have complained that in recent years more education does not guarantee higher wages. Use the following information to evaluate that claim. Level of education refers to the number of years in school after high school, and salaries are expressed in thousands of dollars per year:

| Education | Salary | Education | Salary |
|---|---|---|---|
| 4 | 16.8 | 1 | 15.2 |
| 0 | 11.2 | 0 | 11.6 |
| 6 | 22.9 | 2 | 14.2 |
| 0 | 17.2 | 3 | 15.6 |
| 1 | 12.2 | 5 | 18.2 |
| 4 | 16.2 | 3 | 14.5 |
| 1 | 13.2 | 1 | 12.4 |
| 4 | 16.8 | 4 | 17.8 |
| 4 | 14.9 | 2 | 13.5 |
| 1 | 12.4 | 6 | 20.1 |
| 3 | 17.2 | 4 | 19.2 |
| 2 | 14.2 | 1 | 10.2 |
| 0 | 9.3 | 4 | 16.4 |
| 5 | 17.2 | 6 | 21.0 |
| 0 | 14.2 | 1 | 12.4 |
| 1 | 10.2 | 1 | 17.2 |
| 4 | 18.0 | 5 | 19.1 |
| 4 | 17.2 | 1 | 12.2 |
| 2 | 13.1 | 2 | 12.6 |
| 4 | 15.9 | 1 | 12.4 |
| 5 | 18.7 | 2 | 16.8 |
| 1 | 14.2 | 1 | 10.6 |
| 2 | 15.6 | 4 | 15.9 |
| 1 | 12.1 | 4 | 18.2 |
| 0 | 12.2 | 1 | 16.2 |
| 5 | 17.5 | 0 | 16.2 |
| 0 | 13.4 | 1 | 11.2 |
| 2 | 12.4 | 4 | 19.2 |
| 4 | 14.4 | 3 | 16.2 |
| 4 | 18.7 | 0 | 12.2 |
| 1 | 11.2 | 4 | 17.8 |
| 4 | 21.0 | 0 | 14.2 |
| 1 | 13.2 | 4 | 19.2 |
| 4 | 17.9 | 0 | 17.2 |
| 1 | 15.4 | 5 | 18.9 |
| 4 | 17.5 | 1 | 12.2 |
| 0 | 10.0 | 4 | 16.8 |
| 0 | 12.1 | 1 | 11.8 |
| 5 | 15.2 | 6 | 19.4 |
| 1 | 12.1 | | |

# PROBLEMS    231

Test to see whether level of education and starting salary are related. Use an alpha of .025.

**9-13** Ashley is entering the Miss United States beauty pageant. Her father has a hunch that a contestant's height and score are related. The following are data (height in feet) from previous competitions:

| Height | Score | Height | Score | Height | Score | Height | Score |
|---|---|---|---|---|---|---|---|
| 6.3 | 80 | 5.7 | 94 | 6.1 | 88 | 5.7 | 88 |
| 6.0 | 81 | 6.4 | 92 | 5.9 | 92 | 5.8 | 85 |
| 5.7 | 85 | 5.9 | 85 | 5.9 | 82 | 6.3 | 90 |
| 6.4 | 92 | 6.1 | 88 | 6.3 | 82 | 6.4 | 82 |
| 6.0 | 84 | 6.0 | 92 | 5.8 | 91 | 6.0 | 90 |
| 5.7 | 93 | 5.7 | 81 | 5.9 | 83 | 6.4 | 87 |
| 6.4 | 81 | 6.0 | 91 | 6.2 | 90 | 5.9 | 83 |
| 5.9 | 92 | 5.9 | 86 | 5.7 | 88 | 6.3 | 91 |
| 5.7 | 85 | 6.3 | 89 | 6.2 | 81 | 5.8 | 90 |
| 6.3 | 87 | 6.4 | 90 | 6.0 | 82 | 6.3 | 94 |
| 5.7 | 85 | 5.9 | 85 | 5.9 | 90 | 5.7 | 91 |
| 5.8 | 94 | 6.3 | 83 | 6.1 | 90 | 6.3 | 81 |
| 6.3 | 84 | 5.7 | 82 | 5.7 | 80 | 6.2 | 85 |
| 5.9 | 86 | 6.2 | 92 | 6.2 | 87 | 5.9 | 93 |
| 6.4 | 81 | 5.6 | 91 | 5.9 | 91 | 5.8 | 90 |
| 6.2 | 92 | 5.9 | 80 | 5.8 | 86 | 5.8 | 88 |
| 6.3 | 86 | 6.2 | 91 | 6.2 | 85 | 6.4 | 87 |
| 5.7 | 91 | 6.2 | 83 | 5.8 | 82 | 6.0 | 90 |
| 5.7 | 80 | 5.7 | 86 | 6.2 | 83 | 5.9 | 81 |
| 5.9 | 84 | 5.8 | 91 | 6.3 | 90 | 6.0 | 85 |
| 5.9 | 89 | | | | | | |

**a)** Using a 90% confidence level, test the dependence of the two variables height and score.
**b)** Is Ashley's father's assumption correct?

**9-14** Most advertisements about weight loss show big individuals who have lost a large amount of weight. You are just slightly overweight. You believe that a large weight loss in the first two weeks in a weight loss program is related to being extremely overweight. Following is weight loss data on a random sample of overweight individuals in a weight loss program after two weeks:

| Pounds Overweight | Weight Loss | Pounds Overweight | Weight Loss | Pounds Overweight | Weight Loss | Pounds Overweight | Weight Loss |
|---|---|---|---|---|---|---|---|
| 50 | 12 | 60 | 15 | 99 | 15 | 89 | 19 |
| 100 | 18 | 77 | 18 | 47 | 13 | 60 | 12 |
| 40 | 14 | 50 | 13 | 120 | 25 | 58 | 20 |
| 60 | 11 | 112 | 22 | 54 | 10 | 89 | 21 |
| 92 | 20 | 52 | 12 | 51 | 15 | 105 | 19 |
| 90 | 20 | 87 | 22 | 60 | 10 | 116 | 23 |

*(continued on next page)*

*(Continued)*

| Pounds Overweight | Weight Loss | Pounds Overweight | Weight Loss | Pounds Overweight | Weight Loss | Pounds Overweight | Weight Loss |
|---|---|---|---|---|---|---|---|
| 66 | 13 | 104 | 25 | 56 | 11 | 90 | 25 |
| 70 | 18 | 120 | 22 | 114 | 23 | 72 | 16 |
| 96 | 20 | 55 | 21 | 91 | 12 | 86 | 20 |
| 69 | 11 | 92 | 15 | 66 | 14 | 110 | 17 |
| 123 | 24 | 88 | 24 | 100 | 18 | 52 | 18 |
| 64 | 16 | 57 | 18 | 77 | 15 | 102 | 22 |
| 55 | 14 | 65 | 12 | 98 | 24 | 75 | 14 |
| 62 | 13 | 102 | 24 | 55 | 9 | 90 | 22 |
| 99 | 19 | 54 | 21 | 62 | 12 | 67 | 15 |
| 63 | 12 | 89 | 17 | 95 | 23 | 104 | 21 |
| 82 | 18 | 61 | 14 | 75 | 12 | 102 | 9 |
| 66 | 10 | 79 | 21 | 108 | 19 | 97 | 22 |
| 58 | 13 | 90 | 24 | 65 | 12 | | |

Are the two variables related? Use a 95% confidence level.

**9-15** Mom and Dad love to play golf. Earlier in the day you heard them quarreling. Mom said that age and score for golfers were not related. Dad believed that the two variables were related. Following is data on a random sample of golfers:

| Age | Score | Age | Score | Age | Score | Age | Score |
|---|---|---|---|---|---|---|---|
| 25 | 75 | 43 | 70 | 61 | 76 | 47 | 84 |
| 45 | 80 | 25 | 85 | 35 | 74 | 59 | 82 |
| 52 | 85 | 27 | 71 | 59 | 87 | 62 | 84 |
| 48 | 77 | 42 | 79 | 32 | 77 | 33 | 72 |
| 29 | 80 | 35 | 74 | 62 | 84 | 59 | 87 |
| 24 | 76 | 64 | 70 | 25 | 82 | 34 | 77 |
| 33 | 79 | 36 | 82 | 44 | 82 | 59 | 87 |
| 56 | 79 | 55 | 71 | 34 | 72 | 42 | 84 |
| 49 | 89 | 39 | 73 | 57 | 75 | 29 | 70 |
| 28 | 70 | 27 | 71 | 51 | 79 | 62 | 75 |
| 61 | 85 | 65 | 70 | 41 | 74 | 59 | 79 |
| 58 | 83 | 38 | 78 | 42 | 78 | 60 | 72 |
| 32 | 71 | 61 | 73 | 65 | 72 | 33 | 70 |
| 31 | 89 | 38 | 85 | 58 | 80 | 48 | 76 |
| 64 | 84 | 57 | 86 | 28 | 77 | 37 | 76 |
| 26 | 74 | 32 | 70 | 61 | 77 | 58 | 84 |
| 54 | 70 | 31 | 72 | 52 | 75 | | |
| 60 | 86 | 62 | 71 | 31 | 77 | | |

Using an alpha of .025, test to see if age and score are independent or dependent.

**9-16** In endurance running, a new sport sweeping the country, individuals attempt to run as far as they can. As an avid spectator of the sport, you feel that females run longer distances than males. The following data gives distance in miles, with a female indicated by a 1 and a male by a 0:

# PROBLEMS

| Sex | Distance | Sex | Distance | Sex | Distance | Sex | Distance |
|---|---|---|---|---|---|---|---|
| 1 | 42 | 1 | 29 | 0 | 47 | 1 | 45 |
| 0 | 43 | 0 | 28 | 1 | 45 | 0 | 47 |
| 1 | 50 | 0 | 27 | 1 | 39 | 1 | 39 |
| 1 | 49 | 0 | 33 | 0 | 40 | 0 | 34 |
| 0 | 33 | 1 | 58 | 0 | 54 | 1 | 48 |
| 0 | 27 | 1 | 34 | 0 | 31 | 1 | 47 |
| 1 | 25 | 1 | 50 | 1 | 46 | 0 | 47 |
| 0 | 28 | 1 | 48 | 1 | 51 | 1 | 52 |
| 1 | 49 | 0 | 47 | 0 | 31 | 0 | 48 |
| 0 | 47 | 0 | 32 | 1 | 48 | 1 | 32 |
| 1 | 43 | 1 | 46 | 0 | 33 | 1 | 30 |
| 1 | 50 | 0 | 44 | 1 | 32 | 0 | 47 |
| 0 | 52 | 0 | 40 | 0 | 32 | 0 | 35 |
| 0 | 29 | 1 | 30 | 1 | 46 | 0 | 37 |
| 1 | 39 | 0 | 38 | 1 | 53 | 1 | 32 |
| 1 | 35 | 1 | 44 | 0 | 28 | 1 | 48 |
| 1 | 42 | 1 | 48 | 0 | 35 | 0 | 45 |
| 0 | 33 | 1 | 32 | 1 | 49 | 1 | 47 |
| 1 | 49 | 0 | 52 | 1 | 50 | 1 | 40 |
| 0 | 36 | 0 | 31 | 0 | 59 | 0 | 42 |
| 0 | 32 | 0 | 30 | 1 | 55 | 0 | 40 |
| 0 | 54 | 1 | 49 | 0 | 32 | 1 | 39 |
| 1 | 44 | 0 | 50 | 0 | 34 | 1 | 45 |

Test whether sex and distance are dependent. Use an alpha of .10.

**9-17** In the endurance race described in Problem 9-16, you believe that speed and distance are closely related. You have the following data on speed (in miles per hour) and distance (in miles):

| Speed | Distance | Speed | Distance | Speed | Distance | Speed | Distance |
|---|---|---|---|---|---|---|---|
| 9 | 45 | 14 | 41 | 9 | 37 | 9 | 31 |
| 7 | 32 | 12 | 47 | 8 | 51 | 8 | 35 |
| 12 | 40 | 8 | 40 | 8 | 46 | 14 | 48 |
| 11 | 50 | 13 | 43 | 12 | 47 | 12 | 44 |
| 9 | 42 | 9 | 44 | 14 | 51 | 10 | 39 |
| 15 | 53 | 15 | 57 | 13 | 35 | 10 | 42 |
| 10 | 41 | 12 | 40 | 7 | 30 | 11 | 47 |
| 13 | 49 | 8 | 39 | 9 | 40 | 14 | 52 |
| 16 | 52 | 11 | 48 | 14 | 52 | 8 | 50 |
| 9 | 50 | 10 | 32 | 13 | 48 | 9 | 29 |
| 11 | 35 | 14 | 45 | 10 | 38 | 14 | 49 |
| 10 | 41 | 13 | 39 | 11 | 31 | 12 | 46 |
| 15 | 36 | 9 | 30 | 14 | 49 | 8 | 49 |
| 16 | 60 | 15 | 51 | 8 | 38 | 10 | 42 |
| 12 | 52 | 14 | 50 | 7 | 44 | 13 | 48 |
| 10 | 45 | 8 | 36 | 15 | 50 | 9 | 41 |
| 14 | 51 | 11 | 43 | 12 | 49 | 11 | 47 |
| 8 | 34 | 12 | 40 | 8 | 32 | 11 | 48 |
| 10 | 30 | 11 | 41 | 12 | 40 | 9 | 48 |
| 11 | 40 | 15 | 49 | 15 | 49 | | |

Using an alpha of .10, determine if speed and distance are related.

**9-18** As the typing teacher of Westwood High School, you want to prove to students that speed is related to accuracy. The following is a sample from students in your class. Speed is given in words per minute and errors in errors per page.

| Speed | Errors | Speed | Errors | Speed | Errors | Speed | Errors |
|---|---|---|---|---|---|---|---|
| 45 | 7 | 56 | 4 | 58 | 2 | 61 | 1 |
| 32 | 5 | 63 | 2 | 30 | 4 | 38 | 4 |
| 58 | 3 | 36 | 1 | 46 | 6 | 34 | 7 |
| 42 | 5 | 39 | 11 | 31 | 1 | 31 | 6 |
| 60 | 2 | 45 | 3 | 35 | 2 | 58 | 0 |
| 65 | 2 | 47 | 2 | 61 | 1 | 56 | 3 |
| 43 | 5 | 39 | 2 | 32 | 6 | 44 | 3 |
| 42 | 7 | 57 | 1 | 59 | 2 | 61 | 0 |
| 45 | 1 | 62 | 1 | 55 | 3 | 31 | 1 |
| 61 | 1 | 39 | 4 | 57 | 1 | 56 | 7 |
| 33 | 4 | 55 | 3 | 32 | 7 | 39 | 6 |
| 29 | 8 | 30 | 2 | 61 | 3 | 62 | 1 |
| 34 | 6 | 48 | 3 | 31 | 4 | 51 | 2 |
| 30 | 2 | 56 | 8 | 31 | 1 | 60 | 2 |
| 57 | 2 | 31 | 0 | 55 | 8 | 61 | 2 |
| 60 | 1 | 52 | 3 | 32 | 7 | 39 | 6 |
| 34 | 6 | 46 | 5 | 39 | 4 | 57 | 2 |
| 44 | 5 | 31 | 5 | 49 | 2 | 59 | 4 |
| 61 | 3 | 42 | 5 | 59 | 1 | 58 | 3 |
| 50 | 2 | 61 | 0 | 57 | 3 | 64 | 3 |
| 61 | 0 | 62 | 0 | 58 | 1 | 35 | 2 |
| 33 | 2 | 47 | 8 | 49 | 4 | 46 | 2 |
| 56 | 2 | 58 | 3 | 60 | 4 | 66 | 3 |
| 49 | 0 | 55 | 2 | 43 | 5 | 53 | 6 |

Using an alpha of .05, perform a chi-square test.

**9-19** In the home stereo market, price seems to be inversely related to size: The smaller the product, the more expensive it is. Test this theory. Size is given in inches and price in dollars.

| Size | Price | Size | Price | Size | Price | Size | Price |
|---|---|---|---|---|---|---|---|
| 15 | 320 | 14 | 300 | 18 | 135 | 14 | 360 |
| 12 | 525 | 12 | 369 | 16 | 423 | 9 | 341 |
| 14 | 312 | 9 | 229 | 19 | 112 | 18 | 300 |
| 20 | 225 | 17 | 259 | 12 | 259 | 10 | 523 |
| 10 | 315 | 10 | 520 | 11 | 369 | 19 | 356 |
| 22 | 365 | 12 | 249 | 8 | 345 | 17 | 430 |
| 18 | 326 | 18 | 519 | 9 | 358 | 21 | 312 |
| 11 | 259 | 15 | 160 | 16 | 435 | 12 | 310 |
| 20 | 350 | 11 | 215 | 10 | 261 | 18 | 400 |

*(continued on next page)*

## PROBLEMS

*(Continued)*

| Size | Price | Size | Price | Size | Price | Size | Price |
|------|-------|------|-------|------|-------|------|-------|
| 18 | 525 | 19 | 380 | 10 | 254 | 11 | 525 |
| 16 | 423 | 17 | 340 | 17 | 345 | 17 | 412 |
| 18 | 460 | 14 | 360 | 10 | 260 | 14 | 256 |
| 12 | 200 | 17 | 356 | 19 | 279 | 12 | 298 |
| 9  | 425 | 9  | 450 | 17 | 460 | 19 | 425 |
| 16 | 500 | 17 | 439 | 19 | 425 | 10 | 450 |
| 16 | 425 | 20 | 420 | 10 | 316 | 12 | 340 |
| 18 | 420 | 8  | 412 | 20 | 423 |    |    |
| 10 | 490 | 19 | 300 | 12 | 480 |    |    |

Perform a chi-square test using an alpha of .10.

**9-20** In your company, you believe that older minority workers' average pay is less than that of older nonminority workers. In the following data, salary is given in dollars per month (nonminority = 1; minority = 0):

| Worker | Salary | Worker | Salary | Worker | Salary | Worker | Salary |
|--------|--------|--------|--------|--------|--------|--------|--------|
| 1 | 1200 | 1 | 1265 | 1 | 1295 | 0 | 1321 |
| 1 | 1450 | 1 | 965  | 0 | 1500 | 1 | 1400 |
| 0 | 1050 | 0 | 1520 | 0 | 950  | 0 | 1461 |
| 1 | 1295 | 1 | 987  | 0 | 1460 | 1 | 1495 |
| 1 | 1230 | 0 | 1250 | 1 | 1420 | 1 | 1122 |
| 0 | 1340 | 0 | 1501 | 1 | 950  | 0 | 850  |
| 0 | 989  | 1 | 1212 | 0 | 1420 | 1 | 1263 |
| 0 | 1275 | 0 | 1542 | 1 | 947  | 0 | 1520 |
| 1 | 1501 | 0 | 1125 | 0 | 1492 | 1 | 964  |
| 0 | 1032 | 1 | 1430 | 1 | 975  | 0 | 1253 |
| 1 | 1105 | 0 | 1120 | 0 | 1240 | 0 | 942  |
| 0 | 1245 | 1 | 1435 | 0 | 974  | 0 | 1245 |
| 1 | 1469 | 1 | 1158 | 1 | 1152 | 0 | 1400 |
| 0 | 1326 | 0 | 1236 | 1 | 1310 | 1 | 900  |
| 0 | 1180 | 1 | 941  | 1 | 1522 | 1 | 1500 |
| 0 | 1415 | 0 | 1423 | 1 | 1362 | 1 | 1354 |
| 0 | 899  | 0 | 1250 | 1 | 1250 |   |      |
| 1 | 1421 | 1 | 1321 | 0 | 1429 |   |      |
| 0 | 1300 | 0 | 998  | 0 | 1350 |   |      |

Perform a chi-square test. Use an alpha of .01.

# Correlation and Regression Analysis

*The following Minitab commands will be introduced in this chapter:*

```
CORRelation coefficient of C,...,C
```
A command to compute the linear correlation coefficient between columns using the Pearson product-moment correlation coefficient

```
REGRess y-value in C on K predictors in C,...,C;
  DW.
```
A command to produce simple linear regression or multiple regression equations; the following optional subcommand computes predicted $Y$ for a data set:

```
  PREDict y-values for x-value set E,...,E
```

## SECTION 10·1 Correlation Analysis

Until now we have worked with statistical techniques that have generally required independent observations (except for the paired-sample *t* test); thus items drawn from one sample did not affect the probabilities of items drawn from other samples. In this chapter, we will discuss techniques for examining the relationship *between* two or more variables. The primary statistical techniques to be used are correlation and regression.

## 10·1 CORRELATION ANALYSIS

If you look back at Chapter 9 and our study of chi-square, you can get an understanding of how *all* of our test statistics are calculated. In the contingency table of Chapter 9, we calculated an actual value for the mean and compared that value with an expected mean value. The difference between these values was then weighted by a measure of variation. The *t*, *F*, and chi-square are all calculated in basically the same way. Sometimes a single standard deviation value is used for the weighting process; at other times a pooled standard deviation is used.

In correlation and regression analysis, we will continue this general procedure. Our first test statistic is the coefficient of determination (notated as $r^2$). The coefficient of determination is the square of the Pearson product-moment coefficient of correlation (notated as $r$). Although $r^2$ is simply the square of $r$, we will use these two values in slightly different ways.

The Pearson coefficient of correlation measures the strength of the relationship between two variables. For example, we generally find that taller people weigh more. Thus, if we measured the height and weight of 20 people, we would expect as one variable increased the other variable would also increase. This would be reflected in a high correlation coefficient.

If two variables move in the same direction (that is, as one increases the other increases), then the coefficient of correlation is positive. If two variables move in opposite directions (that is, as one increases the other decreases), then the coefficient of correlation is negative. Besides measuring the direction of the relationship, we will also measure the strength of the relationship. If two variables move in perfect unison with each other then the strength of the relationship is *at its maximum* and is calculated to be +1 or −1. If there is no relationship, then the relationship is calculated to be 0. Thus the possible degrees of relationship range from +1 to 0 to −1. These relationships are shown in Figure 10-1.

A perfect relationship is not necessarily a one-to-one relationship. For example, if for every 2-unit movement in one variable, there was a 5-unit move in the second, then the coefficient of correlation would still be +1 because the pattern of movement between the two variables would be constant. Thus, if everyone gained 5 pounds of weight as they grew 2 inches in height, this relationship would have a correlation coefficient of +1. In Minitab, we can calculate the correlation coefficient by the command

```
CORRelation coefficient of C,...,C
```

For instance, using the Go-Fast data, we might compute the relationship between age (data in column 1) and schooling (data in column 2) by either

```
MTB >CORR C1 C2
```

or

```
MTB >CORR C2 C1
```

**FIGURE 10-1** Range of Possible Values for the Coefficient of Correlation

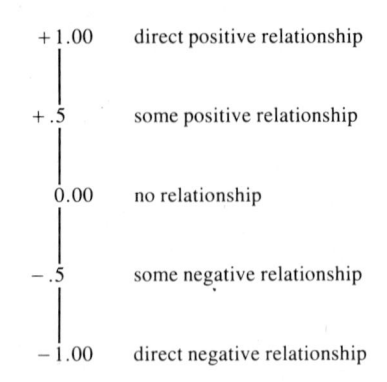

```
+1.00      direct positive relationship
 |
+ .5       some positive relationship
 |
0.00       no relationship
 |
- .5       some negative relationship
 |
-1.00      direct negative relationship
```

The `CORRelation` command is one of the few in which you can enter your columns in any order. The output from this command will take one of two forms. If only two columns are specified, then the output will be printed on a single line. If more than two columns are specified, then the output will be printed in a matrix with the value of the correlation coefficient located at the intersection of each row and column.

**FIGURE 10-2** Correlation between Age and Schooling for Go-Fast Employees

```
MTB >CORR C1 C2

    CORRELATION OF    AGE AND SCHOOL  =  .015
```

Figure 10-2 shows the results of a `CORRelation` command on `C1` and `C2`. Figure 10-3 shows the results of the command

`MTB >CORR C1, C2, C3, C4`

or

`MTB >CORR C1-C4`

**FIGURE 10-3** Correlation Matrix for Variables in Go-Fast Data

```
MTB >CORR C1-C4

            AGE     SCHOOL   PRODUCT
SCHOOL     .015
PRODUCT    .865     .007
HEIGHT    -.915    -.136    -.773
```

Notice that the correlation coefficient for age and schooling is .015. This value is specifically printed for the single command (Figure 10-2) and is found at the intersection of `SCHOOL` and `AGE` in Figure 10-3. The following items are of interest in this correlation matrix:

**(1)** Schooling and age have almost no relationship.

**(2)** Productivity and age have a strong positive relationship. It appears that older workers produce more. (You may recall that this finding is consistent with our paired-sample *t* test.)

**(3)** Height and age have a strong negative relationship. We will discuss this relationship in greater detail in a moment.

### SECTION 10-2 Regression Analysis

(4) Productivity and schooling have a weak negative relationship. Better-educated people do not seem to produce more in our "low-tech" shop.

(5) Height and schooling have a weak negative relationship. The correlation coefficient is really too small to infer a meaningful relationship.

(6) Height and productivity have a negative relationship.

In point 3, we noted a strong negative relationship between height and age. Does that mean that the longer people work at Go-Fast, the shorter they become? Probably not. It is generally true that younger people are taller than older people. This is a good illustration of the need for sound managerial thinking along with statistical skills. For example, if there was a strong positive relationship between years of employment at Go-Fast and hearing loss and if the plant was very noisy, then both managerial instincts and statistical analysis might lead to the conclusion that working conditions were *causing* hearing damage.

The same type of comments also apply to point 6. Does it make sense that taller workers are less productive? As managers, we might want to check the height of workbenches and machinery to see if there was some cause in the working environment. However, there is also a reasonable chance that the relationship is simply a spurious result. (In this case, it could be a side effect of the relationship between age and height or age and productivity.)

## 10-2 REGRESSION ANALYSIS

Regression analysis is probably the most popular statistical technique used today for prediction. At Go-Fast, we might be interested in trying to predict total production level in a future time period. Or a company might be preparing to open a new store and would like an estimate of expected sales. Such a prediction would help in establishing the initial inventory and the number of people to hire.

There are two general forms of regression analysis. The simplest form is called *simple linear regression*. In this form, one independent variable is used to predict the variable of interest. For example, Go-Fast might use age to predict productivity. Here age is the *independent variable* since it is the variable over which we have control; that is, we can choose an employee of any particular age. The *dependent variable* is the one we are trying to predict. In this case, we are trying to predict the level of productivity. In the prediction process, we build a linear model of the relationship between the independent and dependent variable.

The second form of regression is *multiple linear regression*. In this form, we use *more than one* independent variable to predict our single dependent variable. For example, in predicting sales for the new store, we might collect data on the local population, the average income of that population, its average age, traffic patterns, and so on.

Although multiple regression is more powerful, it is also more complex. We will begin our discussion by analyzing simple linear regression. Once these concepts are understood, it is an easy step to appreciating multiple regression. For the sake of simplicity, we will refer to simple linear regression as simply regression analysis.

Regression analysis allows us to elaborate upon the results of correlation analysis and provides

**240** ◂ **CHAPTER 10** Correlation and Regression Analysis

**(1)** a linear model of the relationship between the independent and dependent variables,
**(2)** indicators of the significance of the relationship, and
**(3)** indicators of the validity of the linear model.

Let's walk through a regression analysis. We will examine the relationship between our dependent variable (productivity) and our independent variable (age) at Go-Fast. Our first step is simply to plot the relationship and inspect the result. The **PLOT** command discussed in Chapter 2 will accomplish this result. If we have read age data into column 1 and productivity data into column 3, then we would use the command

```
MTB >PLOT C3 C1
```

When using the **PLOT** command, remember to specify your variables properly. The first variable specified (typically the dependent variable) will be the Y-axis variable; the second (typically the independent variable) will be the X-axis variable.

Figure 10-4 is the result of our **PLOT** command. Notice that the plot (technically called a *scatter diagram*) forms a fairly narrow diagonal band from the lower left to the upper right. The more this band approaches a straight line, the closer the correlation coefficient approaches a value of 1. The less the pattern approaches a straight line, the closer the correlation coefficient is to 0. In Figure 10-4, the pattern moves upward from left to right. This is known as a *positive slope*, since the values of one variable become larger as the values of the other variable become larger. This positive slope means that the correlation coefficient will also be positive. From our correlation analysis (Figure 10-3), we know that the correlation coefficient between productivity and age is +.865. This is certainly consistent with the visual analysis of the scatter diagram.

Later we will plot height and productivity. In that figure, you will see that the band of plotted points moves downward from left to right across the graph. This is a *negative slope* and indicates a negative correlation coefficient. Referring back to our correlation matrix in Figure 10-3, you can see that the correlation between height and productivity is −.915.

**FIGURE 10-4** Minitab Plot of Relationship between Age (Independent Variable) and Productivity (Dependent Variable) of Go-Fast Employees

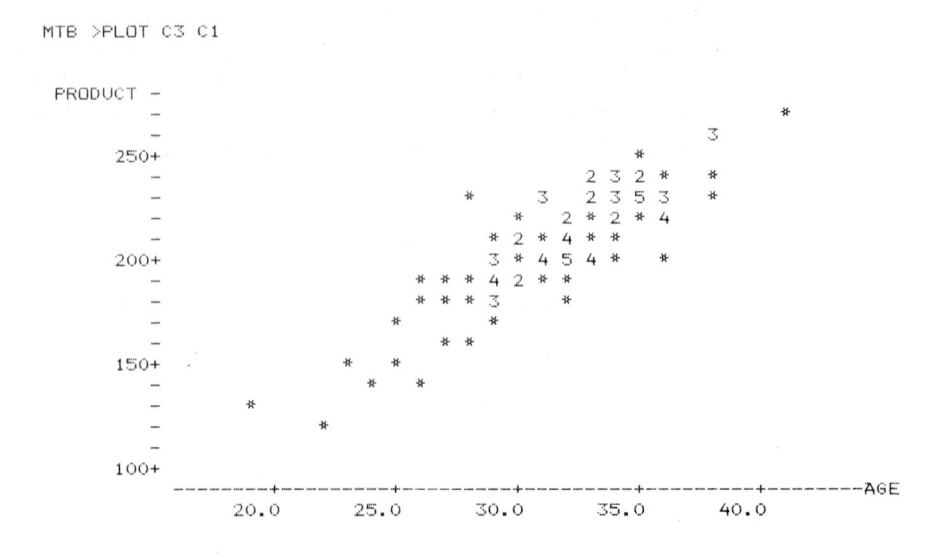

```
       MTB >PLOT C3 C1

    PRODUCT -
            -                                              *
            -                                         3
       250+ -                                    *
            -                               2 3 2 *    *
            -                    *      3   2 3 5 3    *
            -                         *   2 * 2 * 4
            -              * 2 * 4 * *
       200+ -                       3 * 4 5 4 *     *
            -              * * * 4 2 * *
            -              * * * 3       *
            -         *         *
            -              * *
       150+ -  .     *    *
            -        *  *
            -   *
            -        *
            -
       100+
            ---------+---------+---------+---------+---------+--------AGE
                   20.0      25.0      30.0      35.0      40.0
```

## SECTION 10-2 Regression Analysis

From our visual inspection of Figure 10-4, we know that there is a strong relationship between the two variables productivity and age. If you had not already calculated the correlation coefficient, you would certainly do so as your next step in the analysis. If the correlation coefficient was fairly strong (greater than +.5 or less than −.5), you could proceed with calculating a regression equation.

The regression analysis is computed with the Minitab command `REGRess` and takes the following form:

```
REGRess y-value in C on K predictors in C,...,C;
DW.
```

In our illustration, we would use the command

```
MTB >REGRess y-value in C3 on 1 predictor x in C1;
SUBC> DW.
```

This command may look awkward. Since we have only two variables, specifying one predictor seems a bit redundant. In simple linear regression the number of predictors is always one. However, in multiple regression the same command format is used, and it is necessary to specify the number of predictors. For example, if you had six columns of data, you might obtain a regression of `C1` on `5` predictors in `C2-C6`. In both of these examples, the underlying principles are the same. You first specify the column that contains the $Y$ (the dependent variable). You then specify how many predictors will be used to build the regression model. Finally, you inform the computer where the columns of the $X$ (independent) variable(s) are located. Because there is only one predictor in simple linear regression, only a single column must be specified as the independent variable.

It is important to be very careful about which variables you specify as your $X$ and $Y$. The first column specified is your $Y$ variable, and the last one is your $X$. There is a world of difference in the results obtained if you reverse your specification. If you have the time and inclination, do a regression of `C1` on `1` in `C3` and examine the results. Instead of using age to predict productivity you would be using productivity to predict ages. Although this would not be correct for our problem, Minitab would calculate the results.

Returning to our present problem, we issue the Minitab command

```
MTB >REGRess y-value in C3 on 1 predictor in C1;
SUBC> DW.
```

The output shown in Figure 10-5 is obtained.

We should examine Figure 10-5 carefully. First, look at the regression equation,

$$\text{PRODUCT} = 0.9 + 6.45 \text{ AGE}$$

found at the top of the Minitab printout for the `REGRess` command. This equation assumes that the relationship between the two variables is a straight line (thus the name *linear regression*). We can thus predict an expected level of $Y$ (productivity) if we know an employee's age.

Let's assume we had an employee who was 30 years of age. We would substitute the 30 into the equation as follows:

$$\begin{aligned} Y &= 0.9 + 6.45(\text{age}) \\ &= 0.9 + 6.45(30) \\ &= 0.9 + 193.5 \\ &= 194.4 \end{aligned}$$

**FIGURE 10-5** Minitab Regression Analysis of Age (Independent Variable) and Productivity (Dependent Variable) of Go-Fast Employees

```
MTB >REGR C3 1 C1;
SUBC> DW.

The regression equation is
PRODUCT = 0.9 + 6.45 AGE

Predictor        Coef         Stdev        t-ratio
Constant         0.85         12.07        0.07
AGE              6.4517       0.3778       17.08

s = 14.28        R-sq = 74.8%     R-sq(adj) = 74.6%

Analysis of Variance

SOURCE           DF           SS           MS
Regression       1            59446        59446
Error            98           19979        204
Total            99           79425

Unusual Observations
Obs.    AGE     PRODUCT      Fit     Stdev.Fit   Residual   St.Resid
37      31.0    230.00       200.85  1.45        29.15      2.05R
41      22.0    120.00       142.79  3.94        -22.79     -1.66 X
47      19.0    130.00       123.43  5.01        6.57       0.49 X
56      26.0    140.00       168.60  2.59        -28.60     -2.04R
60      36.0    200.00       233.11  2.16        -33.11     -2.35R
61      28.0    230.00       181.50  2.00        48.50      3.43R
68      41.0    265.00       265.37  3.79        -0.37      -0.03 X
99      23.0    150.00       149.24  3.59        0.76       0.05 X

R denotes an obs. with a large st. resid.
X denotes an obs. whose X value gives it large influence.

Durbin-Watson statistic = 2.22
```

Thus for an employee who was 30, we would predict productivity at just under 195 pairs of shoes per week. For an employee who was 40, we would predict

$$Y = 0.9 + 6.45(40)$$
$$= 258.9$$

In this way we can use our regression model to predict the productivity level of any employee at any age.

We now must consider the implications of our model. First, would it make sense for Go-Fast to hire several employees who are 100 years old? Our regression model would predict very high levels of productivity! However, such a recommendation is nonsense. The reason for the obtained relationship is that the model was built with almost all ages between 20 to 50. If you substituted a value for $X$ that was dramatically out of the range of our $X$ variables, this type of nonsensical result would occur.

Second, in our original case description, we stated that the work force at Go-Fast was very stable; thus, age would correlate very highly with experience. Note that the correlation is not perfect because a new employee who is 30 would not produce at the same level as a seasoned 30-year-old employee who had been with the firm since high school.

Third, our prediction will seldom be exactly right. This fact should make sense as well. We know that age is correlated with productivity at only +.865; thus something else

## SECTION 10-2 Regression Analysis

influences productivity besides age. The difference between an actual value and a predicted value from a regression equation is known as a *residual*. For example, in Appendix A, worker 7 is 30 years of age and had a productivity level of 190 units. The difference between the actual productivity of 190 and the predicted productivity of just under 195 (about 5 pairs) is the residual.

Notice that in Figure 10-4, the axes created by the `PLOT` command do not start at zero. That allows the graph to be concentrated in a smaller space. However, if you redrew Figure 10-4 with both axes starting at zero and then plotted the regression line, it would intersect the Y-axis at 0.9 and have a positive slope of 6.45, as shown in Figure 10-6.

The section of Figure 10-5 below the regression equation indicates the significance of the coefficients in the equation. The tests to determine significance are *t* tests, exactly as you performed in Chapter 8. In our tests of significance for the regression equation, we want to answer the question "Is the intercept (0.9) significantly different from 0?" We also want to ask "Is the slope of the regression line (6.45) significantly different from 0?" Both of these questions can be answered with a *t* test. From the laws of probability, we know that the slope and the intercept may simply be due to chance. Our tests of significance will allow us to determine how much confidence we wish to place in each variable.

In the first few lines of Figure 10-5 after the regression equation, these tests are performed. In the first row of values, under the heading `Predictor` the first row is marked `Constant`, which represents the intercept variable. The next value in that row is `0.85` (the intercept value). The third value in the row is the standard deviation of the coefficient, which is a measure of its variation. Finally, if you divide the coefficient value (0.85) by this second value (12.07), you obtain the *t* ratio, shown in the last column as `0.07`. This *t*-value is interpreted like all other *t*-values. Since the intercept is close to 0 and its standard deviation is large, we obtain a very small *t*-value. Using Table 2 of Appendix B, we see that with 98 degrees of freedom, such a *t*-value is not significant. Thus the intercept value is not significantly different from 0.

**FIGURE 10-6** Minitab Plot of Productivity and Age with Both Axes Specified to Start at Zero

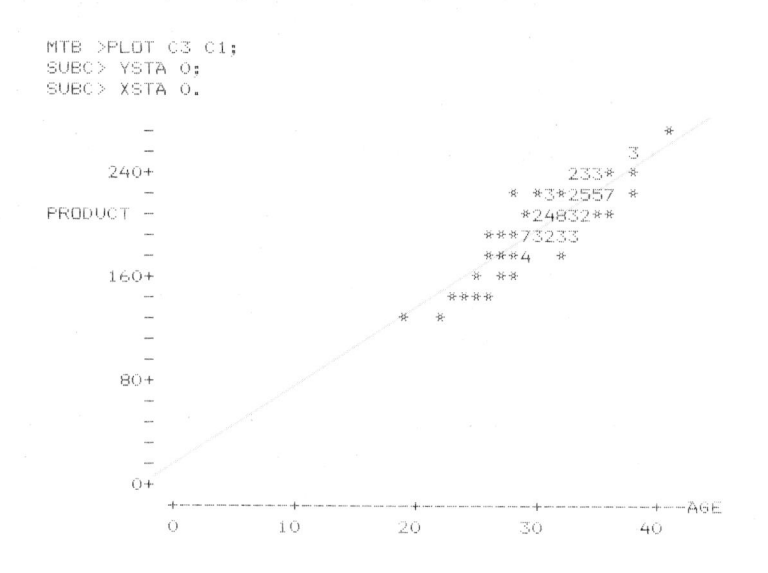

```
MTB >PLOT C3 C1;
SUBC> YSTA 0;
SUBC> XSTA 0.

                                                           *
                                                        3
          240+                                       233*  *
                                                  *  *3*2557  *
        PRODUCT -                                    *24832**
                                                    ***73233
                                                     ***4   *
          160+                                      *  **
                                                   ****
                                              *   *
           80+



            0+
            +---------+---------+---------+---------+----AGE
            0        10        20        30        40
```

Next we evaluate the values of the second row for age, which is the slope of the regression equation. This slope is also called *beta one*. In simple linear regression with only one independent variable, there is only one slope coefficient. In multiple regression we must think of a multidimensional space, and for each independent variable this is a unique slope. Thus, there will be a beta two, beta three, and so on, one for each independent variable.

In the present case, the slope is 6.4517 and the standard deviation is relatively small (0.3778). This produces a very large *t*-value of 17.08, which is very significant. Based upon the *t*-values, we would conclude that the intercept value is of little importance in predicting productivity but that the slope is very important.

With at least one significant coefficient, it is appropriate to question the strength of the relationship. For this analysis we need to find the value of $r^2$. This value is called the *coefficient of determination*, and you will note that the unadjusted value of .748 is the square of the coefficient of correlation 0.865 (found with the command `CORR C3 C1`). The coefficient of determination must be adjusted for degrees of freedom and is then converted by Minitab into a percentage figure of 74.6. An $r^2$ of 74.6% means that 74.6% of the total variation in the dependent variable (productivity) is explained by changes in the independent variable (age). In other words, this equation predicts about 75% of the total change in productivity among employees. The remaining variation, specifically 25.4%, remains unexplained.

Typically, $r^2$ is the first value one examines when doing regression analysis. A high $r^2$ means strong predictability, and a low $r^2$ means weak predictability. The question now becomes "What is high and what is low?" There is no absolute answer. In some applications in business and economics, a typical $r^2$ for regression analysis might be as high as .90 (90%). In such a case, our value of .75 would be low. In other applications, a typical $r^2$ might only be .50.

A word of caution is very important here. The value of $r^2$ cannot be the only measure of the goodness of your regression equation, because it can be artificially inflated by various types of problems in the data. Although the reasons for such problems are well beyond the scope of this text, it is essential that you conduct your regression analysis systematically.

So far we have evaluated the actual regression equation, examined the significance of each coefficient (with *t*-values), and explained the meaning of $r^2$. We will now skip the analysis of variance section in Figure 10-5 and look at the list of `Unusual Observations` in our data set. In this section, Minitab lists observations that may contain errors. Some of the rows are followed by an `R`, which indicates that the predicted value of our equation is more than 1.96 standard deviations from the actual value. An interval of 1.96 standard deviations above and below an expected value should include 95% of the sample observations. Whenever an actual value falls outside this range, Minitab will indicate this. You should then check to make sure that the value is in fact accurate. In our example, 4 out of 100 entries—or 4%—fall outside the 95% interval. This is about what we would expect and is quite acceptable.

An `X` in this column indicates observations that the computer determines to have a large influence on the regression equation. An observation receiving both an `R` and an `X` should be checked for accuracy and appropriateness. Sometimes a simple recording error (data entry error) significantly influences your results. The `R`'s and `X`'s do not mean something is necessarily wrong; they simply mean that the observations are worth checking to ensure their accuracy.

### SECTION 10-3 An Example of No Correlation

Finally, we have the *Durbin–Watson* statistic. This statistic measures the degree to which a variable is correlated to its own previous values (called *autocorrelation*). Whole chapters are devoted to this problem in advanced econometrics texts. For our purposes, we can consider Durbin–Watson values between 1.5 and 2.5 to be acceptable. Values outside of this range indicate unacceptable levels of autocorrelation. In those cases, the regression model should be considered invalid. (Proper treatment of autocorrelation is well beyond the level of this text.) In our case the Durbin–Watson statistic of 2.22 is within our boundaries and indicates no problem.

## 10-3 AN EXAMPLE OF NO CORRELATION

In our world, more things are unrelated than related (at least in our limited view). It would be unrealistic to leave you with the impression that all things can be explained through regression analysis. For example, a comparison of productivity and schooling is a good illustration. In a low-technology production process, we would not expect these two variables to be related. To investigate whether there is a relationship, we would first run a plot of the two variables. For the Go-Fast data with schooling in column 2 and productivity in column 3, the first command would be

```
MTB >PLOT C3 C2
```

This command produces the output shown in Figure 10-7. You can see a great deal of scatter in these points.

The next thing to do in a regression analysis is to calculate the correlation coefficient. Here we can refer to Figure 10-3 and determine that the correlation coefficient ($r$) between schooling and productivity is .007. This is very close to 0 and confirms our initial hypothesis

**FIGURE 10-7** Minitab Plot of Relationship between Schooling and Productivity of Go-Fast Employees

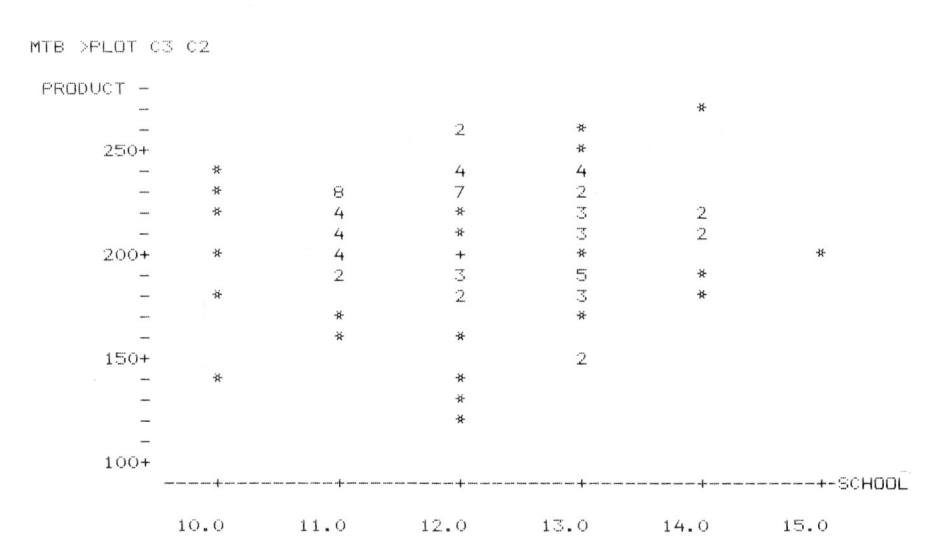

```
MTB >PLOT C3 C2

 PRODUCT -
         -                                              *
         -                        2            *
   250+                                        *
         -       *                4            4
         -       *       8        7            2
         -       *       4        *            3        2
         -               4        *            3        2
   200+          *       4        +            *                        *
         -               2        3            5        *
         -       *                2            3        *
         -               *        *                     *
         -               *        *
   150+                                        2
         -       *                *
         -                        *
         -                        *
         -
   100+
         ----+---------+---------+---------+---------+---------+-SCHOOL
            10.0      11.0      12.0      13.0      14.0      15.0
```

**FIGURE 10-8** Minitab Regression Analysis Showing No Correlation: Schooling and Productivity of Go-Fast Employees

```
MTB >REGR C3 1 C2;
SUBC> DW.

The regression equation is
PRODUCT = 203 + 0.19 SCHOOL

Predictor        Coef       Stdev     t-ratio
Constant       203.15       32.81        6.19
SCHOOL           0.195       2.708       0.07

s = 28.47      R-sq = 0.0%      R-sq(adj) = 0.0%

Analysis of Variance

SOURCE        DF          SS           MS
Regression     1          4.2          4.2
Error         98      79420.8        810.4
Total         99      79425.0

Unusual Observations
Obs.   SCHOOL   PRODUCT       Fit  Stdev.Fit   Residual   St.Resid
 41     12.0    120.00     205.49      2.85     -85.49     -3.02R
 47     12.0    130.00     205.49      2.85     -75.49     -2.67R
 55     12.0    140.00     205.49      2.85     -65.49     -2.31R
 56     10.0    140.00     205.10      6.29     -65.10     -2.34R
 57     13.0    145.00     205.68      3.80     -60.68     -2.15R
 68     14.0    265.00     205.88      5.95      59.12      2.12R
 85     15.0    200.00     206.07      8.43      -6.07     -0.22 X

R denotes an obs. with a large st. resid.
X denotes an obs. whose X value gives it large influence.

Durbin-Watson statistic = 1.62
```

of no relationship between the variables. However, to complete the analysis, let's calculate the regression. Our Minitab command would be

**MTB >REGRess y-value in C3 on 1 predictor in C2;**
**SUBC> DW.**

This command produces Figure 10-8. A quick examination of $r^2$ confirms our hypothesis. The independent variable, schooling, simply does not have *any* power to predict productivity. In fact, the calculation breaks down when we adjust for degrees of freedom; the value then becomes an imaginary number.

As a final point, examine the *t*-values for the intercept (6.19) and the slope (0.07). It appears that the intercept alone is sufficient to describe the data. Because the significance of the slope value is 0, there is no reason to examine the relationship between these two variables. Thus education does not predict the productivity of Go-Fast employees.

## 10-4 AN EXAMPLE OF SPURIOUS CORRELATION

Thus far we have not stated any requirements for regression analysis. For our analysis, we need only the first three below, but we might add one commonsensical assumption as a fourth:

## SECTION 10-4 An Example of Spurious Correlation

**(1)** There are randomly selected samples for each variable.
**(2)** Variables other than the dependent variable are independent of each other.
**(3)** There is no autocorrelation.
**(4)** The relationship must be logical from a managerial viewpoint; that is, it must make sense.

In our Go-Fast data we can study this fourth "requirement" by considering the height of employees. Figure 10-3 shows that productivity and age are highly correlated (.865) and that there is a strong inverse relationship between height and age ($-.915$). We also find a weaker inverse relationship ($-.773$) between height and productivity, which is not surprising. Thus, this information suggests that height predicts age and that age predicts productivity. The statistical relationship between height and productivity, however, would be spurious.

It is important to remember that correlation and regression analysis are mathematical tools, not substitutes for good judgment. As long as there is the proper number of data points in each column, Minitab will calculate the requested analysis, even if the analysis does not make any sense. For example, in this case we could plot the relationship between height and productivity as follows:

```
MTB >PLOT C3 C4
```

This produces the scatter diagram found in Figure 10-9. This is a nice tidy plot showing the strong inverse relationship that was expected.

Since the calculations for correlation analysis have already been completed in Figure 10-3, we can proceed to the regression analysis. In this case, our command is

```
MTB >REGRess y-value in C3 on 1 predictor in C4;
SUBC> DW.
```

**FIGURE 10-9** Minitab Plot of Relationship between Height and Productivity of Go-Fast Employees

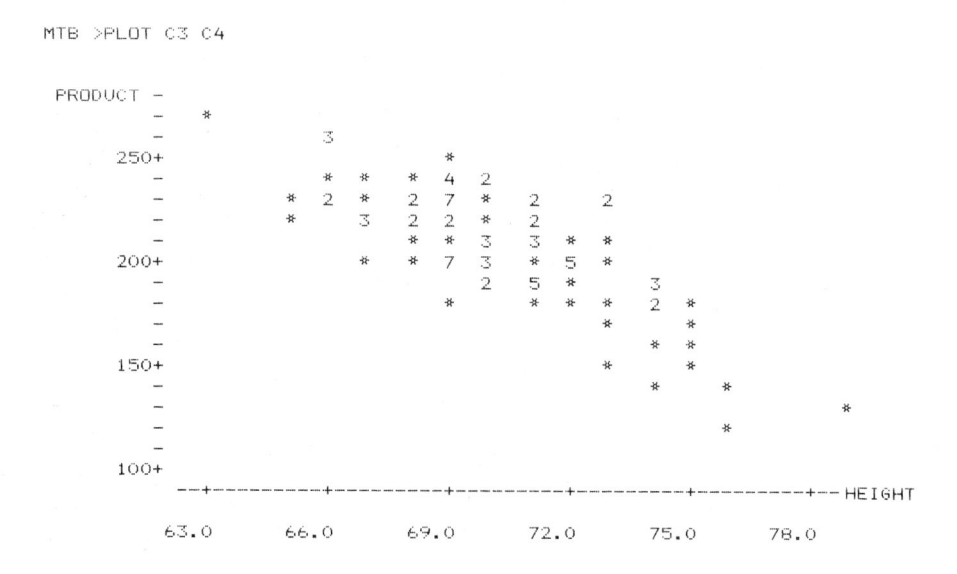

**FIGURE 10-10** Minitab Regression Analysis of a Spurious Relationship: Height and Productivity of Go-Fast Employees

```
MTB >REGR C3 1 C4;
SUBC> DW.

The regression equation is
PRODUCT = 757 - 7.86 HEIGHT

Predictor        Coef        Stdev       t-ratio
Constant        757.46       45.82        16.53
HEIGHT           -7.8582      0.6518     -12.06

s = 18.07       R-sq = 59.7%    R-sq(adj) = 59.3%

Analysis of Variance

SOURCE          DF           SS           MS
Regression       1         47440         47440
Error           98         31985           326
Total           99         79425

Unusual Observations
Obs.   HEIGHT   PRODUCT     Fit  Stdev.Fit  Residual   St.Resid
 13     73.0    225.00   183.81    2.55      41.19      2.30R
 41     76.0    120.00   160.24    4.17     -40.24     -2.29R
 47     79.0    130.00   136.66    5.99      -6.66     -0.39 X
 56     74.0    140.00   175.95    3.04     -35.95     -2.02R
 57     73.0    145.00   183.81    2.55     -38.81     -2.17R
 61     73.0    230.00   183.81    2.55      46.19      2.58R
 68     63.0    265.00   262.39    5.05       2.61      0.15 X

R denotes an obs. with a large st. resid.
X denotes an obs. whose X value gives it large influence.

Durbin-Watson statistic = 1.76
```

Figure 10-10 shows the results of this command. It would be easy to get caught up in an analysis of the numbers without stopping and asking the proper managerial question: "Do the numbers make sense?" In Figure 10-10 the *t*-values are good, $r^2$ is respectable, and the Durbin–Watson is acceptable.

If you carried this analysis to its statistical conclusion, you would recommend to the personnel department that it hire only short people. This is clearly absurd! Thus, you must beware of spurious relationships. No sophisticated mathematical shenanigans can take the place of good managerial judgment.

## 10-5 MULTIPLE REGRESSION

So far we have discussed two types of forecasting models. In the simpler type, we use only one independent variable. For example, if you know your mean score for the first three quizzes in your statistics course, you might predict that your grade on the upcoming quiz will be equal to the mean of your past scores. The second type of forecasting, simple linear regression, involves one independent variable and one dependent variable.

There is no reason to limit ourselves to such simple models, since there is no conceptual difference between models with one independent variable and models with several. In

### SECTION 10-5 Multiple Regression

addition, Minitab performs multiple regression as easily as it does simple regression. However, before looking at the Minitab command format for multiple regression, let's consider an example to illustrate the use of multiple regression.

Suppose that a big shoe retailing chain to which Go-Fast sells its running shoes is planning on opening a new store. It presently has 15 stores around the country. Both its own experience and the experience of Go-Fast indicate that the demand for running shoes in a given area is dependent on several variables. The major determinants include the population density, the average household income, and the average age of people who live near the store. Experience indicates that a two-square-mile area around the store is the most effective area to use. The management of the chain has hired us to measure each of these variables for each of its present store locations. Thus store sales is the dependent variable and population, income, and age are our independent variables. If we can build an effective multiple regression model based on the 15 present stores, we can use it to predict sales at the new store.

The choice to obtain data from a two-square-mile area around each store is a management decision based on previous experience. We could have obtained data from areas of various sizes and then compared the results of each analysis. The problem with that approach would have been its high cost! Gathering that much data would be very expensive and would not be justified managerially. We also could have added other variables to the analysis, but for each additional variable, we would add to the costs of the study.

So, after trips to the chamber of commerce in each city and after studying Census Bureau statistical profiles, we obtain the necessary data on our three variables. This data has been entered into Minitab as shown in Figure 10-11. Our strategy is to determine the relationship between our three independent variables and sales for each of the present stores. Once that relationship is determined, we can obtain the data on the population density, income, and age in the area around the new store and use it to predict the probable sales level.

As a first step, we might plot each of the variables against sales. This has been done in Figures 10-12, 10-13, and 10-14 for population density, income, and age, respectively.

**FIGURE 10-11** Data on 15 Shoe Stores as Entered in Minitab

```
MTB >PRIN 'POP.DEN' 'INCOME' 'AGE' 'SALES'

COLUMN    POP.DEN       INCOME         AGE          SALES
COUNT       15            15            15            15
ROW
  1       1951.00       19388.0        29.         1669.00
  2       2037.00       21712.0        33.         1862.00
  3       2012.00       17556.0        27.         1593.00
  4       2149.00       20409.0        33.         1714.00
  5       1921.00       19966.0        29.         1676.00
  6       2016.00       20712.0        34.         1689.00
  7       2004.00       24282.0        34.         1888.00
  8       2026.00       20004.0        33.         1699.00
  9       1972.00       22487.0        36.         1762.00
 10       1735.00       23397.0        23.         1978.00
 11       1977.00       19983.0        32.         1741.00
 12       1912.00       21987.0        30.         1815.00
 13       2068.00       23499.0        30.         2025.00
 14       2214.00       20521.0        29.         1915.00
 15       2053.00       22010.0        30.         1789.00
```

**FIGURE 10-12** Minitab Plot of Relationship between Sales and Population Density for 15 Shoe Stores

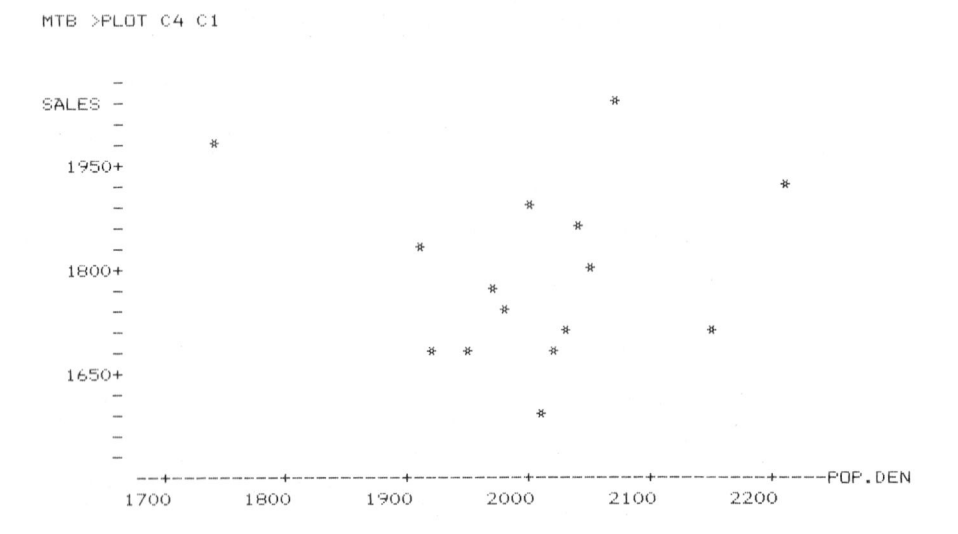

```
MTB >PLOT C4 C1

      -
SALES -                                        *
      -
1950+       *                                              *
      -
      -                        *
      -                                *
      -                *
1800+                                 *
      -                     *
      -                     *
      -                           *        *
      -              *  *     *
1650+
      -
      -                       *
      -
      --+---------+---------+---------+---------+---------+-----POP.DEN
       1700      1800      1900      2000      2100      2200
```

**FIGURE 10-13** Minitab Plot of Relationship between Sales and Income for 15 Shoe Stores

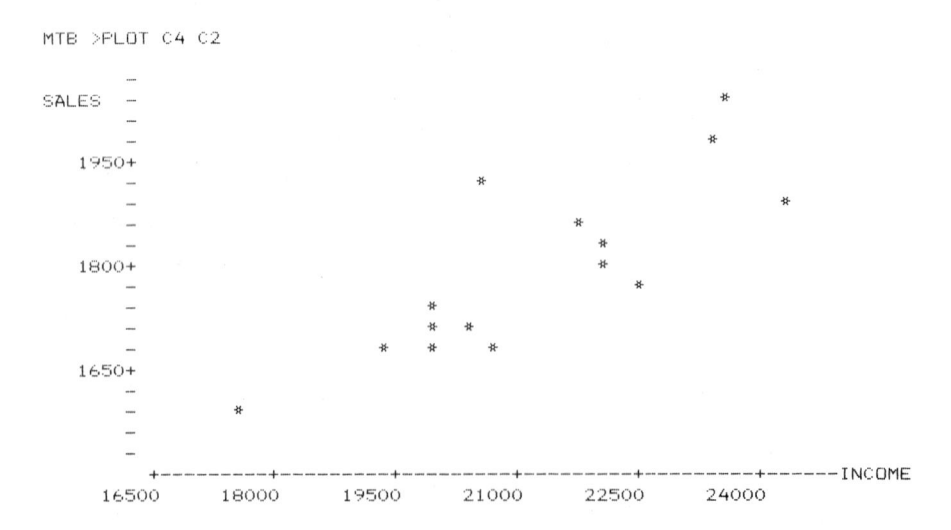

```
MTB >PLOT C4 C2

      -
SALES -                                                   *
      -                                              *
1950+                           *
      -                                                 *
      -                              *
      -                                 *
1800+                                 *
      -                                     *
      -            *
      -                       *  *
      -                    *  *   *
1650+
      -
      -      *
      -
      +---------+---------+---------+---------+---------+------INCOME
      16500     18000     19500     21000     22500     24000
```

The best looking of the graphs is probably Figure 10-13, the relationship between household income and sales in the geographic area. Notice also that in Figures 10-12 and 10-13 there is one value in each graph that is relatively isolated. We will not do anything about those observations now, but it is good to know that such observations exist.

After a brief look at the plots, we should check for multicollinearity between the independent variables. *Multicollinearity* is correlation between the independent variables, and one of our basic assumptions is that the independent variables are not correlated—they

## SECTION 10-5 Multiple Regression

**FIGURE 10-14** Minitab Plot of Relationship between Sales and Age for 15 Shoe Stores

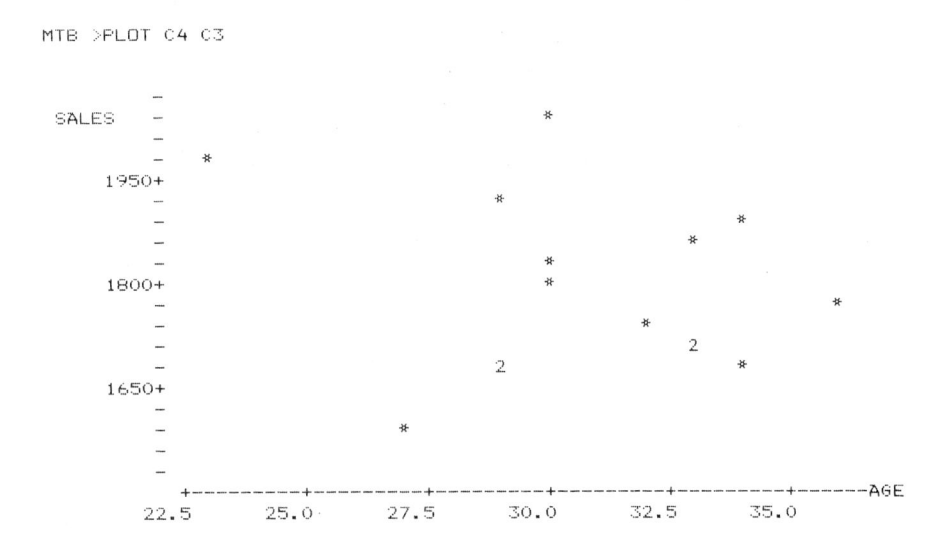

```
MTB >PLOT C4 C3

    SALES  -
           -                              *
           -    *
      1950+
           -                       *
           -                                          *
           -                                  *
           -                          *
      1800+                           *
           -                                          *
           -                              *     2
           -                      2               *
      1650+
           -            *
           -
           -
           +---------+---------+---------+---------+---------+-------AGE
          22.5      25.0      27.5      30.0      32.5      35.0
```

must be independent of each other. We make this check with the `CORRelation` command as follows:

`MTB >CORRelation coefficient of C1-C4`

Figure 10-15 displays the results of this command. As can be seen, there are relatively low correlations between population density, income, and age. The higher correlation between income and sales is acceptable, since sales is our dependent variable and the value that we are trying to predict.

With these steps completed, we can try to develop a multiple regression model. As mentioned at the start of the discussion on simple linear regression, the Minitab command for all regression is the same. For multiple regression you simply need to specify the proper number of predictors (independent variables) and the location of each variable. For this particular problem, the command is

`MTB >REGRess y-value in C4 on 3 predictors in C1, C2, C3;`
`SUBC> DW.`

This command line could be shortened to

`MTB >REGR C4 3 C1-C3;`
`SUBC> DW.`

**FIGURE 10-15** Correlation Matrix for Multiple Regression Analysis of Data on 15 Shoe Stores

```
          MTB >CORR C1-C4

                 POP.DEN   INCOME    AGE
       INCOME    -.215
       AGE        .440     .129
       SALES     -.037     .816    -.202
```

**FIGURE 10-16** Multiple Regression Analysis by Minitab for 15 Shoe Stores

```
MTB >REGR C4 3 C1-C3;
SUBC> DW.

The regression equation is
SALES = 66 + 0.444 POP.DEN + 0.0666 INCOME - 18.8 AGE

Predictor        Coef        Stdev       t-ratio
Constant         66.3        328.4        0.20
POP.DEN          0.4439      0.1432       3.10
INCOME           0.066613    0.007854     8.48
AGE             -18.821      4.679       -4.02

s = 49.82       R-sq = 87.3%      R-sq(adj) = 83.9%

Analysis of Variance

SOURCE           DF          SS           MS
Regression        3       188109        62703
Error            11        27307         2482
Total            14       215415

SOURCE           DF       SEQ SS
POP.DEN           1          296
INCOME            1       147646
AGE               1        40167

Durbin-Watson statistic = 1.79
```

This command produces Figure 10-16. In most cases we will want to examine the value of $r^2$ first to see if we have any predictive power. Since $r^2$ is 83.9%, we have done a reasonably good job of selecting variables, and we apparently have a regression model of some merit. Since the Durbin–Watson statistic is within the acceptable range of 1.5 to 2.5, we can pursue our analysis further.

The $t$ ratios indicate that the intercept coefficient is probably not significant but that all three slope coefficients are.

After examining these preliminary factors, we can develop a model for predicting the sales level at the new store. Our regression equation would look as follows:

$$\text{sales} = 66 + 0.444(\text{population density}) + 0.666(\text{income}) - 18.8(\text{age})$$

This equation reveals several interesting relationships. Notice that as the population density increases, sales should increase. Also, a rise in income should have the same effect. Both of these positive effects are suggested by the plus signs before the respective factors. In contrast, increasing age appears to have a negative effect on running shoe sales. This suggests that as people grow older, they buy fewer running shoes. None of our relationships appear spurious, and they all seem to make good managerial sense.

The next step is to predict probable sales for the new store. If we have the population density, income level, and age for any given location in town, we can predict sales. Let's assume that in the new city the shoe store chain is looking at two possible store locations. At location 1 the population density of the two-square-mile surrounding area is 2,025, family income level is $24,200, and the average age is 29. We would calculate the initial sales level for this location to be

## SECTION 10-5 Multiple Regression

$$\text{sales} = 66 + 0.444(\text{population density}) + 0.0666(\text{income}) - 18.8(\text{age})$$
$$= 66 + 0.444(2{,}025) + 0.0666(24{,}200) - 18.8(29)$$
$$= 66 + 899.1 + 1{,}611.72 - 545.2$$
$$= 2{,}031.92$$

Thus sales at the new store are predicted to be 2,032 pairs per year.

At location 2, the population density is 1,950, the family income level is $23,500, and the average age is 25. With this information, we calculate as follows:

$$\text{sales} = 66 + 0.444(1{,}950) + 0.666(23{,}500) - 18.8(25)$$
$$= 66 + 865.8 + 1565.1 - 470$$
$$= 2{,}027.2$$

Note that Minitab will generate these values with the subcommand

```
PREDict y-value for x-value set E,...,E
```

The actual form of the command set would be

```
MTB >REGR C4 3 C1-C3;
SUBC> DW;
SUBC> PREDict y-value for x-value set 2025, 24200, 29;
SUBC> PRED 1950, 23500, 25.
```

These commands produce all the material shown in Figure 10-16 plus the additional output shown in Figure 10-17. There are minor rounding differences in the numbers.

As you can see, the difference in expected sales at these two store locations is very small. It would be a mistake to choose the location with the larger value, which is only five pairs of shoes per year. All other things being equal, that might be the better location, but seldom are all other things equal. Remember that the $r^2$ is only 83.9%. Thus, other factors also influence the sales level. At this point, such managerial concerns as growth potential, traffic patterns, ease of access by customers and employees, and location of other services (such as lunch facilities for employees) would enter into the final decision. The point is simple: Don't get carried away with just the larger number without considering the whole managerial picture.

It is important to realize that we do not expect the actual sales level to match this prediction exactly. First, the $r^2$ is only 83.9%, so other variables also affect sales. There will also be statistical error and residuals in each of our independent variables. In fact, a much more justifiable prediction would be that sales *at either store* will be in the range between 1,900 and 2,160 pairs per year.

**FIGURE 10-17** Additional Minitab Output to Figure 10-16 When PREDict Subcommand Is Used

```
MTB >REGR C4 3 C1-C3;
SUBC> DW;
SUBC> PRED 2025, 24200, 29;
SUBC> PRED 1950, 23500, 25.

     Fit   Stdev.Fit       95% C.I.          95% P.I.
   2031.5      31.2    ( 1962.7, 2100.2)  ( 1902.0, 2160.9)

   2026.8      35.4    ( 1948.9, 2104.8)  ( 1892.3, 2161.4)
```

The proper managerial question at this time is whether our predictive model is good enough. We could go back and add one or more new variables to our model and see if our $r^2$ might increase. Of course, each new variable would also have to be examined for multicollinearity, and most important, each new variable would add to the cost of the study. Only an effective managerial procedure can decide whether the possible increase in the predictive ability of the model would be worth the cost of obtaining more information.

As more independent variables are added to a multiple regression model, the chances of not meeting the assumption of independence among the variables increase. For example, we might also have collected tax data for the 15 cities in which the stores are located. Let's assume that the tax rate in all these cities is a flat 15% of income. In other words, we could take 15% of column 2 in Figure 10-11 and create a new variable called "tax." Even if you had collected income information from one source and tax information from another, a fundamental relationship would still exist between the two variables.

We can illustrate the consequence of a lack of independence among all the variables. We can take column 2 in Figure 10-11, multiply it by 0.15, and place the results in column 6, which we will name `'TAX'`. The command would be:

```
MTB >LET C6 = C2 * .15
MTB >NAME C6 'TAX'
```

If we then issued `PLOT` and `CORRelation` commands on the new variables, the problem would become apparent. The correlation between `C2` and `C6` would be 1.00, which should set off an alarm that these two variables are definitely not independent. Further, a plot of `C2` and `C6` would show a very nice straight-line relationship between our supposedly independent variables. If we still attempted to enter variables `C1`, `C2`, `C3`, and `C6` into the Minitab multiple regression model with the command

```
MTB >REGR C4 4 C1, C2, C3, C6;
SUBC> DW.
```

we would receive the following message:

- C6 IS HIGHLY CORRELATED WITH OTHER PREDICTOR VARIABLES
- C6 HAS BEEN OMITTED FROM THE EQUATION

Once this type of warning has been issued, Minitab will simply omit the column of data in question and complete the analysis. Thus the output from the Minitab command for the four predictor variables including tax would look exactly like the output in Figure 10-16. If you do not receive this message, it does *not* mean that you have no correlations that are too high. The correlation problem has to be above .99 for this warning to be issued automatically. As you can see, Minitab only issues this warning in extreme cases.

Whenever the correlation coefficients among your independent variables are .60 or above, the possibility of multicollinearity arises. The simplest way to handle this problem is to eliminate one of the two independent variables that are correlated, because if two independent variables are highly correlated, they predict largely the same movement in the dependent variable. If they predict the same movement, then eliminating one of them will not significantly reduce the ability of the regression model to predict the dependent variable.

With the abilities you have available with the computer, it is simple enough to run a regression analysis more than once. If you suspect that multicollinearity is present, you can

**SECTION 10-6** Formulas and Hand Calculations ▶**255**

run the regression with both variables and then rerun the analysis and eliminate one of the two variables.

## 10-6 FORMULAS AND HAND CALCULATIONS

In this section, we will present the formulas and hand calculations for Pearson's *r* and linear regression equations. We will not show the calculation of a multiple regression equation or the tests for the significance of the slope and the *y*-intercept. If you want to calculate multiple regression by hand, numerous basic and advanced statistical texts can show you how.

It is appropriate here to mention that we are quickly approaching the point when computations without the use of a computer are nearly impossible. A multiple regression with 10 variables and 250 observations for each variable would take a tremendous number of hours to calculate and thousands of calculations, all of which must be correct. Not so long ago most businesses simply did not make such calculations. Now, with about $2,000 worth of computer equipment and a few hundred dollars invested in a statistical package such as Minitab (or numerous other packages, including some available in the public domain), we can make these complex calculations. For a single problem such as the projected sales at the new store, it is almost cheaper to buy the equipment and perform the calculations by computer analysis than to pay a statistician to perform the calculations manually.

### Pearson's *r* and the Coefficient of Determination

The correlation coefficient (Pearson's *r*) can be computed by

$$r = \frac{SS(XY)}{\sqrt{[SS(X)][SS(Y)]}}$$

where

$$SS(X) = \sum X^2 - \frac{(\sum X)^2}{n}$$

$$SS(Y) = \sum Y^2 - \frac{(\sum Y)^2}{n}$$

$$SS(XY) = \sum XY - \frac{(\sum X)(\sum Y)}{n}$$

To demonstrate the use of these formulas, let's compute Pearson's *r* for the following data, which might be part of a typical simple linear regression analysis:

| X | Y |
|---|---|
| 1 | 10 |
| 2 | 8 |
| 3 | 7 |
| 4 | 6 |
| 5 | 4 |

The simplest way to handle this computation is to create a table as follows:

|       | X  | Y  | X² | Y²  | XY |
|-------|----|----|----|-----|----|
|       | 1  | 10 | 1  | 100 | 10 |
|       | 2  | 8  | 4  | 64  | 16 |
|       | 3  | 7  | 9  | 49  | 21 |
|       | 4  | 6  | 16 | 36  | 24 |
|       | 5  | 4  | 25 | 16  | 20 |
| Total | 15 | 35 | 55 | 255 | 91 |

$$SS(X) = 55 - \frac{225}{5} = 10$$

$$SS(Y) = 265 - \frac{1{,}225}{5} = 20$$

$$SS(XY) = 91 - \frac{525}{5} = -14$$

$$r = \frac{-14}{\sqrt{(10)(20)}}$$

$$= -.989949$$

We can compute the coefficient of determination simply by squaring Pearson's $r$. Thus, for the $r$ just computed we get

$$r^2 = (-.989949)^2$$
$$= .98$$

# The Regression Equation

The general form of the simple linear regression equation is

$$Y = a + bX + \text{error}$$

where $a$ is the Y-intercept and $b$ is the slope of the line. Using the data from above, we can compute $b$ by

$$b = \frac{SS(XY)}{SS(X)} = \frac{-14}{10}$$
$$= -1.4$$

and

$$a = \frac{1}{n}[\sum Y - (b\sum X)]$$
$$= \frac{1}{5}[35 - (-1.4)(15)]$$
$$= 11.2$$

This gives us a regression equation of

$$Y = 11.2 - 1.4X + \text{error}$$

This relationship can be verified with Minitab by entering our data into columns 1 and 2 and executing the `REGR C2 1 C1` command. The results of this command are shown in Figure 10-18.

**FIGURE 10-18** Check of Hand Calculation Example by Minitab

```
MTB >REGR C2 1 C1;
SUBC> DW.

The regression equation is
C2 = 11.2 - 1.40 C1

Predictor        Coef        Stdev      t-ratio
Constant      11.2000       0.3830        29.25
C1            -1.4000       0.1155       -12.12

s = 0.3651       R-sq = 98.0%     R-sq(adj) = 97.3%

Analysis of Variance

SOURCE        DF           SS           MS
Regression     1       19.600       19.600
Error          3        0.400        0.133
Total          4       20.000


Durbin-Watson statistic = 2.60
```

# SUMMARY

In this chapter, we have studied regression analysis, a powerful statistical tool. Regression is probably the single most important statistical concept in business applications. Whole textbooks have been written just about regression.

This introduction has been necessarily sketchy. You should not get in trouble using regression, however, if you always verify that the necessary assumptions have been observed and that your model makes sense. When using a regression analysis, first check the correlations, review some simple plots of the data, and make sure that the ratios and Durbin–Watson statistic are appropriate. Then look at $r^2$ and the predictive equations with managerial considerations in mind.

# DISCUSSION QUESTIONS

1. What are the major assumptions used in regression analysis?
2. What is multicollinearity?
3. What can be done to correct for multicollinearity?
4. What is autocorrelation?

## PROBLEMS

**10-1** To help determine the factors motivating sales, a manager gathered a random sample of data on sales per month (in dollars) and a random sample of the commission for sales. The data is shown in the following chart:

| Sales | Commission Rate | Sales | Commission Rate | Sales | Commission Rate |
|---|---|---|---|---|---|
| 1,200 | 10 | 1,240 | 11 | 1,500 | 16 |
| 1,100 | 12 | 1,300 | 14 | 1,235 | 10 |
| 1,550 | 18 | 1,400 | 10 | 1,578 | 16 |
| 1,254 | 9 | 1,165 | 9 | 1,650 | 10 |
| 1,300 | 13 | 1,578 | 14 | 1,325 | 16 |
| 1,450 | 12 | 1,150 | 15 | 1,430 | 15 |
| 1,100 | 8 | 1,520 | 15 | 1,587 | 13 |
| 1,421 | 13 | 1,425 | 13 | 1,240 | 13 |
| 1,201 | 15 | 1,600 | 16 | 1,470 | 14 |
| 1,400 | 12 | 1,712 | 14 | 1,600 | 11 |
| 1,122 | 12 | 1,200 | 11 | 1,342 | 13 |

a) Perform a correlation analysis. How well are the two variables correlated?
b) Perform a regression analysis.
c) Test the significance of the coefficients using an alpha of .05.
d) Is there evidence of autocorrelation?
e) How much of the variation in the dependent variable is explained by the regression equation?
f) Is the regression equation adequate? What can be done to improve the equation's predictive ability?

**10-2** A young apprentice once remarked, in a smug tone of voice, "Younger carpenters are much more productive than older carpenters." The following table shows a carpenter's productivity as represented by the number of days required to complete a kitchen and the carpenter's age:

| Productivity | Age | Productivity | Age | Productivity | Age |
|---|---|---|---|---|---|
| 12 | 54 | 14 | 48 | 22 | 51 |
| 15 | 49 | 15 | 28 | 13 | 49 |
| 16 | 50 | 19 | 37 | 15 | 34 |
| 21 | 28 | 23 | 22 | 22 | 26 |
| 24 | 31 | 16 | 47 | 13 | 55 |
| 13 | 56 | 13 | 58 | 21 | 30 |
| 14 | 26 | 22 | 36 | 12 | 61 |
| 12 | 59 | 14 | 57 | 17 | 38 |
| 26 | 27 | 20 | 35 | 18 | 44 |
| 15 | 49 | 22 | 21 | 13 | 59 |
| 20 | 33 | 16 | 30 | 19 | 31 |
| 12 | 57 | 14 | 59 | 16 | 33 |

a) Is productivity highly correlated with age in carpentry? Is it a positive or negative correlation?

# PROBLEMS

b) Create a scatter diagram. Compare the scatter diagram to the correlation coefficient.
c) Perform a regression analysis using Minitab.
d) Using an alpha of .01, test the significance of the intercept and the slope.
e) Does the regression equation have much predictive ability?
f) Is autocorrelation present?
g) Three rows of data were singled out by Minitab. Check the accuracy of the figures.
h) Is it appropriate to have three rows of the data set fall outside of the 95% expected value interval?
i) Was the young apprentice's remark justified?

**10-3** Students in the sixth grade have just completed a series of activities to determine the winners of the school's physical fitness award. Given in the following is a random sample of data on the students' scores, height, weight, and aptitude.

| Score | Height | Weight | Aptitude | Score | Height | Weight | Aptitude |
|-------|--------|--------|----------|-------|--------|--------|----------|
| 88 | 47 | 55 | 8 | 76 | 44 | 59 | 7 |
| 76 | 37 | 50 | 7 | 70 | 47 | 54 | 6 |
| 89 | 50 | 58 | 8 | 90 | 48 | 61 | 8 |
| 69 | 36 | 57 | 6 | 84 | 46 | 56 | 9 |
| 93 | 52 | 57 | 9 | 96 | 50 | 62 | 8 |
| 90 | 49 | 57 | 7 | 74 | 47 | 60 | 8 |
| 70 | 39 | 55 | 8 | 78 | 45 | 56 | 7 |
| 82 | 40 | 44 | 8 | 93 | 49 | 58 | 9 |
| 92 | 51 | 61 | 8 | 90 | 47 | 57 | 8 |
| 94 | 49 | 58 | 7 | 80 | 46 | 60 | 6 |
| 71 | 42 | 58 | 8 | 92 | 50 | 57 | 8 |
| 85 | 48 | 62 | 9 | 79 | 36 | 59 | 8 |
| 80 | 41 | 52 | 7 | 88 | 49 | 58 | 9 |
| 67 | 35 | 54 | 6 | 87 | 48 | 54 | 10 |
| 72 | 38 | 57 | 8 | 91 | 46 | 52 | 8 |
| 93 | 48 | 58 | 8 | 77 | 36 | 49 | 7 |
| 84 | 47 | 55 | 9 | | | | |

a) Perform a correlation analysis. Is there evidence of multicollinearity?
b) Perform a regression analysis.
c) Using an alpha of .10, test the significance of the intercept and the slopes.
d) Is the coefficient of determination acceptable?
e) Is autocorrelation present?
f) The computer suspects errors in a few rows of data; check the accuracy of these inputs.
g) What do the R's and X's indicate?
h) Is the regression equation adequate? Why or why not?

**10-4** During much of the year, 38 employees in the production department of the local telephone company proofread each line of the telephone directory and the Yellow Pages. Productivity has slipped recently, and the manager feels that it is because of the age of some of these workers. Following is a random sample of the number of lines read during a 15-minute period and the age of employees:

## CHAPTER 10 Correlation and Regression Analysis

| Lines Read | Age | Lines Read | Age | Lines Read | Age |
|---|---|---|---|---|---|
| 11 | 57 | 12 | 42 | 14 | 58 |
| 15 | 34 | 14 | 31 | 16 | 27 |
| 15 | 30 | 11 | 57 | 11 | 62 |
| 10 | 62 | 13 | 52 | 15 | 22 |
| 14 | 29 | 15 | 21 | 14 | 29 |
| 11 | 59 | 14 | 39 | 14 | 49 |
| 12 | 61 | 13 | 53 | 11 | 57 |
| 14 | 26 | 15 | 19 | 14 | 24 |
| 17 | 22 | 12 | 51 | 16 | 32 |
| 33 | 16 | 16 | 30 | 11 | 60 |
| 12 | 52 | 13 | 57 | 12 | 54 |
| 13 | 56 | 15 | 33 | 16 | 27 |
| 14 | 27 | 12 | 55 | 13 | 51 |

a) What degree of correlation do you expect between age and productivity?
b) Using Minitab, compute the coefficient of correlation. Are your suspicions correct?
c) Plot the independent variable (productivity) against the dependent variable (age). Does the scatter diagram support the coefficient of correlation computed in (b)?
d) Calculate a regression equation for the dependent variable and the independent variable.
e) Refer to the $t$-values. Are the coefficients significant? Use an alpha of .05.
f) How much variation in the dependent variable is explained by the regression equation? Is this an acceptable level?
g) A data point appears to fall outside the 95% range of predicted values. Is there an error in your inputs or is this acceptable? Explain.

10-5 Refer to Problem 10-4. Assume that the 10th entry was recorded incorrectly. The true values are 16 lines proofread and 33 years of age.
a) What is the coefficient of correlation? Is it different from that found in Problem 10-4?
b) Create a scatter diagram. How does it differ from the one created in Problem 10-4?
c) Perform a regression analysis.
d) Using an alpha of .05, test the significance of the intercept and the slope.
e) How much of the variation in productivity is explained by the independent variable? Is it different from that found in Problem 10-4? Is this an acceptable level?
f) If the coefficient of determination is not acceptable, what could you do to make the equation a better predictor of productivity?
g) Is there autocorrelation in this analysis?
h) One entry falls outside the 95% range of predicted values. Is this acceptable? Why or why not?

10-6 Refer to Problem 10-5. The manager rethought her analysis. Another variable, the number of years of schooling, was thought to influence the productivity of the proofreaders. Add this independent variable to your regression analysis. The data is shown below.

| Lines | Age | School | Lines | Age | School | Lines | Age | School |
|---|---|---|---|---|---|---|---|---|
| 11 | 57 | 14 | 10 | 62 | 15 | 12 | 61 | 15 |
| 15 | 34 | 11 | 14 | 29 | 13 | 14 | 26 | 12 |
| 15 | 30 | 12 | 11 | 59 | 13 | 17 | 22 | 13 |

*(continued on next page)*

# PROBLEMS

*(Continued)*

| Lines | Age | School | Lines | Age | School | Lines | Age | School |
|---|---|---|---|---|---|---|---|---|
| 16 | 33 | 11 | 13 | 53 | 12 | 15 | 22 | 13 |
| 12 | 52 | 14 | 15 | 19 | 11 | 14 | 29 | 12 |
| 13 | 56 | 14 | 12 | 51 | 16 | 14 | 49 | 11 |
| 14 | 27 | 12 | 16 | 30 | 13 | 11 | 57 | 16 |
| 12 | 42 | 15 | 13 | 57 | 13 | 14 | 24 | 16 |
| 14 | 31 | 13 | 15 | 33 | 12 | 16 | 32 | 11 |
| 11 | 57 | 14 | 12 | 55 | 14 | 11 | 60 | 12 |
| 13 | 52 | 14 | 14 | 58 | 11 | 12 | 54 | 14 |
| 15 | 21 | 13 | 16 | 27 | 11 | 16 | 27 | 16 |
| 14 | 39 | 13 | 11 | 62 | 15 | 13 | 51 | 16 |

**a)** Plot the independent variables against productivity. Is there a relationship between the independent variables and the dependent variable?
**b)** Check for multicollinearity between the independent variables.
**c)** Perform a regression analysis using Minitab.
**d)** Evaluate the intercept and the slopes. Are they significant? Use an alpha of .05.
**e)** Does the equation derived from (d) have any predictive value? Does this equation have more predictive ability than the equation obtained in Problem 10-5?
**f)** Is there evidence of autocorrelation in this analysis?
**g)** What is the correlation between the number of years of schooling and productivity? What reasons can you give to explain this relationship?

**10-7** Since its sales seem to be lower than the sales of most other dealers in the state, a local automobile dealership wants to move to another location. Your job is to determine which variables pertinent to location affect sales. The following is a random sample of data on annual sales per dealership, mean population per square mile, mean annual income per household, mean age of head of household, and the number of dealerships within a 20-mile radius.

| Sales | Population | Income | Age | Number |
|---|---|---|---|---|
| 3.3 | 1,010 | 33 | 42 | 4 |
| 4.1 | 1,324 | 37 | 36 | 3 |
| 3.0 | 1,000 | 32 | 47 | 5 |
| 4.2 | 1,429 | 38 | 34 | 3 |
| 3.7 | 1,225 | 39 | 37 | 4 |
| 3.9 | 1,433 | 31 | 40 | 4 |
| 2.9 | 1,022 | 23 | 39 | 3 |
| 4.0 | 1,400 | 37 | 39 | 5 |
| 4.4 | 1,420 | 33 | 33 | 4 |
| 3.1 | 1,065 | 30 | 40 | 4 |
| 4.2 | 1,453 | 31 | 39 | 3 |
| 3.0 | 1,210 | 34 | 41 | 4 |
| 3.6 | 1,324 | 31 | 33 | 4 |
| 3.8 | 1,412 | 35 | 35 | 3 |
| 3.2 | 1,256 | 33 | 32 | 4 |
| 4.4 | 1,500 | 34 | 38 | 2 |
| 4.6 | 1,423 | 38 | 37 | 4 |

*(continued on next page)*

(Continued)

| Sales | Population | Income | Age | Number |
|-------|------------|--------|-----|--------|
| 3.6 | 1,200 | 30 | 37 | 4 |
| 3.4 | 1,314 | 34 | 34 | 3 |
| 4.0 | 1,465 | 34 | 41 | 4 |
| 3.2 | 1,100 | 32 | 34 | 3 |
| 3.9 | 1,460 | 36 | 38 | 3 |
| 3.0 | 1,234 | 30 | 36 | 5 |
| 4.3 | 1,400 | 32 | 37 | 3 |
| 3.3 | 1,378 | 32 | 40 | 3 |
| 3.1 | 1,200 | 33 | 40 | 4 |
| 4.1 | 1,439 | 32 | 37 | 3 |
| 4.2 | 1,419 | 39 | 34 | 4 |
| 3.4 | 1,370 | 35 | 39 | 3 |
| 3.0 | 1,420 | 30 | 41 | 2 |
| 4.2 | 1,475 | 35 | 36 | 3 |
| 3.9 | 1,490 | 32 | 38 | 3 |

**a)** Is there evidence of multicollinearity between the independent variables? If so, drop one of the highly correlated variables.
**b)** Perform a regression analysis using Minitab.
**c)** Using an alpha of .01, test the significance of the intercept and the slopes.
**d)** Are the variables correlated to their own previous values?
**e)** If the owner of the dealership wants a very high coefficient of determination, is this equation acceptable?

**10-8** As part of your economics course, you decide to study the variables affecting enrollment at private postsecondary schools. The following data is a random sample on enrollment per year, tuition per year per student, mean per capita income within the state, average yearly governmental aid to the school, and the number of postsecondary private institutions within the state.

| Enrollment | Tuition | Income | Aid | Institutions |
|------------|---------|--------|-----|--------------|
| 7,200 | $ 8,700 | $3,472 | $22,000 | 3 |
| 8,750 | 5,700 | 3,700 | 26,420 | 2 |
| 9,000 | 10,000 | 3,655 | 25,000 | 1 |
| 6,200 | 7,500 | 3,339 | 29,560 | 4 |
| 8,570 | 10,200 | 3,400 | 30,000 | 3 |
| 8,600 | 5,005 | 3,600 | 21,000 | 4 |
| 5,700 | 9,200 | 3,500 | 22,000 | 5 |
| 7,800 | 10,100 | 3,670 | 22,000 | 3 |
| 9,900 | 5,670 | 3,520 | 26,000 | 2 |
| 10,000 | 8,400 | 3,670 | 24,000 | 3 |
| 5,500 | 9,000 | 3,590 | 25,000 | 4 |
| 9,800 | 8,750 | 3,790 | 20,000 | 2 |
| 8,633 | 9,670 | 3,600 | 21,200 | 3 |
| 6,890 | 7,050 | 3,549 | 23,210 | 2 |
| 8,200 | 7,580 | 3,421 | 19,800 | 2 |

*(continued on next page)*

# PROBLEMS

*(Continued)*

| Enrollment | Tuition | Income | Aid | Institutions |
|---|---|---|---|---|
| 7,000 | $ 8,576 | $3,620 | $23,090 | 1 |
| 6,980 | 9,990 | 3,456 | 24,000 | 4 |
| 8,890 | 10,000 | 3,780 | 21,000 | 2 |
| 9,010 | 7,800 | 3,540 | 30,000 | 1 |
| 6,055 | 8,000 | 3,600 | 20,400 | 3 |
| 6,500 | 6,890 | 3,475 | 21,330 | 5 |
| 5,050 | 7,200 | 3,526 | 24,300 | 5 |
| 9,020 | 10,000 | 3,780 | 26,000 | 2 |
| 5,600 | 8,500 | 3,421 | 19,000 | 4 |
| 8,750 | 7,900 | 3,512 | 24,500 | 2 |
| 7,645 | 9,100 | 3,650 | 29,000 | 1 |
| 8,069 | 5,000 | 3,587 | 28,200 | 2 |
| 5,891 | 5,900 | 3,423 | 20,000 | 4 |
| 9,245 | 6,000 | 3,300 | 29,000 | 2 |
| 7,640 | 7,900 | 3,563 | 22,800 | 3 |
| 8,500 | 6,900 | 3,500 | 25,000 | 2 |
| 9,000 | 6,820 | 3,420 | 28,643 | 2 |

**a)** Perform a correlation analysis on the data.
**b)** Given the correlation coefficients just computed, what will each scatter diagram for the independent and dependent variable look like?
**c)** Plot the dependent variable against each of the independent variables. Compare the scatter diagrams to your response in (b).
**d)** Perform a regression analysis using Minitab.
**e)** Using an alpha of .05, test the significance of the coefficients.
**f)** Is there evidence of autocorrelation?
**g)** How much of the variation in the dependent variable is explained by the regression equation? Does the equation have much predictive ability?

**10-9** Because you were unsuccessful at finding an adequate regression equation in Problem 10-8, you decide that the variables you chose might be more applicable to public postsecondary institutions. A similar random sample of data for public postsecondary institutions follows:

| Enrollment | Tuition | Income | Aid | Institutions |
|---|---|---|---|---|
| 16,000 | $1,012 | $3,401 | $70,000 | 3 |
| 17,600 | 983 | 3,306 | 72,200 | 1 |
| 20,000 | 900 | 3,294 | 72,400 | 1 |
| 16,572 | 1,009 | 3,389 | 70,100 | 2 |
| 16,100 | 1,016 | 3,390 | 70,316 | 3 |
| 20,200 | 901 | 3,310 | 72,560 | 2 |
| 16,116 | 977 | 3,412 | 70,220 | 2 |
| 16,981 | 1,015 | 3,382 | 71,000 | 2 |
| 19,200 | 950 | 3,299 | 72,600 | 3 |
| 16,240 | 1,000 | 3,412 | 70,400 | 2 |
| 16,790 | 1,011 | 3,410 | 71,700 | 2 |
| 16,200 | 1,012 | 3,392 | 72,520 | 3 |

*(continued on next page)*

(Continued)

| Enrollment | Tuition | Income | Aid | Institutions |
|---|---|---|---|---|
| 18,741 | $ 910 | $3,287 | $72,860 | 2 |
| 17,830 | 980 | 3,370 | 72,720 | 1 |
| 19,987 | 916 | 3,301 | 71,000 | 1 |
| 20,112 | 904 | 3,290 | 74,400 | 1 |
| 19,360 | 912 | 3,310 | 71,940 | 2 |
| 18,000 | 967 | 3,402 | 72,310 | 2 |
| 16,500 | 1,002 | 3,411 | 70,617 | 3 |
| 16,722 | 1,005 | 3,419 | 70,644 | 3 |
| 17,982 | 976 | 3,375 | 72,000 | 2 |
| 16,943 | 1,002 | 3,414 | 71,500 | 3 |
| 19,256 | 901 | 3,316 | 72,960 | 1 |
| 18,200 | 924 | 3,381 | 71,420 | 2 |
| 21,630 | 876 | 3,280 | 72,877 | 1 |
| 17,620 | 945 | 3,397 | 71,830 | 2 |
| 19,600 | 912 | 3,282 | 70,870 | 1 |
| 15,900 | 1,016 | 3,416 | 70,012 | 3 |
| 21,777 | 857 | 3,289 | 72,050 | 2 |
| 18,880 | 976 | 3,354 | 72,800 | 1 |

a) Enter the data and perform a correlation analysis.
b) If the independent variables are too highly correlated, eliminate the appropriate variables. Why is this an acceptable procedure?
c) Perform a regression analysis.
d) Use an alpha of .05. Are the coefficients significant?
e) Is autocorrelation at an acceptable level?
f) How much of the variation in the dependent variable can be explained by the regression equation?
g) Were your initial feelings correct? Do the independent variables chosen better explain the enrollment at public postsecondary institutions than at private?

**10-10** As part owner of a service station, you were extremely interested when your daughter remarked that she was working on an equation for predicting gasoline consumption. In the following chart is her randomly sampled data on gallons of gasoline purchased in your city weekly, the number of cars within the city, the population of the city (in millions), and the cost of gasoline per gallon:

| Gallons | Cars | Population | Cost | Gallons | Cars | Population | Cost |
|---|---|---|---|---|---|---|---|
| 72,000 | 667,000 | 1.23 | $0.84 | 70,400 | 659,000 | 0.80 | $0.79 |
| 71,600 | 660,000 | 0.99 | 0.79 | 72,319 | 668,440 | 0.96 | 0.84 |
| 72,010 | 668,505 | 0.97 | 0.82 | 71,612 | 660,450 | 0.87 | 0.83 |
| 71,212 | 660,960 | 0.85 | 0.80 | 72,300 | 668,419 | 0.90 | 0.82 |
| 70,304 | 658,708 | 0.68 | 0.80 | 72,442 | 673,415 | 1.09 | 0.80 |
| 70,000 | 656,200 | 0.72 | 0.81 | 70,540 | 659,275 | 0.92 | 0.82 |
| 72,366 | 669,754 | 1.10 | 0.83 | 70,600 | 659,300 | 0.90 | 0.83 |
| 71,707 | 662,013 | 0.84 | 0.84 | 71,676 | 662,512 | 0.96 | 0.84 |
| 71,340 | 660,012 | 0.76 | 0.80 | 72,500 | 675,400 | 0.99 | 0.81 |

*(continued on next page)*

## PROBLEMS

*(Continued)*

| Gallons | Cars | Population | Cost | Gallons | Cars | Population | Cost |
|---|---|---|---|---|---|---|---|
| 72,399 | 676,000 | 1.11 | $0.81 | 70,002 | 656,314 | 1.00 | $0.80 |
| 70,430 | 650,322 | 0.90 | 0.80 | 72,650 | 669,000 | 1.14 | 0.84 |
| 71,782 | 664,100 | 0.88 | 0.79 | 72,089 | 668,629 | 0.83 | 0.82 |
| 70,860 | 661,200 | 0.82 | 0.82 | 72,115 | 668,700 | 0.85 | 0.80 |
| 70,600 | 649,000 | 0.89 | 0.81 | 70,140 | 657,117 | 0.88 | 0.83 |
| 71,650 | 660,600 | 0.87 | 0.83 | 71,260 | 662,790 | 0.92 | 0.82 |
| 72,340 | 669,000 | 1.22 | 0.82 | 71,232 | 663,040 | 0.89 | 0.80 |
| 70,101 | 652,119 | 0.90 | 0.79 | 70,360 | 658,900 | 0.79 | 0.81 |
| 71,360 | 661,110 | 0.92 | 0.81 | 72,430 | 668,200 | 1.20 | 0.83 |
| 72,111 | 667,400 | 1.06 | 0.80 | 72,422 | 669,000 | 1.18 | 0.80 |

**a)** Enter the data and perform a correlation analysis.
**b)** Refer to the correlation matrix. Before plotting the independent variables against the dependent variable, describe the appearance of the scatter diagrams.
**c)** Create the scatter diagrams and compare them to your answer in (b).
**d)** If the correlation between any of the independent variables is unacceptable, eliminate one of them. Perform a regression analysis.
**e)** Use a 90% confidence level to test the significance of the coefficients.
**f)** Is there evidence of autocorrelation?
**g)** How much variation in the dependent variable is explained by the regression equation?
**h)** How useful would this equation be in determining gasoline consumption?

**10-11** The reproductive success of mosquitoes was studied in a biology class. Careful records were kept on the number of offspring, humidity, amount of food, amount of water for the larvae, and the number of adult mosquitoes within an experimental environment. Randomly sampled data for each variable appears in the following chart (data for food and water are in ounces):

| Offspring | Humidity | Food | Water | Adults | Offspring | Humidity | Food | Water | Adults |
|---|---|---|---|---|---|---|---|---|---|
| 220 | 0.61 | 1.7 | 2.4 | 19 | 250 | 0.50 | 1.9 | 2.6 | 16 |
| 190 | 0.53 | 1.6 | 1.7 | 23 | 210 | 0.60 | 1.9 | 2.1 | 21 |
| 200 | 0.58 | 1.6 | 2.0 | 21 | 170 | 0.64 | 1.9 | 2.0 | 19 |
| 260 | 0.80 | 1.4 | 2.9 | 11 | 160 | 0.49 | 1.7 | 1.8 | 23 |
| 170 | 0.55 | 1.8 | 2.2 | 20 | 250 | 0.76 | 1.5 | 2.8 | 15 |
| 200 | 0.57 | 1.9 | 2.0 | 22 | 260 | 0.70 | 1.2 | 2.9 | 10 |
| 230 | 0.62 | 1.7 | 2.6 | 17 | 200 | 0.53 | 1.9 | 2.4 | 21 |
| 260 | 0.69 | 1.4 | 3.0 | 12 | 180 | 0.54 | 1.7 | 2.2 | 24 |
| 250 | 0.75 | 1.4 | 2.8 | 15 | 200 | 0.49 | 1.9 | 1.8 | 17 |
| 250 | 0.67 | 1.5 | 2.5 | 19 | 220 | 0.71 | 1.8 | 1.9 | 20 |
| 180 | 0.49 | 1.8 | 1.9 | 20 | 230 | 0.63 | 1.6 | 2.5 | 21 |
| 200 | 0.53 | 1.6 | 1.8 | 21 | 230 | 0.72 | 2.0 | 2.3 | 18 |
| 210 | 0.59 | 1.8 | 2.2 | 21 | 210 | 0.60 | 1.6 | 2.2 | 22 |
| 190 | 0.69 | 1.4 | 1.9 | 17 | 210 | 0.61 | 1.8 | 2.0 | 21 |
| 190 | 0.57 | 1.4 | 2.1 | 21 | 190 | 0.58 | 1.9 | 2.1 | 22 |
| 220 | 0.57 | 1.8 | 2.5 | 18 | 260 | 0.65 | 1.7 | 2.8 | 13 |

*(continued on next page)*

## CHAPTER 10 Correlation and Regression Analysis

*(Continued)*

| Offspring | Humidity | Food | Water | Adults | Offspring | Humidity | Food | Water | Adults |
|---|---|---|---|---|---|---|---|---|---|
| 170 | 0.59 | 1.7 | 1.7 | 13 | 170 | 0.52 | 1.9 | 2.0 | 24 |
| 200 | 0.65 | 1.5 | 2.0 | 21 | 190 | 0.56 | 1.9 | 2.0 | 24 |
| 240 | 0.61 | 1.6 | 2.5 | 13 | 170 | 0.50 | 1.9 | 1.9 | 23 |
| 250 | 0.70 | 1.3 | 2.8 | 15 | 200 | 0.57 | 2.0 | 2.1 | 19 |
| 180 | 0.50 | 1.9 | 1.9 | 17 | 250 | 0.66 | 1.3 | 2.7 | 13 |
| 240 | 0.64 | 1.5 | 2.0 | 16 | 240 | 0.69 | 1.4 | 2.8 | 13 |
| 240 | 0.62 | 1.7 | 2.7 | 22 | 200 | 0.61 | 1.8 | 1.7 | 16 |
| 170 | 0.58 | 2.1 | 2.8 | 24 | 230 | 0.63 | 1.9 | 2.8 | 21 |
| 260 | 0.76 | 1.3 | 2.8 | 20 | 180 | 0.60 | 1.6 | 1.8 | 24 |
| 220 | 0.64 | 1.9 | 2.2 | 19 | 210 | 0.62 | 1.8 | 2.2 | 21 |
| 180 | 0.60 | 2.0 | 1.7 | 23 | 190 | 0.58 | 2.0 | 1.9 | 22 |
| 190 | 0.58 | 1.8 | 1.9 | 22 | 260 | 0.70 | 1.8 | 3.0 | 14 |
| 210 | 0.57 | 1.8 | 2.0 | 22 | 240 | 0.65 | 1.7 | 2.6 | 10 |
| 210 | 0.65 | 1.9 | 1.8 | 23 | 180 | 0.52 | 2.0 | 2.0 | 21 |

**a)** Plot the independent variables against the dependent variable (offspring).
**b)** Examine the scatter diagrams. Does there appear to be some correlation between the dependent and independent variables?
**c)** Perform a correlation analysis. Does it confirm your answer in (b)?
**d)** Regress offspring against the independent variables.
**e)** Using a confidence level of 95%, test the significance of the coefficients.
**f)** What do the R's indicate?
**g)** Is it appropriate to have five rows noted with R's? If not, what should you do?
**h)** Is autocorrelation at an acceptable level?
**i)** How much of the variation in the dependent variable, offspring, is explained by the regression equation?

**10-12** Your town's major employer is Big Ben, a clock manufacturer. Recently Big Ben laid off a quarter of its work force. Part of the reason is the recent influx of cheap imported clocks. Following is randomly sampled data on the number of clocks imported annually, the profit per imported clock, the average cost of materials in an imported clock, and the number of imported clock manufacturers:

| Imports | Profit | Cost | Number | Imports | Profit | Cost | Number |
|---|---|---|---|---|---|---|---|
| 20,102 | $1.80 | $0.98 | 10 | 30,000 | $2.19 | $0.88 | 10 |
| 24,612 | 1.96 | 0.92 | 12 | 23,404 | 1.92 | 0.94 | 12 |
| 29,300 | 2.11 | 0.82 | 16 | 22,620 | 2.12 | 0.96 | 10 |
| 21,250 | 1.83 | 0.94 | 10 | 27,397 | 2.08 | 0.82 | 11 |
| 23,975 | 1.91 | 0.98 | 11 | 27,412 | 2.07 | 0.84 | 14 |
| 30,200 | 2.12 | 0.80 | 11 | 28,159 | 1.89 | 0.87 | 15 |
| 20,175 | 2.04 | 1.00 | 12 | 24,919 | 1.94 | 0.92 | 13 |
| 22,367 | 1.88 | 0.89 | 9 | 26,711 | 2.05 | 0.96 | 13 |
| 26,249 | 1.82 | 0.96 | 13 | 21,350 | 1.97 | 0.89 | 12 |
| 20,780 | 1.81 | 1.01 | 9 | 24,829 | 1.80 | 0.93 | 16 |
| 24,367 | 1.95 | 0.82 | 14 | 25,750 | 2.00 | 0.87 | 14 |
| 29,110 | 2.10 | 0.91 | 15 | 20,419 | 1.83 | 0.97 | 12 |

*(continued on next page)*

# PROBLEMS

*(Continued)*

| Imports | Profit | Cost | Number | Imports | Profit | Cost | Number |
|---|---|---|---|---|---|---|---|
| 21,660 | $1.80 | $0.99 | 10 | 24,391 | $1.97 | $0.96 | 11 |
| 26,314 | 2.07 | 0.90 | 12 | 23,005 | 1.94 | 0.93 | 11 |
| 28,212 | 2.08 | 0.97 | 15 | 30,600 | 2.15 | 0.88 | 12 |
| 30,416 | 2.16 | 0.85 | 16 | 29,431 | 2.11 | 0.82 | 17 |
| 22,900 | 1.94 | 0.90 | 11 | 22,900 | 1.87 | 0.92 | 10 |
| 21,315 | 1.85 | 0.96 | 12 | 30,888 | 2.09 | 0.86 | 15 |
| 26,711 | 2.09 | 0.80 | 14 | 23,720 | 1.99 | 0.95 | 14 |
| 20,782 | 1.88 | 0.99 | 16 | 20,840 | 1.85 | 0.98 | 10 |
| 23,900 | 1.90 | 0.92 | 11 | 21,200 | 1.85 | 0.87 | 13 |
| 25,667 | 2.01 | 0.96 | 10 | 25,800 | 2.01 | 0.90 | 12 |
| 24,316 | 1.95 | 0.92 | 15 | 20,677 | 1.83 | 0.84 | 9 |
| 25,319 | 2.00 | 0.91 | 12 | 27,879 | 2.07 | 0.97 | 12 |
| 28,500 | 2.07 | 0.84 | 15 | 30,101 | 2.02 | 0.79 | 10 |
| 29,000 | 2.09 | 0.87 | 16 | 22,412 | 1.83 | 0.91 | 12 |
| 20,249 | 1.82 | 1.01 | 9 | 21,070 | 1.88 | 0.97 | 11 |
| 26,000 | 2.05 | 0.87 | 13 | 28,387 | 2.10 | 0.81 | 15 |

**a)** Plot the independent variables against the dependent variable of imports. Is there reason to perform a regression analysis?
**b)** Perform a correlation analysis. Should any of the variables be removed?
**c)** Using Minitab, perform a regression analysis.
**d)** Use a 99% confidence interval to test the significance of the coefficients.
**e)** How many values fall outside the 95% interval of predicted values? Is this acceptable?
**f)** Is autocorrelation at an acceptable level?
**g)** What is the calculated coefficient of determination? Will this analysis help the local manufacturer understand his market?

**10-13** As the newly appointed city planning commissioner, you wish to determine the factors that affect home purchases within your state. Research indicates that some important factors are interest rates, the number of newlyweds, and the price of homes. Randomly sampled data appears in the following chart:

| Homes Sold | Interest Rate | Newlyweds | Price | Homes Sold | Interest Rate | Newlyweds | Price |
|---|---|---|---|---|---|---|---|
| 10,705 | 14.2% | 8,300 | $71,920 | 11,200 | 13.9% | 8,404 | $72,412 |
| 12,417 | 13.6 | 8,462 | 71,319 | 11,683 | 13.8 | 7,960 | 72,360 |
| 13,626 | 12.7 | 8,423 | 69,120 | 13,457 | 12.0 | 8,900 | 69,590 |
| 11,479 | 14.0 | 8,210 | 72,789 | 13,790 | 13.1 | 8,679 | 69,321 |
| 10,645 | 13.7 | 7,890 | 72,900 | 12,862 | 13.3 | 8,216 | 71,299 |
| 12,500 | 12.1 | 8,810 | 71,300 | 10,286 | 13.8 | 7,916 | 72,840 |
| 13,798 | 11.6 | 9,400 | 75,200 | 10,100 | 14.6 | 7,885 | 69,510 |
| 13,320 | 12.5 | 8,980 | 69,400 | 11,587 | 14.4 | 8,105 | 69,980 |
| 12,170 | 14.2 | 8,549 | 71,479 | 12,222 | 13.3 | 8,445 | 71,333 |
| 11,627 | 12.7 | 8,767 | 72,684 | 13,754 | 12.0 | 8,100 | 69,900 |
| 11,550 | 13.0 | 8,140 | 72,820 | 11,987 | 12.9 | 8,400 | 69,700 |
| 10,314 | 13.9 | 7,910 | 72,316 | 12,350 | 12.8 | 8,419 | 71,300 |

*(continued on next page)*

## CHAPTER 10 Correlation and Regression Analysis

*(Continued)*

| Homes Sold | Interest Rate | Newlyweds | Price | Homes Sold | Interest Rate | Newlyweds | Price |
|---|---|---|---|---|---|---|---|
| 13,200 | 13.0% | 8,897 | $69,449 | 13,153 | 13.0% | 8,979 | $70,230 |
| 12,989 | 14.2 | 8,890 | 72,210 | 13,620 | 12.2 | 8,412 | 69,451 |
| 10,680 | 13.0 | 7,870 | 72,980 | 11,910 | 13.8 | 8,157 | 70,333 |
| 10,901 | 14.6 | 7,989 | 72,760 | 12,780 | 13.8 | 8,200 | 71,400 |
| 11,201 | 13.5 | 8,500 | 70,310 | 12,000 | 13.2 | 8,267 | 70,987 |
| 12,800 | 13.7 | 8,525 | 71,455 | 13,242 | 13.7 | 8,904 | 69,430 |

**a)** Perform a correlation analysis. If the correlation coefficient for two independent variables is unacceptable, eliminate one of those variables from the analysis.
**b)** Using Minitab, perform a regression analysis.
**c)** Use an alpha of .10 to test the significance of the coefficients.
**d)** Is autocorrelation at an acceptable level?
**e)** What percent of the variation in the dependent variable is explained by the regression equation?
**f)** A row of data was singled out with an **X**. What does the **X** indicate and what should you do?

**10-14** It's strawberry season again, but for some reason there don't seem to be as many strawberries in the markets this year. Wondering what went wrong, you gather data from several regions on the amount of strawberries available (in pounds), the number of hours of sunshine, the number of inches of rain, the number of growers, and the price. Randomly sampled data is as follows:

| Strawberries | Sunshine | Rain | Growers | Price | Strawberries | Sunshine | Rain | Growers | Price |
|---|---|---|---|---|---|---|---|---|---|
| 66,000 | 690 | 3 | 184 | $0.66 | 66,500 | 699 | 4 | 184 | $0.68 |
| 72,400 | 720 | 7 | 199 | 0.75 | 72,303 | 710 | 4 | 206 | 0.75 |
| 65,312 | 687 | 7 | 180 | 0.64 | 68,225 | 700 | 4 | 201 | 0.72 |
| 67,900 | 707 | 6 | 186 | 0.65 | 71,040 | 711 | 6 | 206 | 0.74 |
| 70,240 | 710 | 7 | 200 | 0.70 | 70,000 | 709 | 5 | 202 | 0.72 |
| 66,990 | 692 | 3 | 190 | 0.69 | 72,960 | 721 | 6 | 200 | 0.70 |
| 71,300 | 712 | 5 | 202 | 0.72 | 69,344 | 702 | 8 | 204 | 0.71 |
| 72,209 | 719 | 8 | 212 | 0.76 | 70,100 | 717 | 3 | 207 | 0.74 |
| 70,000 | 697 | 5 | 216 | 0.72 | 66,240 | 714 | 6 | 189 | 0.67 |
| 65,790 | 689 | 3 | 197 | 0.67 | 68,937 | 701 | 4 | 190 | 0.69 |
| 68,241 | 703 | 7 | 196 | 0.70 | 70,200 | 709 | 8 | 201 | 0.71 |
| 68,790 | 710 | 5 | 192 | 0.71 | 72,309 | 720 | 7 | 212 | 0.76 |
| 72,300 | 708 | 7 | 195 | 0.70 | 67,304 | 705 | 7 | 200 | 0.68 |
| 72,202 | 724 | 7 | 204 | 0.74 | 70,190 | 711 | 4 | 208 | 0.72 |
| 68,970 | 711 | 4 | 192 | 0.69 | 71,300 | 693 | 6 | 201 | 0.73 |
| 67,425 | 684 | 7 | 201 | 0.71 | 69,240 | 704 | 5 | 210 | 0.74 |
| 71,312 | 710 | 4 | 201 | 0.71 | 66,783 | 694 | 3 | 187 | 0.66 |
| 70,600 | 720 | 5 | 202 | 0.69 | 67,850 | 696 | 4 | 180 | 0.68 |
| 71,922 | 701 | 6 | 199 | 0.70 | 72,987 | 720 | 5 | 210 | 0.75 |
| 67,212 | 700 | 4 | 181 | 0.68 | 70,309 | 707 | 6 | 200 | 0.73 |
| 68,490 | 700 | 4 | 189 | 0.69 | 69,000 | 704 | 4 | 189 | 0.70 |

# PROBLEMS

a) Perform a correlation analysis on the variables.
b) Is there evidence of multicollinearity? If so, eliminate a variable.
c) Perform a regression analysis.
d) Test the significance of the coefficients using an alpha of .05.
e) Check the accuracy of the data singled out by Minitab. Are there any errors? Is it acceptable for these values to fall outside of the 95% range of predicted values?
f) Is there evidence of autocorrelation?
g) Can this equation be of use in predicting sales? Explain.

**10-15** In your state there are two rival discount grocery chains. The owner of one of these chains believes that the key to success is to derive a formula that predicts annual sales. He believes the key variables to be per capita income, annual advertising expenditures, and average waiting time (in minutes) in the checkout lane. Following is randomly sampled data:

| Sales | Income | Advertising | Waiting Time | Sales | Income | Advertising | Waiting Time |
|---|---|---|---|---|---|---|---|
| $822,340 | $3,787 | $32,000 | 4.6 | $823,119 | $3,780 | $41,000 | 4.3 |
| 864,960 | 3,731 | 76,000 | 3.7 | 829,000 | 3,739 | 36,000 | 4.7 |
| 848,420 | 3,702 | 57,000 | 4.0 | 851,472 | 3,679 | 60,000 | 4.3 |
| 832,340 | 3,752 | 39,000 | 4.4 | 836,012 | 3,762 | 47,000 | 4.6 |
| 839,670 | 3,612 | 54,000 | 4.2 | 825,312 | 3,773 | 40,000 | 4.2 |
| 840,010 | 3,730 | 56,000 | 4.4 | 856,019 | 3,663 | 71,000 | 4.1 |
| 821,306 | 3,789 | 39,000 | 4.1 | 820,000 | 3,801 | 34,000 | 5.0 |
| 833,725 | 3,750 | 38,000 | 4.0 | 853,700 | 3,677 | 75,000 | 4.4 |
| 856,910 | 3,739 | 58,000 | 4.1 | 860,800 | 3,400 | 75,000 | 3.7 |
| 857,460 | 3,660 | 74,000 | 4.6 | 859,320 | 3,682 | 67,000 | 3.9 |
| 830,000 | 3,758 | 37,000 | 4.5 | 827,916 | 3,770 | 42,000 | 4.6 |
| 824,306 | 3,781 | 36,000 | 4.2 | 821,600 | 3,687 | 52,000 | 4.4 |
| 839,024 | 3,665 | 50,000 | 4.5 | 844,100 | 3,721 | 45,000 | 4.7 |
| 853,700 | 3,670 | 68,000 | 4.2 | 850,612 | 3,795 | 62,000 | 4.2 |
| 849,090 | 3,690 | 74,000 | 4.2 | 836,020 | 3,739 | 52,000 | 4.5 |
| 824,036 | 3,778 | 39,000 | 4.6 | 832,400 | 3,750 | 37,000 | 4.7 |
| 827,033 | 3,655 | 44,000 | 4.8 | 829,000 | 3,754 | 62,000 | 3.9 |
| 845,060 | 3,718 | 54,000 | 3.7 | 847,970 | 3,712 | 55,000 | 4.1 |
| 859,510 | 3,764 | 64,000 | 3.6 | 852,004 | 3,692 | 42,000 | 4.2 |
| 862,346 | 3,611 | 78,000 | 3.9 | 823,700 | 3,780 | 37,000 | 4.8 |
| 839,108 | 3,730 | 50,000 | 3.8 | 826,940 | 3,771 | 35,000 | 4.8 |
| 824,612 | 3,662 | 39,000 | 4.7 | 825,820 | 3,769 | 51,000 | 4.2 |
| 820,000 | 3,798 | 30,000 | 4.0 | 852,705 | 3,748 | 60,000 | 4.3 |
| 822,019 | 3,785 | 45,000 | 4.6 | 833,000 | 3,682 | 39,000 | 4.6 |
| 857,349 | 3,777 | 74,000 | 4.2 | 839,904 | 3,730 | 39,000 | 4.5 |
| 843,212 | 3,725 | 47,000 | 4.4 | 831,516 | 3,763 | 38,000 | 4.9 |
| 860,019 | 3,649 | 75,000 | 3.7 | 847,827 | 3,719 | 57,000 | 3.9 |
| 822,341 | 3,784 | 32,000 | 4.1 | 851,009 | 3,702 | 60,000 | 4.3 |
| 847,060 | 3,708 | 55,000 | 3.9 | 858,600 | 3,657 | 50,000 | 4.0 |
| 840,347 | 3,729 | 55,000 | 4.2 | 841,000 | 3,728 | 50,000 | 4.4 |
| 849,118 | 3,690 | 70,000 | 4.3 | 838,790 | 3,733 | 47,000 | 4.2 |
| 860,201 | 3,661 | 72,000 | 3.9 | 823,040 | 3,775 | 39,000 | 4.9 |

*(continued on next page)*

*(Continued)*

| Sales | Income | Advertising | Waiting Time | Sales | Income | Advertising | Waiting Time |
|---|---|---|---|---|---|---|---|
| $856,319 | $3,674 | $73,000 | 4.0 | $860,004 | $3,622 | $57,000 | 3.8 |
| 867,340 | 3,604 | 77,000 | 3.2 | 855,000 | 3,700 | 64,000 | 3.7 |
| 824,800 | 3,777 | 47,000 | 3.9 | 849,312 | 3,672 | 74,000 | 4.3 |
| 821,620 | 3,797 | 35,000 | 4.7 | 822,009 | 3,652 | 45,000 | 4.9 |
| 837,024 | 3,741 | 50,000 | 4.8 | 820,600 | 3,750 | 33,000 | 4.0 |
| 839,568 | 3,730 | 79,000 | 4.4 | 859,845 | 3,786 | 73,000 | 3.9 |
| 860,011 | 3,634 | 81,000 | 3.0 | 860,212 | 3,644 | 76,000 | 4.0 |
| 834,300 | 3,746 | 36,000 | 4.8 | 847,000 | 3,709 | 55,000 | 3.5 |
| 825,018 | 3,770 | 33,000 | 4.7 | 841,370 | 3,750 | 69,000 | 4.3 |
| 866,012 | 3,605 | 52,000 | 3.6 | 832,400 | 3,735 | 39,000 | 4.6 |
| 820,480 | 3,796 | 34,000 | 4.8 | 855,930 | 3,668 | 50,000 | 4.1 |
| 834,790 | 3,680 | 40,000 | 4.5 | 867,200 | 3,601 | 80,000 | 3.0 |
| 849,000 | 3,756 | 57,000 | 4.4 | 825,340 | 3,775 | 34,000 | 4.0 |
| 831,330 | 3,701 | 56,000 | 4.2 | 820,319 | 3,790 | 32,000 | 4.1 |
| 857,218 | 3,661 | 72,000 | 3.9 | 845,212 | 3,610 | 46,000 | 4.9 |
| 848,790 | 3,704 | 39,000 | 4.4 | 850,090 | 3,687 | 59,000 | 4.3 |
| 852,344 | 3,748 | 64,000 | 4.0 | 833,280 | 3,755 | 42,000 | 4.9 |
| 826,930 | 3,768 | 35,000 | 4.6 | 846,570 | 3,710 | 58,000 | 4.1 |

**a)** Plot the independent variables against the dependent variable. Examine the scatter diagrams. Is a regression analysis justified?

**b)** Perform a correlation analysis on the variables. If independent variables are highly correlated, eliminate the appropriate value from your analysis.

**c)** Using Minitab, perform a regression analysis.

**d)** Using a 90% confidence level, test the significance of the coefficients.

**e)** Minitab suspects errors in seven input rows. Check the accuracy of these values. Is it acceptable to have these values fall outside of the 95% range of predicted values? If not, what can be done?

**f)** Is there evidence of autocorrelation?

**g)** How much of the variation in the dependent variable is explained by the regression equation?

# 11

# FORECASTING

*The following Minitab commands will be introduced in this chapter:*

`TSPLot [with K periods,] data in C`
A command to create a time series plot

`LAG by K, data in C, put in C`
A command to lag the values in a given column by a specified number of periods

`UNSTack C into C,...,C;`
  `SUBScripts in C.`
A command to separate values in a given column to create multiple columns based upon a set of subscripts (codes) found in an associated column

`DIFFerence [of lag K] for data in C, put in C`
A command to calculate the difference between the value for each term and a previous value (lag K)

`ACF for series in C`
A command to calculate the autocorrelation function

`PACF for series in C`
A command to calculate the partial autocorrelation function

`ARIMa p=K, d=K, q=K [seasonal values sp=K, sd=K, sq=K, s=K];`
  `[[FORECast start after period K] forecast length K periods].`
A command to calculate the Box–Jenkins forecasting model

In Chapter 10, we introduced one of the most powerful—and most frequently misused—tools in statistical analysis, regression analysis. Regression analysis, and more specifically multiple regression analysis, can be used to forecast future values of a dependent variable by substituting projected values for each independent variable into the regression equation.

However, a slightly different approach, time series, can be used to project a variable into the future. Occasionally time series analysis provides a better forecast than regression analysis, especially when the data fluctuates systematically over time. For example, if sales fluctuated seasonally (e.g., high in spring and low in winter), time series analysis would be effective in predicting future sales.

## 11-1 TIME SERIES ANALYSIS

Time series analysis can be computed in several ways. However, among the most popular methods today are classical decomposition and Box–Jenkins. We will discuss both methods, but you should know that forecasting is a very sophisticated topic. This chapter will only provide you with some preliminary insights and computational skills. The full complexity of this topic is beyond the scope of an introductory text. We suggest you follow our discussion closely and with pad, pencil, and computer terminal handy.

Regression analysis attempts to combine several variables in a common regression equation, smoothing out fluctuations among the independent variables. However, there are times when the fluctuations of a particular variable follow a predictable pattern. For example, sales of Go-Fast running shoes could be related to the season of the year, being higher in the spring and lower in the winter. If we look at sales data *over time*, we might be able to forecast sales with some accuracy.

Notice that this approach does not take into account such variables as population density, per capita income, or traffic patterns. Those types of variables go into a multiple regression analysis. In time series analysis, the key is to break down the dependent variable, in this case sales, by time intervals.

The time interval is important. Some sales follow a weekly pattern, others a monthly, others a quarterly, and others an annual. If you select an inappropriate time unit, your time series analysis will be ineffective. Typically, the proper time unit is determined by managerial experience.

To make a forecast using time series analysis, we will isolate four elements in our sales data:

(1) a *trend* component, that is, growth or decline over time;
(2) a *cyclical* component, that is, a predictable pattern related to some long-term phenomenon;
(3) a *seasonal* component, that is, a predictable pattern related to some short-term phenomenon (e.g., seasons, months, or days); and
(4) *random fluctuation*, that is, an unpredictable portion in the forecast variable.

In general, we will build the relationship

$$V = T \times C \times S \times R$$

where $V$ is the forecasted value, and $T$, $C$, $S$, and $R$ are the trend, cyclical, seasonal, and random components, respectively. (There are other forms of this general relationship, nota-

**SECTION 11-2** Classical Decomposition  ▶ **273**

bly an additive model where $V = T + C + S + R$, but such forms are beyond the scope of this book.)

Time series analysis can be computed in several different ways. In this brief introduction, we will simply use the raw computational power of Minitab through a series of **LET** commands and other operations, including linear regression. In our first approach we will remove various components—thus the term *decomposition*. We will then look briefly at the approach known as Box–Jenkins.

## 11-2 CLASSICAL DECOMPOSITION

To begin an analysis using classical decomposition, let's look at the following sales history for a competitor's most popular tennis shoe, known as tennis shoe A, over the past 15 years. The sales data is found in Table 11-1.

**TABLE 11-1** Sales of Tennis Shoe A by Quarter, 1972–1986

| Year | Winter   | Spring   | Summer   | Autumn   |
|------|----------|----------|----------|----------|
| 1972 | 2,331.24 | 2,723.61 | 1,704.42 | 1,535.06 |
| 1973 | 2,696.44 | 2,816.47 | 1,815.17 | 1,485.15 |
| 1974 | 2,523.83 | 2,820.05 | 1,720.55 | 1,568.81 |
| 1975 | 2,514.53 | 2,840.89 | 1,784.83 | 1,622.81 |
| 1976 | 2,723.29 | 3,189.37 | 1,879.81 | 1,777.64 |
| 1977 | 2,980.86 | 3,294.40 | 2,088.08 | 1,690.05 |
| 1978 | 2,959.52 | 3,286.46 | 1,903.68 | 1,768.43 |
| 1979 | 2,927.78 | 3,341.63 | 2,094.28 | 1,756.98 |
| 1980 | 3,332.26 | 3,439.41 | 2,217.08 | 1,913.71 |
| 1981 | 3,470.13 | 3,887.83 | 2,373.91 | 2,058.13 |
| 1982 | 3,243.26 | 3,659.29 | 2,317.32 | 1,969.15 |
| 1983 | 3,247.12 | 3,548.10 | 2,292.67 | 2,082.63 |
| 1984 | 3,613.11 | 3,928.44 | 2,401.24 | 2,252.05 |
| 1985 | 3,851.08 | 3,970.40 | 2,512.48 | 2,312.82 |
| 1986 | 3,625.30 | 3,929.24 | 2,389.11 | 2,193.64 |

Whenever you try to understand complex relationships within data, it is always a good idea to examine the data visually if possible. We can use a Minitab command that simply plots one value after another. The command is called the time series plot (**TSPLot**). Assume that we have entered the data in Table 11-1 into column 1 starting with winter of 1972, then spring 1972, and so on through autumn 1986. We can then use the following command:

```
MTB >TSPLot data in C1
```

The results of this command are shown in Figure 11-1.

A quick visual inspection reveals that the data for sales contains some classic time series relationships. For example, notice that in 1972 periods 1 and 2 are high and periods 3 and 4 are low; this pattern is repeated in 1973 for periods 5 through 8. Clearly sales have a seasonal component. Also, there is a general upward trend in the sales over the entire 15-year period. These characteristics make this data a good candidate for time series analysis.

**FIGURE 11-1** Minitab Time Series Plot of Tennis Shoe A Sales by Quarter, 1972–1986

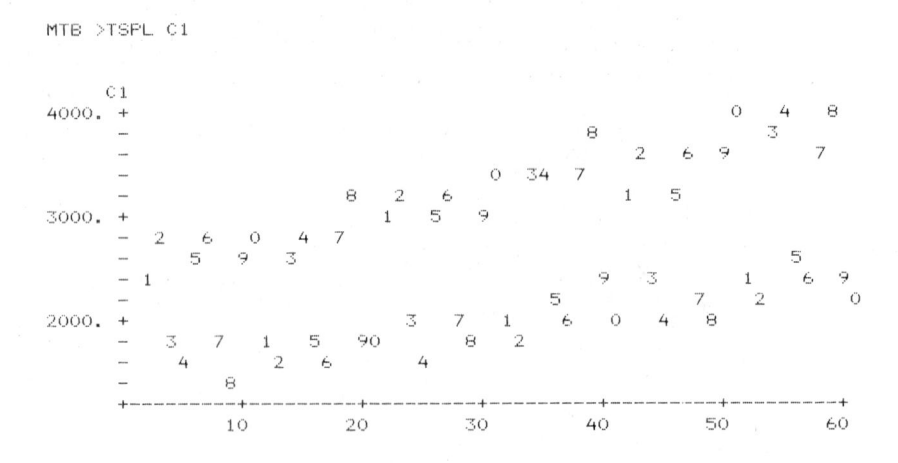

```
MTB >TSPL C1

         C1
4000.  +                                                      0   4   8
       -                                          8                 3
       -                                            2     6 9       7
       -                              0  34  7
       -                 8    2   6              1    5
3000.  +                   1    5     9
       -    2   6  0  4  7
       -       5   9  3                                             5
       -   1                                   9   3      1     6 9
       -                                   5              7  2       0
2000.  +                     3    7  1      6    0   4  8
       -    3  7  1   5   90        8    2
       -     4         2   6         4
       -
       -  8
       +---------+---------+---------+---------+---------+---------+
              10        20        30        40        50        60
```

The seasonal component can be seen even better by slightly modifying the **TSPLot** command. In the default command, we do not specify the number of periods to use in the plotting cycle; all 10 digits from 1 through 0 are used. However, because this data seems to cycle every four periods, we can replot the data with the following version of the **TSPLot** command:

```
MTB >TSPLot with 4 periods, data in C1
```

As you can see in Figure 11-2, we now have a very nice cycle of four periods, which is easily identified. There is clearly nothing new in the data—Figure 11-2 simply relabels the data points in Figure 11-1.

**FIGURE 11-2** Minitab Time Series Plot with a Four-Period Cycle of Tennis Shoe A Sales by Quarter, 1972–1986

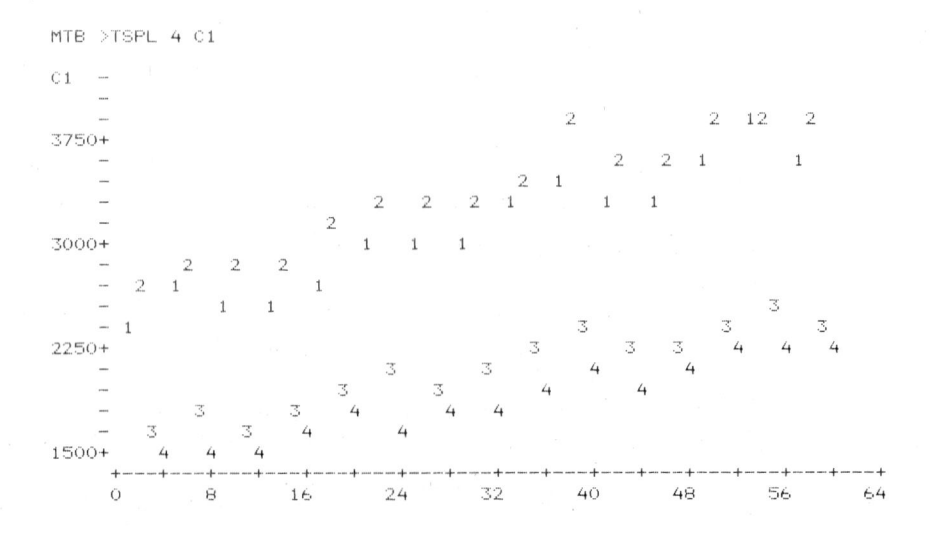

```
MTB >TSPL 4 C1

C1   -
     -
     -                                      2           2  12    2
3750+
     -                                         2    2   1        1
     -                                    2  1
     -                       2    2   2 1        1  1
     -               2
3000+                   1    1    1
     -        2   2   2
     -    2  1              1
     -                1    1                                     3
     -   1                              3             3         3
2250+                                3       3    3    4   4    4
     -                           3       3        4    4
     -           3      3    4        4   4
     -       3      3    4        4
     -  3    4    4    4
1500+
     +---+---+---+---+---+---+---+---+---+---+---+---+---+---+---+
         0       8      16      24      32      40      48      56      64
```

## SECTION 11-2 Classical Decomposition

# Identifying the Trend Component

The first element that we will identify is the straight-line trend component. To do this, we will build a simple straight-line regression equation of the form

$$Y = a + bX + e$$

where $a$ is the $Y$-intercept, $b$ is the slope of the line, and $e$ represents unidentified random components. We can determine this equation quickly by using the **REGRess** command introduced in Chapter 10. However, before we can use the regression command, we must have a second variable to regress sales against. In this case, we want to regress sales against time as represented by the 60 quarters of time in the past 15 years. To do this, we need to enter the numbers 1 to 60 in column 2. This can be done quickly with the **SET** command as follows:

```
MTB >SET C2
MTB >1:60
```

Now, with our data in **C1** and our time intervals in **C2**, we can run our simple regression as follows:

```
MTB >REGRess y-value in C1 on 1 predictor in C2
```

This command will produce a complete regression output. However, because we are interested in only part of the output, Figure 11-3 shows only the sections of interest.

The regression equation is

```
C1 = 2078 + 17.2 C2
```

An examination of $r^2$ shows that about 15% of the variation can be predicted by the trend component. In the next step of our analysis, we remove the unidentified trend component from our forecast variable. To do this we put the trend component in **C3** by multiplying each of our time intervals (**C2**) by the trend equation (regression line). The command is

```
MTB >LET C3 = 2077.9 + (17.224) * C2
```

**FIGURE 11-3** Minitab Time Series Linear Regression Analysis of Tennis Shoe A Sales by Quarter, 1972–1986

```
MTB >REGR C1 1 C2

The regression equation is
C1 = 2078 + 17.2 C2

Predictor      Coef       Stdev     t-ratio
Constant      2077.9      177.3      11.72
C2             17.224      5.056      3.41

s = 678.2      R-sq = 16.7%     R-sq(adj) = 15.2%

Analysis of Variance

SOURCE        DF        SS            MS
Regression     1      5338553       5338553
Error         58     26675910        459929
Total         59     32014464
```

Note that this command uses the coefficients from the second part of the printout. This prevents rounding error in our subsequent calculations.

Column 3 now represents the trend component for each of the 60 observations. Now we can *remove* the trend component from our original forecast variable by dividing each sales value. With the trend value removed, we have a new variable that contains the remaining seasonal, cyclical, and random variation components. We will place this new variable in column 4 with the command

```
MTB >LET C4 = C1/C3
```

To review, we started with

$$C1 = T * C * S * R$$

We then identified the trend component (T) and put the values for that trend component in C3. After removing C3 (trend) from our original relationship, we had

$$C4 = C * S * R$$

## Identifying the Cyclical Component ▼▲▼▲▼▲▼▲▼▲▼▲▼▲▼▲▼▲▼▲

A `TSPLot` of `C4` will provide us with some useful information. Such a plot is shown in Figure 11-4.

Take a minute to compare Figures 11-2 and 11-4. Notice that in Figure 11-4, we still have the zigzag pattern of sales, with values 1 and 2 high and values 3 and 4 low. However, the general upward trend of the data in Figure 11-2 has disappeared in Figure 11-4. We have removed the trend! Now you can see why the term *decomposition* is used in this type of time series analysis. (The trend component is presently isolated in column 3. You might want to plot column 3 and examine the results. What you will find is a straight line that moves upward with a slope of 17.224 units per time period.)

**FIGURE 11-4** Minitab Time Series Plot of Tennis Shoe A Sales by Quarter, 1972–1986, with Trend Component Removed

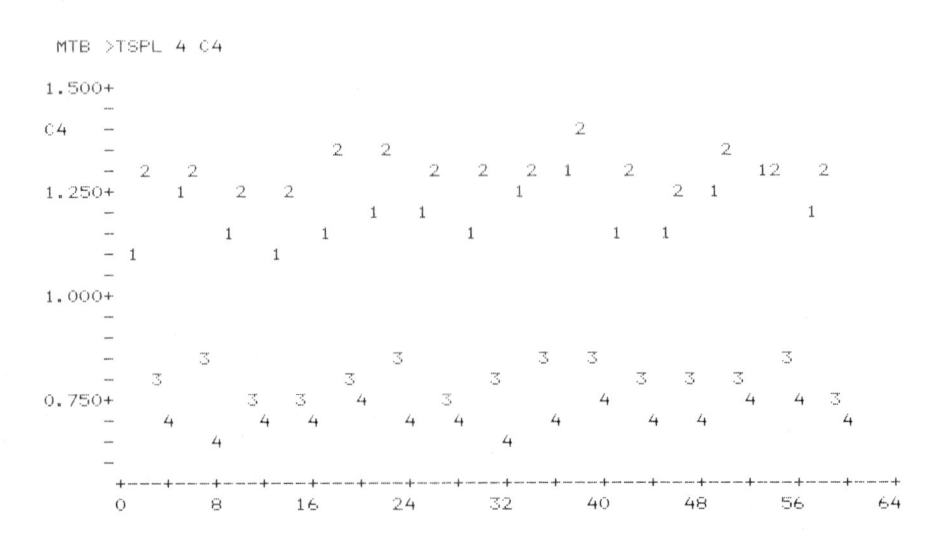

```
MTB >TSPL 4 C4

1.500+
C4    -
      -                                              2
      -                    2   2                               2
      -  2   2                       2   2   1   2            12   2
1.250+     1   2   2                         1       2   1         1
      -                       1   1             1   1
      -        1       1               1
      - 1               1
      -
1.000+
      -
      -        3                    3               3    3              3
      -  3               3   3               3               3   3
0.750+              3   3       4        3               4           4   3
      -        4           4   4            4   4               4       4
      -                4               4
      -
      +---+---+---+---+---+---+---+---+---+---+---+---+---+---+---+---+
      0       8      16      24      32      40      48      56      64
```

### SECTION 11-2 Classical Decomposition

Now that we have removed the straight-line growth over time (note that the trend in other data sets could show a decline rather than growth) we want to remove any nonlinear effects. The first of the remaining nonlinear effects that we wish to remove is the business cycle. *Business cycle* effects usually repeat over years instead of weeks or months. We can isolate the cyclical component in this example by computing an annual, or four-quarter, *moving average*. It is important to include one full set of seasons (for convenience, a year) in each of our calculations. In this way we can identify the cyclical component without removing or distorting the seasonal values.

Since the remaining components of our forecast variable are in `C4`, we will compute a four-quarter moving average of `C4`. This is done by obtaining the mean for quarters 1 through 4, then the mean for quarters 2 through 5, then the mean for quarters 3 through 6, and so on through all the data.

In our problem, the first two such calculations are as follows:

| Moving Average for Periods 1–4 | Moving Average for Periods 2–5 |
| --- | --- |
| 2,331.24 | 2,723.61 |
| 2,723.61 | 1,704.42 |
| 1,704.42 | 1,535.06 |
| +1,535.06 | +2,696.44 |
| 8,294.33 ÷ 4 = 2,073.58 | 8,659.53 ÷ 4 = 2,164.88 |

We are considering a full year at a time to avoid including any seasonal influence. However, notice in our example what the value of 2,073.58 actually represents. Since this is the average of sales for the first year, it represents the sales value midway through the first year, or at a point halfway between period 2 and period 3. As we progress in this calculation, we will need to move this calculated value so that it matches up exactly with one of our time periods (year end rather than mid-year).

If we are just a bit clever, we can have Minitab calculate the moving average for us. First, we need to make a copy of `C4` in a computational area. Let's move `C4` into `C11` to create some room for our computations. The command is

```
MTB >COPY C4 into C11
```

Now we need to lag all the values in `C11` by one period. In other words, we want to create a `C12` in which there is no value in row 1 and all the remaining values in `C11` are found in the following rows. Thus each value in `C11` will be one row sooner in time in `C12`. Thus, the value in the 8th row in `C11` is the value for the 8th period, but the value in the 8th row in `C12` is the value for the 7th period from `C11`; similarly, the value in the 33rd row of `C11` is the value for the 33rd period, but that in `C12` is for the 32nd period; and so on.

To accomplish this process we use the **LAG** command. The form of the **LAG** command is

```
LAG by K, data in C, put in C
```

With this command we can lag the data in `C11` by one period and put the results in `C12` by substituting a `1` for the `K`-value. Since we have a four-season cyclical component, we will also lag our values by two periods and place them in `C13` and by three periods and place them in `C14`.

The commands are

```
MTB >COPY C4 to C11
MTB >LAG by 1, data in C4, put in C12
MTB >LAG by 2, data in C4, put in C13
MTB >LAG by 3, data in C4, put in C14
```

A print command shows the resulting entries (Figure 11-5; `C15` will be described below).

Our data is now established with zero lag in `C11`, one-period lag in `C12`, two-period lag in `C13`, and three-period lag in `C14`. With this done, notice what we have in *row* 4. In `C11` we have our fourth value, in `C12` we have our third, in `C13` our second, and in `C14` our first. These four values can then be added together and divided by 4 to obtain our first four-quarter average. Notice that `C12` loses the original 60th value in the data set, `C13` loses the last two values, and `C14` the last three. Since we will not be using these values, this loss is of no consequence.

After the lags have been created, we can compute the moving averages. Remember, the data we want to average is the row values: We want to add up the values in each row from rows 4 through 60 of `C11`, `C12`, `C13`, and `C14` and find the mean. Since the first few rows have an unequal number of observations, we cannot use the row mean (`RMEAn`) command described in Chapter 6, which requires columns of equal length. We will use a simple **LET** command instead as follows:

```
MTB >LET C15 = (C11 + C12 + C13 + C14)/4
```

Figure 11-5 shows `C15`. Remember that we have already removed the trend component from the forecast variable and are now working with the remaining components.

Looking at Figure 11-5, we see that `C12` through `C14` contain our lags (indicated by the asterisks at the top of the columns). Column `C15` starts with the fourth row and continues to calculate averages through period 60. Since we pushed down the values from `C11` as we entered them into `C12` through `C14`, rows with asterisks have been eliminated in our present calculations.

Next we have to take care of the midpoint problem mentioned earlier. The first value in `C15`, `0.97937`, is the moving average value of the remaining components for the first four quarters. However, this value represents a time halfway between period 2 and period 3. We now need to move this average so that it is centered over a single quarter. We will do this by averaging the first and second values (`0.97937` and `1.01270`) in `C15`. Since the first value is centered between quarters 2 and 3 and the second value is centered between quarters 3 and 4, averaging them together will produce a value corresponding directly to the third quarter.

The logic for what we now need to do is exactly the same as that behind the creation of `C12`, `C13`, and `C14`. We take the values in `C15`, lag them by one period, and put the results in `C16`. We then find the row mean of `C15` and `C16`. The commands to do this, which will produce moving averages that are properly centered over the periods, are as follows:

```
MTB >LAG by 1, data in C15, put in C16
MTB >LET C17 = (C15 + C16)/2
```

This creates the data set shown in Figure 11-6. As can be seen at the top of `C15` in Figure 11-6 there are three asterisks; at the top of `C17` there are four.

## SECTION 11-2 Classical Decomposition ▸ 279

**FIGURE 11-5** Development of Lagged Quarters for Calculation of Moving Average of Tennis Shoe A Sales, 1972–1986

```
MTB >PRIN C11-C15
 ROW       C11        C12        C13        C14        C15

   1    1.11270         *          *          *          *
   2    1.28938     1.11270        *          *          *
   3    0.80036     1.28938     1.11270        *          *
   4    0.71505     0.80036     1.28938     1.11270     0.97937
   5    1.24603     0.71505     0.80036     1.28938     1.01270
   6    1.29122     1.24603     0.71505     0.80036     1.01316
   7    0.82565     1.29122     1.24603     0.71505     1.01949
   8    0.67029     0.82565     1.29122     1.24603     1.00830
   9    1.13028     0.67029     0.82565     1.29122     0.97936
  10    1.25328     1.13028     0.67029     0.82565     0.96988
  11    0.75883     1.25328     1.13028     0.67029     0.95317
  12    0.68669     0.75883     1.25328     1.13028     0.95727
  13    1.09241     0.68669     0.75883     1.25328     0.94780
  14    1.22503     1.09241     0.68669     0.75883     0.94074
  15    0.76397     1.22503     1.09241     0.68669     0.94203
  16    0.68954     0.76397     1.22503     1.09241     0.94274
  17    1.14872     0.68954     0.76397     1.22503     0.95681
  18    1.33562     1.14872     0.68954     0.76397     0.98446
  19    0.78158     1.33562     1.14872     0.68954     0.98886
  20    0.73384     0.78158     1.33562     1.14872     0.99994
  21    1.22186     0.73384     0.78158     1.33562     1.01822
  22    1.34092     1.22186     0.73384     0.78158     1.01955
  23    0.84399     1.34092     1.22186     0.73384     1.03515
  24    0.67839     0.84399     1.34092     1.22186     1.02129
  25    1.17980     0.67839     0.84399     1.34092     1.01077
  26    1.30120     1.17980     0.67839     0.84399     1.00084
  27    0.74861     1.30120     1.17980     0.67839     0.97700
  28    0.69075     0.74861     1.30120     1.17980     0.98009
  29    1.13594     0.69075     0.74861     1.30120     0.96912
  30    1.28791     1.13594     0.69075     0.74861     0.96580
  31    0.80184     1.28791     1.13594     0.69075     0.97911
  32    0.66829     0.80184     1.28791     1.13594     0.97350
  33    1.25922     0.66829     0.80184     1.28791     1.00431
  34    1.29130     1.25922     0.66829     0.80184     1.00516
  35    0.82704     1.29130     1.25922     0.66829     1.01146
  36    0.70932     0.82704     1.29130     1.25922     1.02172
  37    1.27804     0.70932     0.82704     1.29130     1.02643
  38    1.42286     1.27804     0.70932     0.82704     1.05931
  39    0.86335     1.42286     1.27804     0.70932     1.06839
  40    0.74385     0.86335     1.42286     1.27804     1.07703
  41    1.16493     0.74385     0.86335     1.42286     1.04875
  42    1.30628     1.16493     0.74385     0.86335     1.01960
  43    0.82217     1.30628     1.16493     0.74385     1.00931
  44    0.69440     0.82217     1.30628     1.16493     0.99695
  45    1.13815     0.69440     0.82217     1.30628     0.99025
  46    1.23618     1.13815     0.69440     0.82217     0.97273
  47    0.79402     1.23618     1.13815     0.69440     0.96569
  48    0.71700     0.79402     1.23618     1.13815     0.97134
  49    1.23657     0.71700     0.79402     1.23618     0.99594
  50    1.33661     1.23657     0.71700     0.79402     1.02105
  51    0.81224     1.33661     1.23657     0.71700     1.02561
  52    0.75736     0.81224     1.33661     1.23657     1.03570
  53    1.28765     0.75736     0.81224     1.33661     1.04847
  54    1.31995     1.28765     0.75736     0.81224     1.04430
  55    0.83051     1.31995     1.28765     0.75736     1.04887
  56    0.76018     0.83051     1.31995     1.28765     1.04957
  57    1.18487     0.76018     0.83051     1.31995     1.02388
  58    1.27702     1.18487     0.76018     0.83051     1.01314
  59    0.77215     1.27702     1.18487     0.76018     0.99855
  60    0.70505     0.77215     1.27702     1.18487     0.98477
```

**FIGURE 11-6** Development of Lagged Relationships and Moving Averages of Tennis Shoe A Sales by Quarter, 1972–1986

```
MTB >PRIN C11-C17
 ROW      C11        C12        C13        C14        C15        C16        C17

   1   1.11270          *          *          *          *          *          *
   2   1.28938    1.11270          *          *          *          *          *
   3   0.80036    1.28938    1.11270          *          *          *          *
   4   0.71505    0.80036    1.28938    1.11270    0.97937          *          *
   5   1.24603    0.71505    0.80036    1.28938    1.01270    0.97937    0.99604
   6   1.29122    1.24603    0.71505    0.80036    1.01316    1.01270    1.01293
   7   0.82565    1.29122    1.24603    0.71505    1.01949    1.01316    1.01633
   8   0.67029    0.82565    1.29122    1.24603    1.00830    1.01949    1.01389
   9   1.13028    0.67029    0.82565    1.29122    0.97936    1.00830    0.99383
  10   1.25328    1.13028    0.67029    0.82565    0.96988    0.97936    0.97462
  11   0.75883    1.25328    1.13028    0.67029    0.95317    0.96988    0.96152
  12   0.68669    0.75883    1.25328    1.13028    0.95727    0.95317    0.95522
  13   1.09241    0.68669    0.75883    1.25328    0.94780    0.95727    0.95254
  14   1.22503    1.09241    0.68669    0.75883    0.94074    0.94780    0.94427
  15   0.76397    1.22503    1.09241    0.68669    0.94203    0.94074    0.94138
  16   0.68954    0.76397    1.22503    1.09241    0.94274    0.94203    0.94238
  17   1.14872    0.68954    0.76397    1.22503    0.95681    0.94274    0.94978
  18   1.33562    1.14872    0.68954    0.76397    0.98446    0.95681    0.97064
  19   0.78158    1.33562    1.14872    0.68954    0.98886    0.98446    0.98666
  20   0.73384    0.78158    1.33562    1.14872    0.99994    0.98886    0.99440
  21   1.22186    0.73384    0.78158    1.33562    1.01822    0.99994    1.00908
  22   1.34092    1.22186    0.73384    0.78158    1.01955    1.01822    1.01889
  23   0.84399    1.34092    1.22186    0.73384    1.03515    1.01955    1.02735
  24   0.67839    0.84399    1.34092    1.22186    1.02129    1.03515    1.02822
  25   1.17980    0.67839    0.84399    1.34092    1.01077    1.02129    1.01603
  26   1.30120    1.17980    0.67839    0.84399    1.00084    1.01077    1.00581
  27   0.74861    1.30120    1.17980    0.67839    0.97700    1.00084    0.98892
  28   0.69075    0.74861    1.30120    1.17980    0.98009    0.97700    0.97854
  29   1.13594    0.69075    0.74861    1.30120    0.96912    0.98009    0.97461
  30   1.28791    1.13594    0.69075    0.74861    0.96580    0.96912    0.96746
  31   0.80184    1.28791    1.13594    0.69075    0.97911    0.96580    0.97246
  32   0.66829    0.80184    1.28791    1.13594    0.97350    0.97911    0.97630
  33   1.25922    0.66829    0.80184    1.28791    1.00431    0.97350    0.98890
  34   1.29130    1.25922    0.66829    0.80184    1.00516    1.00431    1.00474
  35   0.82704    1.29130    1.25922    0.66829    1.01146    1.00516    1.00831
  36   0.70932    0.82704    1.29130    1.25922    1.02172    1.01146    1.01659
  37   1.27804    0.70932    0.82704    1.29130    1.02643    1.02172    1.02407
  38   1.42286    1.27804    0.70932    0.82704    1.05931    1.02643    1.04287
  39   0.86335    1.42286    1.27804    0.70932    1.06839    1.05931    1.06385
  40   0.74385    0.86335    1.42286    1.27804    1.07703    1.06839    1.07271
  41   1.16493    0.74385    0.86335    1.42286    1.04875    1.07703    1.06289
  42   1.30628    1.16493    0.74385    0.86335    1.01960    1.04875    1.03418
  43   0.82217    1.30628    1.16493    0.74385    1.00931    1.01960    1.01446
  44   0.69440    0.82217    1.30628    1.16493    0.99695    1.00931    1.00313
  45   1.13815    0.69440    0.82217    1.30628    0.99025    0.99695    0.99360
  46   1.23618    1.13815    0.69440    0.82217    0.97273    0.99025    0.98149
  47   0.79402    1.23618    1.13815    0.69440    0.96569    0.97273    0.96921
  48   0.71700    0.79402    1.23618    1.13815    0.97134    0.96569    0.96851
  49   1.23657    0.71700    0.79402    1.23618    0.99594    0.97134    0.98364
  50   1.33661    1.23657    0.71700    0.79402    1.02105    0.99594    1.00850
  51   0.81224    1.33661    1.23657    0.71700    1.02561    1.02105    1.02333
  52   0.75736    0.81224    1.33661    1.23657    1.03570    1.02561    1.03065
  53   1.28765    0.75736    0.81224    1.33661    1.04847    1.03570    1.04208
  54   1.31995    1.28765    0.75736    0.81224    1.04430    1.04847    1.04638
  55   0.83051    1.31995    1.28765    0.75736    1.04887    1.04430    1.04658
  56   0.76018    0.83051    1.31995    1.28765    1.04957    1.04887    1.04922
  57   1.18487    0.76018    0.83051    1.31995    1.02388    1.04957    1.03673
  58   1.27702    1.18487    0.76018    0.83051    1.01314    1.02388    1.01851
  59   0.77215    1.27702    1.18487    0.76018    0.99855    1.01314    1.00585
  60   0.70505    0.77215    1.27702    1.18487    0.98477    0.99855    0.99166
```

## SECTION 11-2 Classical Decomposition

We want our first value in `C17` (which is `0.99604`) to be opposite the third-quarter value in `C4`. Thus we need to eliminate the first two asterisks and move everything up two rows. We can accomplish this with the `DELEte` command as follows:

```
MTB >DELEte rows 1 and 2 in C17
```

`C17` *now contains the four-quarter moving averages that correspond to the values in* `C4`. Let's move `C17` next to `C4` to keep things together. The command is

```
MTB >COPY C17 into C5
```

Now we can remove the cyclical component by dividing `C4` by `C5`. However, to do this we face a minor technical problem: `C4` contains 60 values and `C5` contains 58. To divide these two columns, they must be of equal length. So we add two empty values (`'*'`) to the bottom of `C5`. The commands are

```
MTB >LET C5(59) = '*'
MTB >LET C5(60) = '*'
```

Now we can execute the division with the command

```
MTB >LET C6 = C4/C5
```

We have now isolated the remaining three components as follows:

**(1)** `C3` is the trend component.
**(2)** `C5` is the cyclical component.
**(3)** `C6` contains the seasonal component and random fluctuations.

Expressed as an equation, `C6` is

$$C6 = S * R$$

Our complete set of data is printed in Figure 11-7.

## Identifying the Seasonal Component ▼▲▼▲▼▲▼▲▼▲▼▲▼▲▼▲▼▲▼▲▼▲

A `TSPLot` of `C6` will produce Figure 11-8. With both the trend and cyclical influences omitted, the plot becomes more scattered. However, as can be seen by inspection, there is still a very noticeable pattern of two low periods followed by two high periods.

The seasonal index is very important for short-term forecasts (less than one year). It should be noted that the term *seasonal* is used in a general sense. Some businesses might have seasonal patterns that are quarterly, others weekly, and still others daily. Any short-term pattern is termed "seasonal" and the technique developed below will handle any such pattern. The ability to deal with these short-term fluctuations is essential to business.

The first thing we need to do is to set up a computational area. We will copy `C6` into `C20` for the present. Next we need a set of identification values to represent the four seasons, so we will set the values of 1 through 4 repeatedly in `C21`. The following commands will get us started:

```
MTB >SET C21
DATA>15(1:4)
DATA>END
```

Our computational area now looks like Figure 11-9.

**FIGURE 11-7** Minitab Time Series Analysis of Tennis Shoe A Sales by Quarter, 1972–1986, with Trend (C3), Cyclical (C5), and Seasonal and Random Components (C6) Isolated

```
MTB >PRIN C1-C6
ROW      C1     C2       C3        C4         C5         C6

  1   2331.24    1   2095.12    1.11270         *          *
  2   2723.61    2   2112.35    1.28938         *          *
  3   1704.42    3   2129.57    0.80036    0.99604    0.80354
  4   1535.06    4   2146.80    0.71505    1.01293    0.70592
  5   2696.44    5   2164.02    1.24603    1.01633    1.22602
  6   2816.47    6   2181.24    1.29122    1.01389    1.27353
  7   1815.17    7   2198.47    0.82565    0.99383    0.83078
  8   1485.15    8   2215.69    0.67029    0.97462    0.68774
  9   2523.83    9   2232.92    1.13028    0.96152    1.17551
 10   2820.05   10   2250.14    1.25328    0.95522    1.31203
 11   1720.55   11   2267.36    0.75883    0.95254    0.79664
 12   1568.81   12   2284.59    0.68669    0.94427    0.72722
 13   2514.53   13   2301.81    1.09241    0.94138    1.16043
 14   2840.89   14   2319.04    1.22503    0.94238    1.29993
 15   1784.83   15   2336.26    0.76397    0.94978    0.80437
 16   1622.81   16   2353.48    0.68954    0.97064    0.71039
 17   2723.29   17   2370.71    1.14872    0.98666    1.16425
 18   3189.37   18   2387.93    1.33562    0.99440    1.34314
 19   1879.81   19   2405.16    0.78158    1.00908    0.77454
 20   1777.64   20   2422.38    0.73384    1.01889    0.72024
 21   2980.86   21   2439.60    1.22186    1.02735    1.18933
 22   3294.40   22   2456.83    1.34092    1.02822    1.30411
 23   2088.08   23   2474.05    0.84399    1.01603    0.83068
 24   1690.05   24   2491.28    0.67839    1.00581    0.67447
 25   2959.52   25   2508.50    1.17980    0.98892    1.19302
 26   3286.46   26   2525.72    1.30120    0.97854    1.32973
 27   1903.68   27   2542.95    0.74861    0.97461    0.76812
 28   1768.43   28   2560.17    0.69075    0.96746    0.71398
 29   2927.78   29   2577.40    1.13594    0.97246    1.16812
 30   3341.63   30   2594.62    1.28791    0.97630    1.31917
 31   2094.28   31   2611.84    0.80184    0.98890    0.81084
 32   1756.98   32   2629.07    0.66829    1.00474    0.66514
 33   3332.26   33   2646.29    1.25922    1.00831    1.24884
 34   3439.41   34   2663.52    1.29130    1.01659    1.27023
 35   2217.08   35   2680.74    0.82704    1.02407    0.80760
 36   1913.71   36   2697.96    0.70932    1.04287    0.68016
 37   3470.13   37   2715.19    1.27804    1.06385    1.20133
 38   3887.83   38   2732.41    1.42286    1.07271    1.32641
 39   2373.91   39   2749.64    0.86335    1.06289    0.81227
 40   2058.13   40   2766.86    0.74385    1.03418    0.71927
 41   3243.26   41   2784.08    1.16493    1.01446    1.14833
 42   3659.29   42   2801.31    1.30628    1.00313    1.30221
 43   2317.32   43   2818.53    0.82217    0.99360    0.82747
 44   1969.15   44   2835.76    0.69440    0.98149    0.70750
 45   3247.12   45   2852.98    1.13815    0.96921    1.17431
 46   3548.10   46   2870.20    1.23618    0.96851    1.27637
 47   2292.67   47   2887.43    0.79402    0.98364    0.80722
 48   2082.63   48   2904.65    0.71700    1.00850    0.71096
 49   3613.11   49   2921.88    1.23657    1.02333    1.20838
 50   3928.44   50   2939.10    1.33661    1.03065    1.29686
 51   2401.24   51   2956.32    0.81224    1.04208    0.77944
 52   2252.05   52   2973.55    0.75736    1.04638    0.72379
 53   3851.08   53   2990.77    1.28765    1.04658    1.23034
 54   3970.40   54   3008.00    1.31995    1.04922    1.25803
 55   2512.48   55   3025.22    0.83051    1.03673    0.80109
 56   2312.82   56   3042.44    0.76018    1.01851    0.74637
 57   3625.30   57   3059.67    1.18487    1.00585    1.17798
 58   3929.24   58   3076.89    1.27702    0.99166    1.28775
 59   2389.11   59   3094.12    0.77215         *          *
 60   2193.64   60   3111.34    0.70505         *          *
```

## SECTION 11-2 Classical Decomposition

▶ **283**

**FIGURE 11-8** Minitab Time Series Plot of Seasonal and Random Fluctuation Components in Tennis Shoe A Sales Data

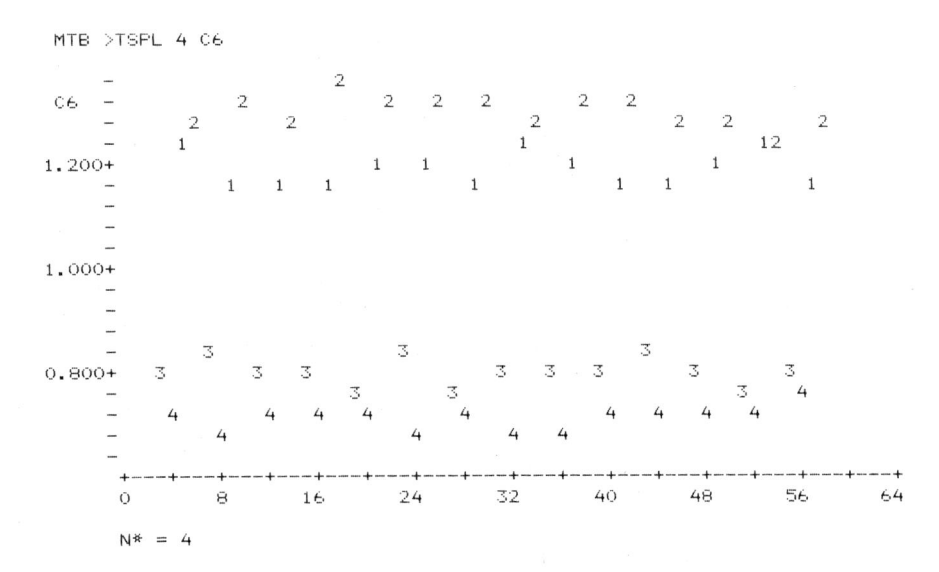

```
MTB >TSPL 4 C6

                           2
C6  -           2                2    2    2           2    2
    -      2         2                      2                    2    2         2
    -  1                                1              1                   12
1.200+                1    1    1              1              1                   1
    -         1    1    1         1              1    1              1
    -
    -
    -
1.000+
    -
    -
    -
    -         3         3              3                   3
0.800+   3         3         3                   3    3         3    3
    -              3         3                        3         3
    -    4         4    4    4         4              4    4    4    4
    -         4                   4         4
       +---+---+---+---+---+---+---+---+---+---+---+---+---+---+---+---+
       0        8        16       24       32       40       48       56       64

N* = 4
```

**FIGURE 11-9** Minitab Worksheet for Isolating Seasonal Component after Setting Seasonal Index

```
MTB >PRIN C20 C21
ROW      C20       C21       ROW      C20       C21
  1       *         1         31     0.81084     3
  2       *         2         32     0.66514     4
  3     0.80354     3         33     1.24884     1
  4     0.70592     4         34     1.27023     2
  5     1.22602     1         35     0.80760     3
  6     1.27353     2         36     0.68016     4
  7     0.83078     3         37     1.20133     1
  8     0.68774     4         38     1.32641     2
  9     1.17551     1         39     0.81227     3
 10     1.31203     2         40     0.71927     4
 11     0.79664     3         41     1.14833     1
 12     0.72722     4         42     1.30221     2
 13     1.16043     1         43     0.82747     3
 14     1.29993     2         44     0.70750     4
 15     0.80437     3         45     1.17431     1
 16     0.71039     4         46     1.27637     2
 17     1.16425     1         47     0.80722     3
 18     1.34314     2         48     0.71096     4
 19     0.77454     3         49     1.20838     1
 20     0.72024     4         50     1.29686     2
 21     1.18933     1         51     0.77944     3
 22     1.30411     2         52     0.72379     4
 23     0.83068     3         53     1.23034     1
 24     0.67447     4         54     1.25803     2
 25     1.19302     1         55     0.80109     3
 26     1.32973     2         56     0.74637     4
 27     0.76812     3         57     1.17798     1
 28     0.71398     4         58     1.28775     2
 29     1.16812     1         59       *         3
 30     1.31917     2         60       *         4
```

## CHAPTER 11 Forecasting

We will now make separate columns for each of our four seasons. To accomplish this task takes two steps. First, we must omit any period for which we do not have a value; this means the first two periods and the last two. We can omit these periods with the **DELEte** command as follows:

```
MTB >DELEte rows 1, 2, 59, 60 from C20, C21
```

The computational area now appears as in Figure 11-10. Notice that our values start with period 3 at the top (row 1 of **C21**) and end with period 2 at the bottom (row 56 of **C21**).

**FIGURE 11-10** Worksheet for Isolating Seasonal Component after Eliminating Blank Data Points

```
MTB >PRIN C20 C21
ROW        C20      C21      ROW        C20      C21
  1     0.80354      3       29      0.81084      3
  2     0.70592      4       30      0.66514      4
  3     1.22602      1       31      1.24884      1
  4     1.27353      2       32      1.27023      2
  5     0.83078      3       33      0.80760      3
  6     0.68774      4       34      0.68016      4
  7     1.17551      1       35      1.20133      1
  8     1.31203      2       36      1.32641      2
  9     0.79664      3       37      0.81227      3
 10     0.72722      4       38      0.71927      4
 11     1.16043      1       39      1.14833      1
 12     1.29993      2       40      1.30221      2
 13     0.80437      3       41      0.82747      3
 14     0.71039      4       42      0.70750      4
 15     1.16425      1       43      1.17431      1
 16     1.34314      2       44      1.27637      2
 17     0.77454      3       45      0.80722      3
 18     0.72024      4       46      0.71096      4
 19     1.18933      1       47      1.20838      1
 20     1.30411      2       48      1.29686      2
 21     0.83068      3       49      0.77944      3
 22     0.67447      4       50      0.72379      4
 23     1.19302      1       51      1.23034      1
 24     1.32973      2       52      1.25803      2
 25     0.76812      3       53      0.80109      3
 26     0.71398      4       54      0.74637      4
 27     1.16812      1       55      1.17798      1
 28     1.31917      2       56      1.28775      2
```

The next step is to sort out all the **1**'s, **2**'s, **3**'s, and **4**'s from **C21** and carry along the values from **C20** to find our seasonal effect. The **UNSTack** command can accomplish this task. The **UNSTack** command is

```
UNSTack C into C,...,C;
  SUBScripts in C.
```

We will unstack all the **1**'s from **C21** and carry along the corresponding values from **C20**. This set of values from **C20** comprises the seasonal fluctuations for winter, and we put them in **C25**. We will unstack all of the **2**'s from **C21** and place the corresponding values from **C20**, which represent seasonal fluctuations from spring, into **C26**. We will do similarly for the summer and autumn data, placing them in **C27** and **C28**, respectively. Note that the values in **C21** are *not* moved; they are just used as identifiers to group values from **C20**.

Our commands to create the data space described above are as follows:

## SECTION 11-2 Classical Decomposition

```
MTB >UNSTack C20 into C25, C26, C27, C28;
SUBC> SUBScripts in C21.
```

We now want to find the mean values of C25 through C28 and put those values in K1 through K4, respectively. The commands are

```
MTB >MEAN of values in C25, put in K1
MTB >MEAN of values in C26, put in K2
MTB >MEAN of values in C27, put in K3
MTB >MEAN of values in C28, put in K4
```

Figure 11-11 lists the results of the two previous sets of commands. *Values K1 through K4 give us the seasonal indices* (stated as a proportion, not a percentage) for seasons 1 (winter) through 4 (autumn). As you will see shortly, these four seasonal index values are the important values that we have been trying to obtain.

**FIGURE 11-11** Seasonal Components, Including Mean Values, of Tennis Shoe A Sales, 1972–1986

```
MTB >PRIN C25-C28, K1-K4
K1      1.19044
K2      1.29996
K3      0.803900
K4      0.706652

ROW        C25        C26        C27        C28
  1    1.22602    1.27353    0.803543    0.705917
  2    1.17551    1.31203    0.830778    0.687743
  3    1.16043    1.29993    0.796643    0.727218
  4    1.16425    1.34314    0.804368    0.710393
  5    1.18933    1.30411    0.774541    0.720238
  6    1.19302    1.32973    0.830675    0.674470
  7    1.16812    1.31917    0.768117    0.713977
  8    1.24884    1.27023    0.810836    0.665138
  9    1.20133    1.32641    0.807599    0.680158
 10    1.14833    1.30221    0.812273    0.719269
 11    1.17431    1.27637    0.827470    0.707497
 12    1.20838    1.29686    0.807224    0.710957
 13    1.23034    1.25803    0.779439    0.723789
 14    1.17798    1.28775    0.801090    0.746369
```

We can now set our four seasonal indices into C7 and remove the seasonal effects from our time series model. The command is

```
MTB >SET C7
DATA>15(K1,K2,K3,K4)
DATA>END
```

With these values set, we can remove the seasonal fluctuation by

```
LET >C8 = C6/C7
```

(Remember that our trend component is in C3, our cyclical component in C5, our seasonal component in C7, and our random component in C8.) Figure 11-12 shows the results of our commands.

We can now compare our predictable portion of the model ($T \times C \times S$) to our original data with the command

```
MTB > LET C9 = C3 * C5 * C7
```

**FIGURE 11-12** Values of Trend (C3), Cyclical (C5), Seasonal (C7), and Random (C8) Components in Tennis Shoe A Sales Data

```
MTB >PRIN C3 C5 C7 C8
 ROW       C3        C5         C7        C8

   1    2095.12       *       1.19044       *
   2    2112.35       *       1.29996       *
   3    2129.57    0.99604    0.80390    0.99956
   4    2146.80    1.01293    0.70665    0.99896
   5    2164.02    1.01633    1.19044    1.02988
   6    2181.24    1.01389    1.29996    0.97966
   7    2198.47    0.99383    0.80390    1.03344
   8    2215.69    0.97462    0.70665    0.97324
   9    2232.92    0.96152    1.19044    0.98746
  10    2250.14    0.95522    1.29996    1.00928
  11    2267.36    0.95254    0.80390    0.99097
  12    2284.59    0.94427    0.70665    1.02910
  13    2301.81    0.94138    1.19044    0.97479
  14    2319.04    0.94238    1.29996    0.99997
  15    2336.26    0.94978    0.80390    1.00058
  16    2353.48    0.97064    0.70665    1.00529
  17    2370.71    0.98666    1.19044    0.97800
  18    2387.93    0.99440    1.29996    1.03321
  19    2405.16    1.00908    0.80390    0.96348
  20    2422.38    1.01889    0.70665    1.01922
  21    2439.60    1.02735    1.19044    0.99907
  22    2456.83    1.02822    1.29996    1.00319
  23    2474.05    1.01603    0.80390    1.03331
  24    2491.28    1.00581    0.70665    0.95446
  25    2508.50    0.98892    1.19044    1.00216
  26    2525.72    0.97854    1.29996    1.02290
  27    2542.95    0.97461    0.80390    0.95549
  28    2560.17    0.96746    0.70665    1.01037
  29    2577.40    0.97246    1.19044    0.98125
  30    2594.62    0.97630    1.29996    1.01477
  31    2611.84    0.98890    0.80390    1.00863
  32    2629.07    1.00474    0.70665    0.94125
  33    2646.29    1.00831    1.19044    1.04905
  34    2663.52    1.01659    1.29996    0.97713
  35    2680.74    1.02407    0.80390    1.00460
  36    2697.96    1.04287    0.70665    0.96251
  37    2715.19    1.06385    1.19044    1.00915
  38    2732.41    1.07271    1.29996    1.02035
  39    2749.64    1.06289    0.80390    1.01042
  40    2766.86    1.03418    0.70665    1.01785
  41    2784.08    1.01446    1.19044    0.96462
  42    2801.31    1.00313    1.29996    1.00173
  43    2818.53    0.99360    0.80390    1.02932
  44    2835.76    0.98149    0.70665    1.00120
  45    2852.98    0.96921    1.19044    0.98645
  46    2870.20    0.96851    1.29996    0.98185
  47    2887.43    0.98364    0.80390    1.00414
  48    2904.65    1.00850    0.70665    1.00609
  49    2921.88    1.02333    1.19044    1.01507
  50    2939.10    1.03065    1.29996    0.99761
  51    2956.32    1.04208    0.80390    0.96957
  52    2973.55    1.04638    0.70665    1.02425
  53    2990.77    1.04658    1.19044    1.03351
  54    3008.00    1.04922    1.29996    0.96774
  55    3025.22    1.03673    0.80390    0.99651
  56    3042.44    1.01851    0.70665    1.05620
  57    3059.67    1.00585    1.19044    0.98953
  58    3076.89    0.99166    1.29996    0.99061
  59    3094.12       *       0.80390       *
  60    3111.34       *       0.70665       *
```

**SECTION 11-2** Classical Decomposition

Both `C1` and `C9` are printed in Figure 11-13. Column `C9` is also plotted using the `TSPLot` command in Figure 11-14.

A glance at `C1` and `C9` in Figure 11-13 indicates that our model seems to fit the actual data reasonably well. Also, in Figure 11-14 we can clearly see that our model produces a plot that resembles the plot of the original data very well (see Figure 11-2).

## Making Projections Using Time Series Analysis

It is now time to make a few sales projections using the components of our time series analysis. However, before we start, we should note that most economists agree that it is practically impossible to predict business cycles. Consequently, when we make forecasts we will not use the cyclical component of our model.

You might well ask why in the #*%$#! did we bother to calculate and isolate the cyclical component in the first place? The reason is that we had to eliminate the cyclical component in order to calculate the seasonal component. Without the cyclical component eliminated, the calculated seasonal component would have been inaccurate.

Thus, we will use only the trend and seasonal components of our model for prediction, lumping the random fluctuation and cyclical components together as error.

**FIGURE 11-13** Actual (C1) and Predicted (C9) Sales of Tennis Shoe A by Quarter Based on Minitab Time Series Analysis

```
MTB > PRIN C1 C9

ROW      C1        C9       ROW      C1        C9
  1   2331.24       *        31    2094.28   2076.37
  2   2723.61       *        32    1756.98   1866.65
  3   1704.42   1705.18      33    3332.26   3176.43
  4   1535.06   1536.66      34    3439.41   3519.93
  5   2696.44   2618.21      35    2217.08   2206.92
  6   2816.47   2874.92      36    1913.71   1988.26
  7   1815.17   1756.45      37    3470.13   3438.65
  8   1485.15   1525.99      38    3887.83   3810.31
  9   2523.83   2555.87      39    2373.91   2349.45
 10   2820.05   2794.12      40    2058.13   2022.04
 11   1720.55   1736.22      41    3243.26   3362.17
 12   1568.81   1524.44      42    3659.29   3653.01
 13   2514.53   2579.54      43    2317.32   2251.32
 14   2840.89   2840.97      44    1969.15   1966.81
 15   1784.83   1783.80      45    3247.12   3291.73
 16   1622.81   1614.27      46    3548.10   3613.67
 17   2723.29   2784.54      47    2292.67   2283.23
 18   3189.37   3086.85      48    2082.63   2070.03
 19   1879.81   1951.06      49    3613.11   3559.47
 20   1777.64   1744.12      50    3928.44   3937.84
 21   2980.86   2983.63      51    2401.24   2476.59
 22   3294.40   3283.93      52    2252.05   2198.73
 23   2088.08   2020.77      53    3851.08   3726.17
 24   1690.05   1770.70      54    3970.40   4102.76
 25   2959.52   2953.13      55    2512.48   2521.30
 26   3286.46   3212.92      56    2312.82   2189.75
 27   1903.68   1992.37      57    3625.30   3663.66
 28   1768.43   1750.28      58    3929.24   3966.49
 29   2927.78   2983.74      59    2389.11      *
 30   3341.63   3292.98      60    2193.64      *
```

**FIGURE 11-14** Minitab Time Series Plot of Predicted Tennis Shoe A Sales, 1972–1986

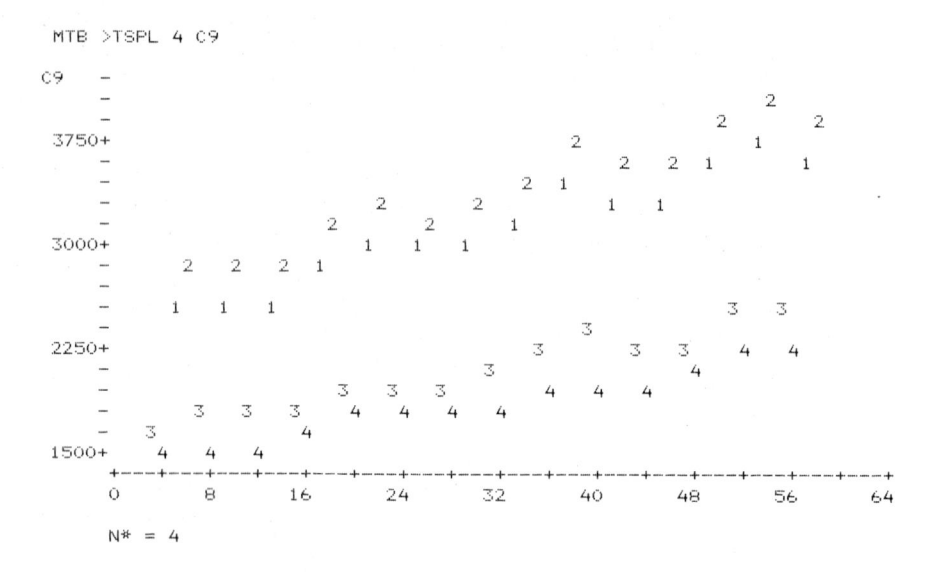

```
MTB >TSPL 4 C9

C9  -
    -                                                          2
    -                                               2       2
3750+                                          2        2
    -                                    2       2 1      1
    -                                2 1
    -                          2           2 1  1
    -                    2           2   1
3000+                  1     1   1
    -        2  2  2 1
    -
    -     1  1  1                                      3  3
2250+                                    3                3 3 4  4
    -                          3                    4
    -              3  3  3          4  4  4
    -        3  3  3    4  4  4  4
    -     3        4
1500+  4  4  4
    +---+---+---+---+---+---+---+---+---+---+---+---+---+---+---+---+
    0      8     16     24     32     40     48     56     64

N* = 4
```

Our trend component was computed before as

$$T = 2{,}077.9 + 17.224(\text{period})$$

Our seasonal components are

$$\begin{aligned}\text{fall} &= 1.1904 \\ \text{spring} &= 1.3000 \\ \text{summer} &= 0.80390 \\ \text{autumn} &= 0.70665\end{aligned}$$

Finally, our overall prediction model is

$$V = T \times S$$

Let's predict sales for period 59. We would first substitute 59 into our trend equation as follows:

$$\begin{aligned}T &= 2{,}077.9 + 17.224(59) \\ &= 3{,}094.12\end{aligned}$$

Next, since period 59 is a summer period, we would use the seasonal index of 0.80390. Thus, our final sales prediction would be

$$\begin{aligned}V &= 3{,}094.12(S) \\ &= 3{,}094.12(0.80390) \\ &= 2{,}487.36\end{aligned}$$

Actual sales for period 59 (summer 1986) were 2,389.11. Our projection using time series is off by less than 100. Notice that if we had simply calculated a simple linear regression and predicted sales from the trend component only, our forecast would have been

### SECTION 11-2 Classical Decomposition

3,094.12, an error of slightly over 700. Clearly the additional weighting of the prediction by the seasonal index is of great value.

As you can see, the seasonal adjustment decreases our trend prediction in periods 3 and 4 of each year and increases our trend prediction in periods 1 and 2. Although the trend will lead to a good estimate of sales for the entire year, it is less accurate in predicting quarterly sales.

Let's now forecast sales for period 60. In this case our trend analysis is

$$T = 2,077.9 + 17.224(60)$$
$$= 3,111.34$$

When this trend projection is weighted by our seasonal index for winter (0.70665), the results are

$$V = 3,111.34(S)$$
$$= 3,111.34(0.70665)$$
$$= 2,198.63$$

This projection is off by about 5. The projection based on the trend alone would have been off by over 900. Here the effect of the weighting factor for period 4 is dramatic.

Now let's forecast sales for the next two years, that is, for the next eight quarters. The general form of our model is

$$V = T \times S$$

For each period, the calculations are as follows:

$$V_{61} = [2,077.9 + (17.224)(61)]1.1904 = 3,724.24$$
$$V_{62} = [2,077.9 + (17.224)(62)]1.3000 = 4,089.52$$
$$V_{63} = [2,077.9 + (17.224)(63)]0.80390 = 2,542.74$$
$$V_{64} = [2,077.9 + (17.224)(64)]0.70665 = 2,247.31$$
$$V_{65} = [2,077.9 + (17.224)(65)]1.1904 = 3,806.26$$
$$V_{66} = [2,077.9 + (17.224)(66)]1.3000 = 4,179.09$$
$$V_{67} = [2,077.9 + (17.224)(67)]0.80390 = 2,598.13$$
$$V_{68} = [2,077.9 + (17.224)(68)]0.70665 = 2,296.00$$

At this point, managerial experience must again take over. We should always ask, "Do these results make sense?" The values seem to follow the trend set in previous years. Go-Fast could use this information to schedule production and distribution more efficiently. Notice that we are predicting a large swing in sales next year between the second and third periods from 4,089.52 to 2,542.74. The more a company can predict such swings, the more effectively it can run its organization.

In our illustration for time series analysis, our problem worked out reasonably well. This will not always be the case. Many times trends and cycles will not be well defined. As stated at the beginning of this chapter, only an elementary overview of the topic is presented here.

Speaking of "elementary overviews," we are now going to turn our attention to Box–Jenkins. It is difficult to decide just how much of theory and terminology to present concerning Box–Jenkins without overburdening the novice. We will attempt to give you some useful tools without baffling you.

## 11-3 BOX–JENKINS FORECASTING

An interesting alternative to classical time series analysis using decomposition was developed by Box and Jenkins. The Box–Jenkins approach uses mathematics beyond the scope of this text, but we can use Minitab to perform the necessary calculations.

The Box–Jenkins approach to forecasting in a Minitab environment requires the assumption that the data set is stationary—that is, it must have no trend. Thus our first task in using Box–Jenkins is to make the data stationary.

### Making the Series Stationary

We can remove any trend component from the series by first examining the series using the `TSPLot` command. If there is a systematic change over time, we can remove any first-order (linear) trend with the `DIFFerence` command. We can then examine the differenced series and remove any remaining second- or higher-order trends by more differencing. We set the variable `d` equal to the number of differences needed to remove the trend component, thus rendering the series stationary.

We can illustrate this process using the data on the sales of tennis shoe A that we have been analyzing. Putting the data into `C1` with the `SET` command, we can evaluate the trend component with the `TSPLot` command. Look back at Figure 11-2.

Because there is an upward trend, we will use the `DIFFerence` command once and see if the trend is removed. Notice that we will difference with a lag of 4 since we wish to compare results for seasons. One of the advantages of Box–Jenkins is that it is very easy to try several different lags by simply entering a different value in the `DIFFerence` command. The general form of the command is

```
DIFFerence [of lag K] for data in C, put in C
```

For this example, we are using a lag of 4, so our command is

```
MTB >DIFFerence of lag 4 for data in C1, put in C2
```

Then we examine the data to see whether the trend is gone. If the data is stationary, there should be no recognizable upward trend overall or by individual seasons. The data has been plotted in Figure 11-15 by the following command:

```
MTB >TSPLot with 4 periods, data in C2
```

### Autocorrelation and Partial Autocorrelation Functions

Once the trend has been removed, we use the `ACF` (autocorrelation function) command to determine the order of the autoregressive function needed to model the series. Then we use the `PACF` (partial autocorrelation function) to determine the order of the moving average function needed. Don't panic, we will explain these terms after some examples.

Some of the most common charts created by the `AFC` and `PACF` commands for various models are shown in Figure 11-16. In the Minitab command for Box–Jenkins we

### SECTION 11-3 Box–Jenkins Forecasting

**FIGURE 11-15** Minitab Times Series Plot of Tennis Shoe A Sales after Removal of First-Order Trend

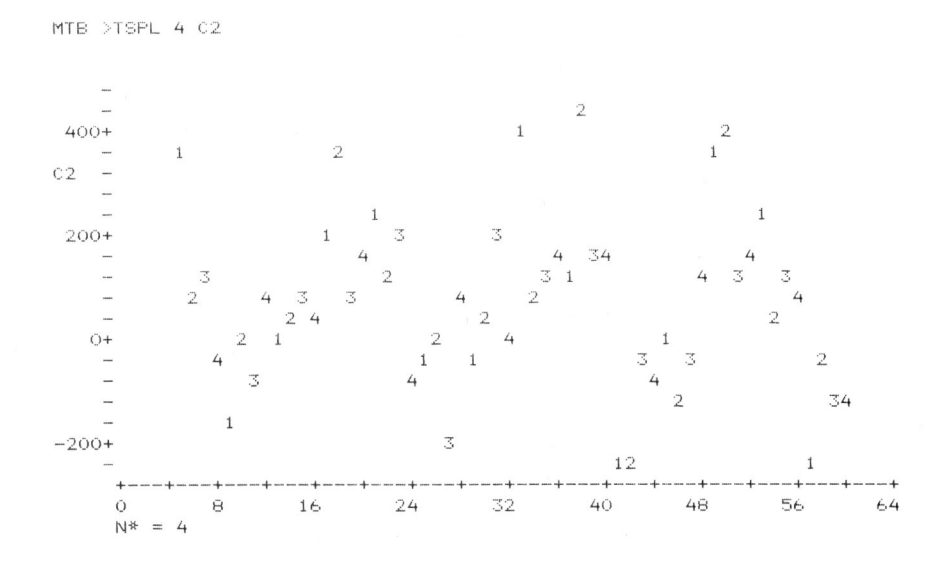

```
MTB >TSPL 4 C2

                                          2
                                    1              2
      400+
           -     1         2                       1
      C2   -
           -
           -                  1                        1
      200+                 1      3      3
           -                   4              4  34         4
           -         3             2           3  1       4   3   3
           -      2     4  3  3        4   2                        4
           -         2  4               2                     2
        0+            2  1          2       4              1
           -     4                1  1           3    3              2
           -        3                  4         4                    34
           -                                          2
           -     1
      -200+                        3
                                              12              1
                 +----+----+----+----+----+----+----+----+----+----+----+----+
                 0        8       16       24       32       40       48       56       64
                N* = 4
```

will input a **p** value (for the autoregression) and a **q** value (for the moving average). The appropriate **p** and **q** values to input for each type of chart are shown in the figure.

You may remember that correlation indicates the degree to which two variables move together. If both move in the same direction, then the correlation is positive; if they move in opposite directions, the correlation is negative. Minitab has a command to print an autocorrelation chart. We first want to calculate the autocorrelation and then compare our results with the charts in Figure 11-16 to determine our **p** and **q** values. We then want to calculate the partial autocorrelation and then compare our chart with the charts for partial autocorrelation.

The **ACF** and **PACF** commands in Minitab are very straightforward. The commands are

```
ACF for series in C
```

and

```
PACF for series in C
```

Both of these commands have options for lags and storing output, but we will not use these options here. We will store the results of the **DIFFerence** command in **C2** and then perform both the **AFC** and **PACF** commands on **C2**. Remember, it is important that **ACF** and **PACF** be used only on stationary data. The first command is

```
MTB >ACF for series in C2
```

The results are shown in Figure 11-17.

**FIGURE 11-16** Common Charts Made by the ACF and PACF Commands

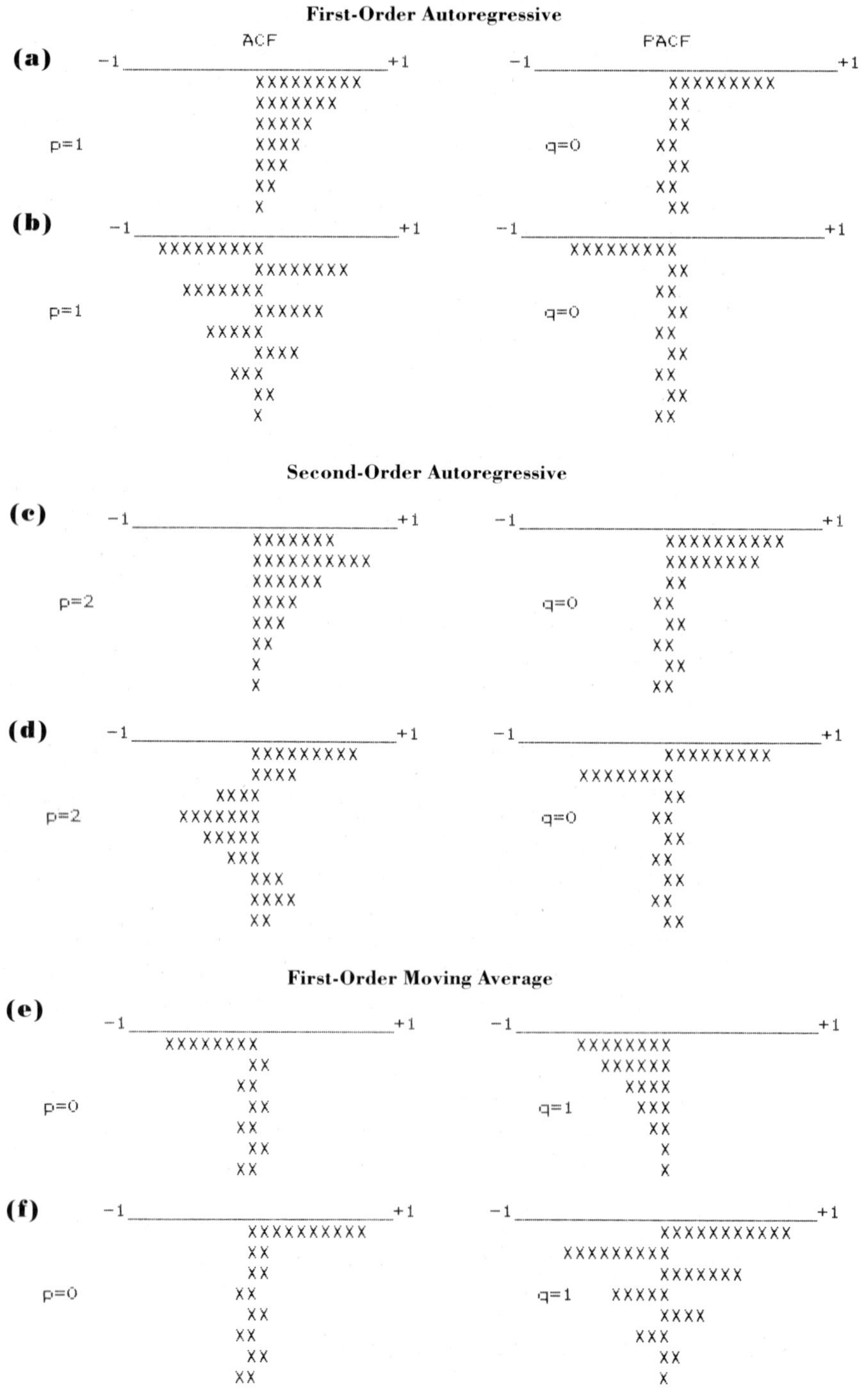

### First-Order Autoregressive

```
                          ACF                              PACF
(a)         -1_____+1      -1_____+1
                      XXXXXXXX                         XXXXXXXX
                      XXXXXX                           XX
                      XXXXX                            XX
         p=1          XXXX                q=0          XX
                      XXX                              XX
                      XX                               XX
                      X                                XX
(b)         -1_____+1      -1_____+1
                 XXXXXXXX                         XXXXXXXX
                   XXXXXXX                             XX
                XXXXXX                                 XX
         p=1      XXXXXX                  q=0          XX
                 XXXXX                                 XX
                  XXXX                                 XX
                XXX                                    XX
                 XX                                    XX
                 X                                     XX
```

### Second-Order Autoregressive

```
(c)         -1_____+1      -1_____+1
                     XXXXXX                           XXXXXXXXX
                     XXXXXXXXX                        XXXXXXX
                     XXXXXX                           XX
         p=2         XXXX                 q=0         XX
                     XXX                              XX
                     XX                               XX
                     X                                XX
                     X                                XX
(d)         -1_____+1      -1_____+1
                     XXXXXXXX                         XXXXXXXXX
                     XXXX                         XXXXXXX
                   XXXX                               XX
         p=2       XXXXXX                 q=0         XX
                    XXXXX                             XX
                    XXX                               XX
                    XXX                               XX
                    XXXX                              XX
                    XX                                XX
```

### First-Order Moving Average

```
(e)         -1_____+1      -1_____+1
                   XXXXXXX                          XXXXXXX
                   XX                               XXXXX
                   XX                               XXXX
         p=0       XX                     q=1       XXX
                   XX                               XX
                   XX                               X
                   XX                               X
(f)         -1_____+1      -1_____+1
                   XXXXXXXXX                        XXXXXXXXXX
                   XX                            XXXXXXXX
                   XX                                    XXXXXXX
         p=0       XX                     q=1        XXXXX
                   XX                                 XXXX
                   XX                                 XXX
                   XX                                 XX
                   XX                                 X
```

## SECTION 11-3 Box–Jenkins Forecasting

**FIGURE 11-17** Results of AFC Command on Tennis Shoe A Sales Data after DIFFerence Command

```
MTB >ACF C2
              -1.0 -0.8 -0.6 -0.4 -0.2  0.0  0.2  0.4  0.6  0.8  1.0
               +----+----+----+----+----+----+----+----+----+----+
   1   0.402                                    XXXXXXXXXXX
   2   0.313                                    XXXXXXXXX
   3   0.102                                    XXXX
   4  -0.200                              XXXXXX
   5  -0.004                                    X
   6  -0.365                        XXXXXXXXXXX
   7  -0.383                        XXXXXXXXXXX
   8  -0.512                  XXXXXXXXXXXXXX
   9  -0.488                   XXXXXXXXXXXXX
  10  -0.175                              XXXXX
  11  -0.125                                XXXX
  12   0.078                                    XXX
  13   0.165                                    XXXXX
  14   0.159                                    XXXXX
  15   0.394                                    XXXXXXXXXXX
  16   0.429                                    XXXXXXXXXXXX
  17   0.350                                    XXXXXXXXXX
```

The autocorrelation function indicates a second-order autoregressive component. We know this because the shape of Figure 11-17 resembles the shape in Figure 11-16(d).

Next we evaluate the shape of our partial autocorrelation function. The command is

**MTB >PACF for series in C2**

The results are shown in Figure 11-18. The strong partial autocorrelation function (it did not die out after one or two periods) indicates a first-order moving average component. This figure is similar to that in Figure 11-16(f).

In the Minitab command to run Box–Jenkins, a value is entered for the order of the autoregressive part of the command structure (the **p** value) and another is entered for the

**FIGURE 11-18** Results of PAFC Command on Tennis Shoe A Sales Data after DIFFerence Command

```
MTB >PACF C2
              -1.0 -0.8 -0.6 -0.4 -0.2  0.0  0.2  0.4  0.6  0.8  1.0
               +----+----+----+----+----+----+----+----+----+----+
   1   0.402                                    XXXXXXXXXXX
   2   0.181                                    XXXXXX
   3  -0.092                                 XXX
   4  -0.329                         XXXXXXXXX
   5   0.217                                    XXXXXX
   6  -0.391                        XXXXXXXXXXX
   7  -0.259                           XXXXXXX
   8  -0.378                        XXXXXXXXXX
   9  -0.043                                   XX
  10  -0.060                                   XXX
  11  -0.041                                   XX
  12  -0.055                                 XXXXX
  13   0.073                                    XXX
  14  -0.182                                XXXXXX
  15   0.120                                    XXXX
  16   0.102                                    XXXX
  17  -0.022                                   XX
```

order of the moving average part of the command structure (the `q` value). Since our `ACF` looked like the chart in Figure 11-16(d), the `p` value used in the Minitab command will be `2`. Since the `PACF` calculation produced an output resembling Figure 11-16(f), we will assign `q` the value of `1` in the Minitab command.

## Minitab Commands

The Minitab command `ARIMa` fits nonseasonal and seasonal models to a time series. The general form of the Minitab command lines is

```
ARIMa p=K, d=K, q=K (seasonal values sp=K, sd=K, sq=K, s=K) for C[;]
   [[FOREcast start after period K] forecast length K periods.]
```

From our discussion above we know that in our example `p = 2` and `q = 1`. Those values will simply be repeated in the seasonal portion of the model; thus `sp = 2` and `sq = 1`. The value for `d` is the number of differences to be applied; since we did one difference calculation (i.e., an annual computation), our `d` value is `1`. That value is also repeated in our seasonal section as `sd = 1`. The final value to determine is the seasonal lag `s`. Our model is built on four seasons per year, thus `s = 4`.

In the subcommand, we will instruct a forecast of 8 periods starting after the 56th period. This will allow us to evaluate the strength of our model and, if feasible, to forecast the next 4 periods of activity. (If you are using a PC *without* an Intel 8087 math coprocessor, stock up on groceries. This computation will take a while! In fact, this computation should be restricted to mainframe computers and PCs with at least an 8087 coprocessor.)

Our commands are thus

```
MTB >ARIMa p=2, d=1, q=1, sp=2, sd=1, sq=1, s=4 for C1;
SUBC> FOREcast start after period 56 forecast length 8 periods.
```

In the following shortened version of this same command group, you can easily see the essential input:

```
MTB >ARIM 2 1 1 2 1 1 4 C1;
SUBC> FORE 56, 8.
```

The output of this command is shown in Figure 11-19.

Our first step in evaluating this output is to evaluate the chi-square values. Looking them up in Appendix B, Table 4, you will find that none are significant. If these statistics were significant, we would reject this model and try again.

Although we will not evaluate all of the printout, we will look at a couple of things. Our primary interest is the forecast. The last section listed in the printout contains the forecasts for periods 57 through 64. For each period the following information is given: the forecasted amount, an estimate of the standard 95% confidence interval (just like any other confidence interval we have studied), and the actual sales level for periods where this information is available.

Clearly, this is not an extensive analysis of Box–Jenkins forecasting. Our purpose here is simply to illustrate another approach to studying time series data. One of the beauties of Minitab is its ability to take us places that our own computational skills cannot. Box–Jenkins is certainly an example of this.

# SUMMARY

**FIGURE 11-19** Minitab Box–Jenkins Analysis with `ARIMa` Command for Tennis Shoe A Sales Data, 1972–1986

```
MTB >ARIM 2, 1, 1, 2, 1, 1, 4 C1;
SUBC> FORE 56, 8.

** Convergence criterion not met after 25 iterations

Final Estimates of Parameters
Type       Estimate    St. Dev.    t-ratio
AR   1     -1.3215     0.1178      -11.21
AR   2     -0.5724     0.1245       -4.60
SAR  4     -0.0541     0.3256       -0.17
SAR  8     -0.3962     0.1601       -2.47
MA   1     -0.9489     0.0696      -13.64
SMA  4      0.2459     0.3785        0.65

Differencing: 1 regular, 1 seasonal of order 4
No. of obs.: Original series 60, after differencing 55
Residuals:   SS =  777650  (backforecasts excluded)
             MS =   15870  DF = 49
```

```
Modified Box-Pierce chisquare statistic
Lag            12           24           36           48
Chisquare   8.1(DF= 6)  24.2(DF=18)  36.0(DF=30)  40.5(DF=42)

Forecasts from period 56
                      95 Percent Limits
Period   Forecast    Lower      Upper      Actual
  57     3747.26    3500.29    3994.22    3625.30
  58     3947.58    3656.02    4239.13    3929.24
  59     2527.87    2206.51    2849.24    2389.11
  60     2350.12    1964.03    2736.21    2193.64
  61     3746.72    3261.42    4232.02
  62     4021.42    3447.07    4595.77
  63     2581.95    1957.93    3205.97
  64     2409.49    1708.22    3110.75
```

However, before we leave this subject, we should issue a word of caution. There is no guarantee that our forecasted sales will prove accurate. Like most forecasts, the predictions for periods closer in time will generally be more accurate than predictions for more distant periods. Thus, in our example, we can be reasonably confident of our prediction for period 61, but much less confident of our prediction for, say, period 68. Always keep in mind the planning horizon of the organization. If a company is in a stable environment, then predictions for two or more years into the future might make sense. On the other hand, in organizations where sales are very erratic, predictions of sales two years hence based upon past sales patterns would clearly be unwise.

# SUMMARY

In many situations, we can use simple linear and multiple regression to forecast business variables with acceptable accuracy. However, when a data set contains predictable patterns of variation, time series analysis will generate more accurate forecasts. It is possible to isolate trend, seasonal, and cyclical components in such data sets.

## DISCUSSION QUESTIONS

1. What is the major difference between regression analysis and time series analysis?
2. Define the four components used in time series analysis.
3. Which component is not used in forecasting? Why?

## PROBLEMS

These problems can be solved by either time series analysis or Box–Jenkins analysis.

**11-1** American Seeds, a well-established company with stable sales growth, sells seeds to the home gardener. Recently it has attempted to manage its inventories better through detailed analysis of past sales. Following are quarterly sales of seed packets for the last 15 years:

| Year | Winter | Spring | Summer | Fall |
|---|---|---|---|---|
| 1972 | 2,792.11 | 5,612.43 | 5,014.49 | 2,610.09 |
| 1973 | 2,600.59 | 5,471.79 | 5,212.67 | 2,897.36 |
| 1974 | 2,942.73 | 5,792.82 | 5,219.30 | 2,800.14 |
| 1975 | 3,022.57 | 5,998.53 | 5,444.16 | 2,879.44 |
| 1976 | 3,306.80 | 5,504.01 | 5,302.02 | 2,990.47 |
| 1977 | 3,117.92 | 5,785.16 | 5,600.93 | 3,021.92 |
| 1978 | 3,250.34 | 5,901.90 | 5,409.82 | 3,340.10 |
| 1979 | 3,451.40 | 6,020.47 | 5,513.37 | 3,116.31 |
| 1980 | 3,660.92 | 6,315.62 | 5,718.33 | 3,523.60 |
| 1981 | 3,512.88 | 6,550.17 | 5,820.81 | 3,590.64 |
| 1982 | 3,800.12 | 6,316.11 | 5,544.77 | 3,410.55 |
| 1983 | 3,927.31 | 6,793.90 | 5,690.90 | 3,709.79 |
| 1984 | 4,240.64 | 6,837.71 | 5,702.62 | 3,911.82 |
| 1985 | 4,010.70 | 6,525.33 | 5,720.39 | 4,001.32 |
| 1986 | 4,390.83 | 6,902.43 | 5,715.17 | 4,014.10 |

a) Perform a time series analysis using `TSPLot`. Describe the time series relationships present in the plot.
b) Using Minitab, regress sales against time.
c) What percent of the variation in sales can be predicted by the trend component?
d) Remove the trend component from the forecast variable.
e) Check to see that the trend component has been removed. Perform a time series analysis on the variable representing seasonal, cyclical, and random variation. Compare this `TSPLot` to the one created in (a). What do you notice?
f) Isolate the cyclical component by computing a moving average.
g) Remove the cyclical component from the forecast variable.

## PROBLEMS

**h)** Perform a time series analysis on the variable representing seasonal and random variation. How does it differ from the figure created in (e)?

**i)** Isolate and remove the seasonal component.

**j)** Compare the predictable portion of the model ($T \times C \times S$) to the original data.

**k)** Create a time series plot for the predictable portion of the model ($T \times C \times S$). Does the figure resemble the one created in (a)? What does this indicate?

**l)** Using the time series model just developed, predict sales of American Seeds for the next eight quarters.

**m)** During the next eight quarters, when should minimum inventory be reordered?

**11-2** Employees of a kite and wind sock manufacturing company set out to develop an equation that would help the manufacturing division improve production planning. The following data, adjusted for inflation, shows quarterly sales:

| Year | Winter | Spring | Summer | Fall |
|------|--------|--------|--------|------|
| 1966 | 1,433.02 | 3,800.23 | 2,433.86 | 2,510.11 |
| 1967 | 1,612.10 | 3,884.00 | 2,592.22 | 2,409.73 |
| 1968 | 1,344.75 | 3,916.12 | 2,401.19 | 2,608.60 |
| 1969 | 1,790.37 | 3,709.31 | 2,733.42 | 2,790.02 |
| 1970 | 1,640.88 | 3,817.92 | 2,850.64 | 2,897.33 |
| 1971 | 1,739.10 | 3,933.85 | 2,697.04 | 2,682.20 |
| 1972 | 1,845.28 | 4,140.90 | 2,900.69 | 3,040.19 |
| 1973 | 1,910.61 | 4,360.13 | 3,104.30 | 3,016.02 |
| 1974 | 1,802.91 | 4,202.10 | 3,331.87 | 3,240.13 |
| 1975 | 1,940.12 | 4,460.39 | 3,212.14 | 3,104.91 |
| 1976 | 1,900.33 | 4,502.79 | 3,310.97 | 3,300.09 |
| 1977 | 2,200.36 | 4,600.84 | 3,377.02 | 3,210.18 |
| 1978 | 2,110.14 | 4,529.07 | 3,112.16 | 3,462.66 |
| 1979 | 2,292.73 | 4,802.21 | 3,300.99 | 3,500.21 |
| 1980 | 2,401.02 | 4,850.18 | 3,412.02 | 3,733.71 |
| 1981 | 2,312.16 | 4,736.01 | 3,550.56 | 3,610.89 |
| 1982 | 2,413.10 | 4,865.60 | 3,600.20 | 3,830.19 |
| 1983 | 2,500.07 | 4,912.02 | 3,522.44 | 3,900.12 |
| 1984 | 2,409.57 | 5,300.10 | 3,610.65 | 3,710.09 |
| 1985 | 2,603.99 | 5,100.03 | 3,822.03 | 3,830.61 |
| 1986 | 2,670.02 | 5,210.60 | 3,970.10 | 3,900.40 |

**a)** Using `TSPLot`, perform a time series analysis on the data. Describe any relationships that are present.

**b)** Using Minitab, regress sales against time.

**c)** What percent of the variation in sales can be predicted by the trend component?

**d)** Remove the trend component from the forecast variable.

**e)** Check to see that the trend component has been removed. Perform a time series analysis on the variable representing seasonal, cyclical, and random variation. Compare this time series plot to the one created in (a). What are your observations?

**f)** Isolate the cyclical component by computing a moving average.

**g)** Remove the cyclical component from the forecast variable.

**h)** Isolate and remove the seasonal component.

**i)** Compare the predictable portion of the model ($T \times C \times S$) to the original data.

**j)** Use a time series plot to analyze the predictable portion of the model. Does the figure resemble the one created in (a)?

**k)** Using the time series model, predict sales for the next eight quarters.

**l)** If you heard rumors that a large distributor of imported kites intended to enter the market in the next year, would it be wise to make sales predictions for the next eight quarters? Why or why not?

**11-3** In recent years farming has become less profitable. As a result, Farmers' Equipment, Inc., foresees a substantial sales drop. Help the company predict sales for the next two years. The following is data on quarterly sales volumes for the last 17 years:

| Year | Winter | Spring | Summer | Fall |
|------|--------|--------|--------|------|
| 1970 | 604.72 | 597.02 | 356.02 | 172.45 |
| 1971 | 544.32 | 577.06 | 340.79 | 160.66 |
| 1972 | 511.16 | 561.23 | 331.19 | 170.11 |
| 1973 | 532.04 | 584.72 | 320.85 | 151.99 |
| 1974 | 529.89 | 560.91 | 345.16 | 143.02 |
| 1975 | 502.01 | 531.07 | 319.01 | 110.11 |
| 1976 | 490.03 | 500.66 | 310.14 | 147.16 |
| 1977 | 471.61 | 510.14 | 302.16 | 130.18 |
| 1978 | 452.04 | 499.16 | 290.04 | 121.49 |
| 1979 | 463.07 | 480.70 | 281.66 | 116.64 |
| 1980 | 440.00 | 410.01 | 280.10 | 112.01 |
| 1981 | 431.79 | 470.34 | 261.92 | 108.95 |
| 1982 | 412.88 | 461.40 | 235.50 | 129.14 |
| 1983 | 407.67 | 452.10 | 251.16 | 100.49 |
| 1984 | 381.30 | 491.16 | 240.19 | 96.12 |
| 1985 | 363.16 | 434.87 | 210.10 | 81.69 |
| 1986 | 371.09 | 430.16 | 201.69 | 90.88 |

**a)** Use `TSPLot` to perform an analysis of the data. Describe the relationships in the time series plot.

**b)** Using Minitab, regress sales against time.

**c)** What percent of the variation in sales can be predicted by the trend component?

**d)** Remove the trend component from the forecast variable.

**e)** Isolate the cyclical component by computing the moving average.

**f)** Remove the cyclical component from the forecast variable.

**g)** Isolate and remove the seasonal component.

**h)** Compare the predictable portion of the model ($T \times C \times S$) to the original data.

**i)** Perform an analysis, using `TSPLot`, on the predictable portion of the model ($T \times C \times S$). Does the figure resemble the one created in (a) of the original data?

**j)** Using the times series model, predict sales for the next eight quarters.

**k)** Do your predicted values make good managerial sense? Why or why not?

**11-4** A national greeting card company needs your help in determining monthly sales volume. Because a quarter of its manufacturing staff is part-time, forecasts of sales volume per month are needed. The following is data on monthly sales volume.

## PROBLEMS

| Year | 1 | 2 | 3 | 4 | 5 | 6 | 7 | 8 | 9 | 10 | 11 | 12 |
|---|---|---|---|---|---|---|---|---|---|---|---|---|
| 1977 | 5,622.01 | 8,006.19 | 7,425.26 | 5,420.22 | 7,360.20 | 7,540.60 | 7,243.80 | 6,710.03 | 5,141.32 | 6,140.33 | 5,970.20 | 8,201.79 |
| 1978 | 5,790.07 | 8,114.37 | 7,591.33 | 5,597.16 | 7,515.63 | 7,711.22 | 7,350.51 | 6,655.90 | 5,371.74 | 6,317.12 | 6,212.67 | 8,345.61 |
| 1979 | 5,912.16 | 8,010.32 | 7,660.10 | 5,710.90 | 7,702.40 | 7,600.02 | 7,510.63 | 6,777.31 | 5,470.61 | 6,404.14 | 6,149.89 | 8,533.11 |
| 1980 | 5,810.91 | 8,260.91 | 7,790.02 | 5,666.55 | 7,600.01 | 7,791.46 | 7,631.93 | 6,820.97 | 5,631.21 | 6,571.04 | 6,241.32 | 8,440.91 |
| 1981 | 6,020.10 | 8,314.82 | 7,803.09 | 5,697.65 | 7,743.82 | 7,903.70 | 7,514.02 | 6,972.29 | 5,670.10 | 6,730.09 | 6,350.71 | 8,730.56 |
| 1982 | 6,140.70 | 8,327.27 | 7,981.92 | 5,708.71 | 7,914.50 | 7,946.89 | 7,709.97 | 7,003.40 | 5,911.12 | 6,610.92 | 6,412.76 | 8,941.61 |
| 1983 | 6,221.11 | 8,490.43 | 7,880.31 | 5,808.34 | 7,830.61 | 8,140.33 | 7,839.75 | 6,900.82 | 5,731.49 | 6,741.82 | 6,609.81 | 9,001.14 |
| 1984 | 6,117.23 | 8,527.81 | 7,999.02 | 5,910.01 | 7,951.71 | 8,220.66 | 7,901.30 | 7,100.30 | 5,833.88 | 6,891.13 | 6,742.20 | 9,210.27 |
| 1985 | 6,431.12 | 8,710.11 | 8,110.11 | 5,871.42 | 8,140.11 | 8,341.87 | 7,810.65 | 7,250.06 | 5,960.01 | 6,910.07 | 6,811.09 | 9,440.30 |
| 1986 | 6,714.57 | 8,694.81 | 8,290.60 | 6,104.12 | 8,240.01 | 8,310.43 | 7,990.62 | 7,390.47 | 6,010.12 | 7,102.45 | 6,901.06 | 9,512.16 |

**a)** Create a time series plot with `TSPLot` to analyze the data. Describe the time series relationships in the figure.
**b)** Using Minitab, regress sales against time.
**c)** What percent of the variation in sales can be predicted by the trend component?
**d)** Remove the trend component from the forecast variable.
**e)** Isolate the cyclical component by computing the moving average.
**f)** Remove the cyclical component from the forecast variable.
**g)** Isolate and remove the seasonal component.
**h)** Compare the predictable portion of the model ($T \times C \times S$) to the original data. What does this tell you?
**i)** Using the time series model, predict sales for the next 12 months.
**j)** How can these predictions help with the management of the part-time work force?

**11-5** Fun World Amusement Park is open year round. As part of its new marketing plan, the park intends to increase advertising during the months of low attendance. Your job is to develop a model that predicts attendance during each month of the year. Monthly attendance data is given in the following:

| Year | 1 | 2 | 3 | 4 | 5 | 6 | 7 | 8 | 9 | 10 | 11 | 12 |
|---|---|---|---|---|---|---|---|---|---|---|---|---|
| 1977 | 7,912.12 | 7,833.02 | 8,651.22 | 7,910.49 | 9,716.67 | 10,810.82 | 10,831.70 | 10,590.16 | 7,789.82 | 7,840.81 | 7,412.70 | 10,491.81 |
| 1978 | 7,794.49 | 7,960.69 | 8,711.42 | 8,010.16 | 9,801.10 | 10,960.16 | 10,901.11 | 10,310.17 | 7,841.16 | 7,990.10 | 7,589.16 | 10,525.28 |
| 1979 | 7,840.33 | 8,010.44 | 8,790.60 | 8,144.32 | 9,990.42 | 11,040.17 | 11,140.21 | 10,455.75 | 7,916.12 | 8,141.71 | 7,697.44 | 10,667.91 |
| 1980 | 8,002.82 | 8,270.06 | 8,880.11 | 8,321.22 | 10,020.16 | 11,249.18 | 10,982.01 | 10,497.16 | 8,001.91 | 8,266.49 | 7,701.80 | 10,701.30 |
| 1981 | 8,114.66 | 8,410.11 | 8,910.16 | 8,497.81 | 10,144.30 | 11,447.16 | 11,197.16 | 10,591.32 | 7,906.16 | 8,390.90 | 7,541.89 | 10,651.11 |
| 1982 | 8,370.12 | 8,296.10 | 8,945.52 | 8,360.11 | 10,220.90 | 11,399.60 | 11,290.33 | 10,671.49 | 8,140.33 | 8,482.30 | 7,641.12 | 10,713.49 |
| 1983 | 8,499.16 | 8,455.61 | 9,125.75 | 8,526.79 | 10,360.41 | 11,459.90 | 11,402.82 | 10,701.77 | 8,210.40 | 8,301.87 | 7,812.13 | 10,921.82 |
| 1984 | 8,510.90 | 8,497.43 | 9,220.41 | 8,690.81 | 10,410.72 | 11,525.75 | 11,311.70 | 10,812.33 | 8,349.67 | 8,517.67 | 7,930.87 | 10,980.82 |
| 1985 | 8,441.01 | 8,591.16 | 9,290.94 | 8,731.14 | 10,260.16 | 11,660.14 | 11,450.31 | 10,760.40 | 8,410.14 | 8,604.40 | 7,842.26 | 11,440.95 |
| 1986 | 8,566.02 | 8,640.89 | 9,361.16 | 8,791.91 | 10,530.91 | 11,791.34 | 11,659.02 | 10,891.11 | 8,397.90 | 8,501.16 | 7,901.51 | 11,260.70 |

**a)** Create a time series plot with `TSPLot`. Describe the time series relationships that are present.
**b)** Using Minitab, regress sales against time.
**c)** What percent of the variation in sales can be predicted by the trend component?
**d)** Remove the trend component from the forecast variable.

**e)** Isolate the cyclical component by computing the moving average.
**f)** Remove the cyclical component from the forecast variable.
**g)** Isolate and remove the seasonal component.
**h)** Compare the predictable portion of the model ($T \times C \times S$) to the original data.
**i)** Using the time series model, predict sales for the next 12 months.
**j)** When will Fun World choose to advertise?

# 12

# PERSPECTIVE

No new tests, no more computer commands. Let's just talk a bit about good sense.

In the previous 11 chapters we discussed several elementary statistical methods. This chapter will be quite different. Here we will discuss morality in the use of statistics—the ethics, if you will, of statistical analysis.

Actually, we have been trying to do that since the very first chapter, and we have frequently paused to discuss the proper and improper use of statistical tests. In this chapter we will refer back to many of those discussions. Although a chapter-by-chapter discussion of the ethical issues in statistical analysis is the best way to approach this topic, we have chosen to devote an entire chapter to this critical issue.

## 12-1 INFORMATION ETHICS

We live in the midst of an information revolution. Because of the avalanche of information, we often find ourselves accepting new ideas uncritically. As individuals with a newly acquired understanding of statistical logic and technique, you face a twofold responsibility:

**(1)** You must not blindly accept the statistical evidence of others.
**(2)** You must not contribute to the statistical inaccuracies already in abundance.

If it is true that a little knowledge is a dangerous thing, then, having completed this course, you are extremely dangerous. When the authors of this text were undergraduates (and we are not exactly ancient now), our teachers had a powerful tool for safeguarding statistical integrity—the pencil. They did not hit or jab with it; they computed with it. To perform statistical computations, long and tedious calculations based on an extensive theoretical background were required.

People did not misuse multiple regression or analysis of variance because, in general, it was computationally too difficult to do so. Like many other modern moral trespasses, our predecessors simply did not have the opportunity to commit such sins. Minitab (and other statistical languages) have changed all that. Suddenly statistical analysis is available to the masses. Some folks, however, are simply not ready. In the course of this text, we have frequently asked you to rerun a particular analysis several different ways. For example, we have asked you to create not only a 90% confidence interval but also a 95% and a 99% interval. In regression we suggested you run a regression command backwards just to see what might happen if you specified your dependent and independent variables in the reverse order. In plotting we suggested you reverse the *X*- and *Y*-axis.

All of these exercises were designed so you could gain a better insight into each statistical technique. However, while we were illustrating why one test was more powerful than another (such as the four different types of *t* test), we were also showing you how to manipulate a statistical analysis. In the good old days, to rerun a multiple regression would have required several hours of pencil pushing. A second or third or fourth confidence interval required numerous additional calculations, and each one of those human calculations was subject to human error. Yes, the pencil was a good tool for statistical integrity. It was simply too time consuming to play with numbers long enough to manipulate them.

Minitab, however, gives you the ability to "shop around" until you get some type of statistic that supports your analysis, even when proper statistical techniques do *not* support your position. If that use of Minitab is all you have learned in this book, we as authors consider ourselves total failures.

In short, the computer and statistical packages like Minitab have allowed for great leaps forward in the teaching and comprehension of statistics. Unfortunately, they have also allowed for a quantum jump in mischief! Whether you are evaluating someone else's analysis or setting up your own, you need to ask several questions.

▶ ▶ ▶ ▶ ▶ **Question 1** *What are you trying to prove?* Have you established your experimental hypothesis so that you will obtain an answer? Remember, if you fail to reject your null hypothesis, *you have not proven anything*. In research, we have a word for a study that accepts the null hypothesis—*inconsequential*. Also remember that the test of your hypothesis must be established *before* you begin the test. You cannot simply collect data and then see what you can prove from it.

You must keep all of our discussion of this point in Chapter 7 in mind. It is easy to try to prove a point and use the wrong test. As the power of a test increases, the number of assumptions required by the test also increases. Parametric and nonparametric tests require different types of data and employ different procedures.

In Chapter 10 we discussed regression analysis. The potential for abuse in this area is immense. For example, with regression and correlation analysis, the performance of the stock market has been related to whether the AFL or the NFL team won the Super Bowl.

### SECTION 12·1 Information Ethics

Clearly, such correlations are spurious! Always ask, "What am I trying to prove and why am I trying to prove it?"

▶ ▶ ▶ ▶ ▶ **Question 2** *Will the statistical test chosen answer the research question adequately?* All too often an improper test is applied in statistical analysis. This problem has become so prevalent since the advent of computer technology and statistical packages that the term "Type III error" has been coined to refer to it. The Minitab program is particularly susceptible to this type of misuse. As long as the proper number of observations is found in each column, Minitab (and any similar statistical language) will run any test commanded.

Early in Chapter 3 we discussed the improper use of data. In our illustration there we noted that there is no 25.5th state of the union (the "average" state), even if such an arithmetic mean can be calculated. If the data is ordinal but you run a metric data test, Minitab will be happy to oblige!

▶ ▶ ▶ ▶ ▶ **Question 3** *Can we meet the assumptions of the statistical test in question?* Remember that assumptions are what the developers of the particular statistical test assumed. As users of their statistical tests, we must take their assumptions as *givens*. If we cannot be certain that we are meeting the assumptions of any particular test, then the results may be invalid.

We have presented only a very small number of statistical techniques in this text; there are hundreds or maybe even thousands in existence. There is essentially a statistical technique for any question. Each of these tests requires certain assumptions about the data. Thus you must make sure that you use a test that fits your data and has assumptions that can be met.

In Chapter 3 we first mentioned the importance of this point. There is a very real danger of subjecting noninterval data to statistical tests designed only for interval data. In Chapter 8, we mentioned numerous times where the use of noninterval data is inappropriate. In particular, a *t* test is frequently used on nominal or ordinal data. In Chapter 9, we completed our discussion of tests for metric data and developed our discussion of nonparametric testing. Chapters 8 and 9 contain several illustrations of improper statistical procedures. We always like to use the most powerful test available, but we must meet the assumptions for that test.

▶ ▶ ▶ ▶ ▶ **Question 4** *Have you randomly selected your data?* This is not simply another assumption. There is no known statistical technique for dealing with biased (that is, nonrandom) data. If data is not randomly selected, you must not use it and you cannot accept statistical inferences based on it. Likewise, you must consider the parent population of a sample. For example, if you could rationalize that class registration is a random process, your classmates would be a random sample. But what are you and your classmates a random sample of? You simply represent the population of students available and wishing to take this course. You do not represent all the students in the world, or in the country, or in your school. If you randomly select students walking past the library at noon, you are not getting a representative sample of all students, just of those students who are walking by the library at noon. This is important because our inferential statements pertain *only* to the population being sampled.

Probably the most common mistake in statistical analysis is misinterpreting which population is being studied. Frequently a sample is carefully drawn with proper statistical techniques and then accurately analyzed, but statistical inferences are made for the wrong population. Unfortunately, this is also one of the easiest ways to purposefully abuse statistics. There are several illustrations of this in the next section of this chapter.

In Chapter 4 we illustrated problems in proper sampling. Improper sampling can be intentional or unintentional. However, in either case, the results are the same: garbage in, garbage out. If you have sampled the wrong group, no statistical procedure will allow you to draw conclusions about another group or population.

▶ ▶ ▶ ▶ ▶ **Question 5** *Have the results been presented so that they clarify rather than distort the issue?* This is a difficult subject. Distortion can take one of two forms. First, sophisticated statistical tools can be used to "do a snow job" on the users of the statistical analysis. These types of abusers are relatively easy to catch, since they tend to be a bit leery of data they do not understand. The second, and more dangerous, type of distortion is when data is presented so that the users think they understand it and are purposefully misled. Such presentations include the use of plots, histograms, and pie charts.

We often see graphs and tables constructed to present a certain point of view rather than an unbiased appraisal of the facts. Do both axes on the chart start at zero, or is it clearly marked that the axes start at another value? Are the increments in the chart or plot consistent, or does the width of the increments change? Whenever you see evidence of manipulations of this type, you should assume the worst and proceed accordingly. Typically, someone is trying to emphasize their point of view.

In our Minitab analysis, we have seen many examples of how the formulation of a problem or the selection of a significance level can have a dramatic effect on the results. Be careful whenever one part of an analysis is done at one level of significance and another part is done at another. Also be careful of unusual significance levels. Technically, a statistician can set the level of significance for a given test at 87%, but we have never seen it done! Most statistical analyses are done with levels of significance of 90%, 95%, 98%, or 99%. Other values might be acceptable, but their use should raise a red flag of warning. You have seen how changing a confidence interval from the 90% level to the 99% level spreads the width sufficiently to include some values that are rejected at the former level. You must not use statistics to manipulate, but rather to clarify information!

We first discussed this issue in Chapter 1. For proper statistical analysis, standards must be established before testing and adhered to throughout the analysis. In Chapter 3 we mentioned the issue of changing the axis of a histogram to alter the appearance of the data. Such manipulation is one of the most prevalent misuses of statistical analysis. We again cautioned you on this issue at the end of Chapter 3. In Chapter 7 we discussed the issue of changing significance levels after seeing the data. In Chapter 8 we further discussed changing a significance level in a hypothesis test so that a value could either be accepted or rejected.

Computer technology has greatly contributed to this problem. It is very easy to experiment with significance levels and confidence intervals until you find one that suits your needs. Nevertheless, by all accepted statistical standards, such a procedure is inappropriate and unethical.

## 12-2 EXAMPLES OF MISUSE

An interesting example of the misuse of statistical analysis occurred on TV in early 1986. A pharmaceutical company that manufactures a name-brand aspirin surveyed 1,000 doctors and asked them, "If you were stranded on a desert island and could have only one type of pain reliever, which would it be?" They were then given a choice of four responses: (1) the name-brand aspirin, (2) a brand of nonaspirin, (3) an extra-strength form of the same nonaspirin, or (4) a second brand of nonaspirin. All three of the nonaspirin products were based on the same chemical group.

Let's look closely at what we know and don't know about the ad. There are two groups of assumptions in a survey of this type. The first group concerns the sample—that is, the doctors themselves. Since a purely random sample of all the doctors in the United States would be very expensive, the sample was probably not purely random. Most likely it was a convenience sample of some type, such as a mail survey, which is biased by the fact that only certain types of people return mail surveys. People who do not return such surveys constitute a completely different population of doctors who may have different opinions.

Another possible way to survey 1,000 doctors is to survey doctors by telephone who were chosen from a city's Yellow Pages. Clearly, all such doctors would be from the same region of the country. The types of doctors attracted to the open spaces of Wyoming and Colorado would probably be different from the types attracted to New York City or Los Angeles.

Another method of surveying is for the company's drug representatives to ask the question as they make their normal sales rounds. This sample would clearly be biased toward doctors who already use this firm's products.

Finally, the TV ad implies that the pain reliever that doctors would take for themselves is the same type that members of the general public should take. Doctors clearly have greater knowledge of the use of a drug than the average layman. The inference the public is asked to draw is that the drug of choice for a doctor is the same drug that the doctor would prescribe for a nonphysician. Notice that we (the authors) are not saying that doctors would not prescribe the same drug for themselves and others—we are only noting that the ad infers that doctors would do so.

To summarize our critique of the first set of assumptions: The ad surveyed 1,000 doctors. Are they representative of all doctors? What are we to assume about their ages, their geographical location, and so on? What is the relationship between the drug that they would take themselves and the drug that they would prescribe to others?

Answers to these questions would prove helpful in understanding the ad. However, the questions raised so far pertain to *unintentional* sample bias. The next group of issues pertains to *intentional* statistical misrepresentation.

Notice how the survey was set up. There was only one aspirin product and three nonaspirin products, all based on the same chemical compound. *Clearly, the survey was designed to split the responses favoring the nonaspirin product, thus allowing the aspirin product (of which there was only one option) to win.*

The aspirin manufacturer proudly announced that its product was preferred by 2 to 1 over the extra-strength nonaspirin product (product 3). But if you look at the data carefully, a nonaspirin easily won over an aspirin. In fact, it was only by splitting the responses favoring the regular-strength nonaspirin product (product 2) and the extra-strength nonaspirin product (product 3) that the aspirin won. Remember, products 2 and 3 were from the same company; the only difference between them was their strength. Remember the first question in this chapter, "What are you trying to prove?" This ad campaign was clearly designed to prove that the name-brand aspirin was more popular.

We can obtain the same type of results on almost any issue. For example, let's assume we want to survey high school students in a given area about their choice of colleges. Let's also assume there is Hometown University, which is a good, reputable school in the local area with an arts and sciences division and a business administration division. We might word our survey question as follows: "Do you want to go to Hometown U. or to one of the top 10 schools in the country?" Notice that this choice is an either/or option. In all likelihood, Hometown U. would come in second in our survey.

However, we might also word the question differently by listing different schools. Our question might read something like: "Do you want to attend Hometown U. or Stanford, Harvard, Berkeley, Chicago, M.I.T., Cal Tech, Colorado School of Mines, or Carnegie–Mellon School of Engineering?"

In response to this second question, Hometown U. would probably come out on top. The reason is simple. A certain group of students would want to stay near home and friends, and they could receive a very good education right at home. Those students who wish to leave home, however, could choose among numerous high-quality options. Hometown U. does not have a school of mines or an engineering school so we would automatically split off those responses. Because different schools have different academic strengths, the students who want to go away to school will have different choices according to their interest. Also, if Hometown U. is in a nice climate and several of the schools are in a cold climate, students' choices could be affected.

The misuse of a survey based on the second question could now begin. Hometown U. might advertise that "more local high school students prefer to attend our university than Stanford or Harvard or M.I.T.!" If you split people who favor something other than what you want, you can almost always get the results you desire.

For one final illustration, assume that we compare a low-priced American sedan with 15 European sports cars all priced over $30,000. If we take a sample from people who live in Detroit (which is very sensitive to U.S. jobs in the auto industry), we could probably get the American car to come out on top in a survey. Also, since the sports cars are low to the ground, hold only two people, and promote high performance, individuals with large families would probably favor the American sedan. Similarly, a sample of all elderly people would have difficulty getting into and out of a sports car and would probably favor the American car.

In these illustrations, the deception is easy to spot. However, when you examine a statistical analysis in the middle of a long report, such biases are not so easy to discern. Also, in a 30-second TV ad, most people do not have the time to fully understand the facts of an ad. You can be overrun by a quick presentation that does not allow enough time for a proper analysis. Such inundation with data in a short time is apparent in the aspirin advertisement. Besides the flaws already mentioned, the response rates shown on TV did not add to 100%. While in printed form, this problem would probably be caught, most people watching TV would not notice it.

However, whether people catch such deception is really only part of the issue. The advertiser bears the responsibility for the unethical presentation. It knew that the numbers did not add properly, and it knew that the "survey" design would split the responses to prove the advertiser's point.

## 12-3 LONG-RANGE EFFECTS OF STATISTICS

Now that you have completed the course, you might well ask if there is any significance to what you have studied (besides passing the final exam). We strongly feel that this material has tremendous signficance for your business and your personal lives.

If you look at the world around you, there are two major trends that this book relates to directly. The first trend is the increase in the use of computers and in their sophistication, and the second is the increased speed of communication. Computers have allowed us to process mountains of data almost instantaneously. The advent of communications satellites, has enabled us to transmit all the material calculated just as quickly.

To illustrate how these two factors affect our lives, consider the recent "Big Bang" in England, the day the London stock market went to a computer-based system. In and of itself, the event is of limited interest. However, it heightens the awareness that we are now on the verge of worldwide, 24-hour-a-day trading. We are not quite at that point yet, but all the major pieces are now in place.

Communications satellites have collapsed the so-called information float. Almost anything known anywhere in the world this second can be known by the rest of the world in the next second. As we become more and more of a data-drenched society, the question becomes, "What is the upper limit of the amount of data we can effectively use?" The logical answer is the capacity of human beings to absorb data. Individuals able to absorb more data will survive and compete more effectively.

Human beings are capable of storing and handling information in two different ways. First, we can hold, understand, and control a number of isolated bits of data at one time. However, this number is relatively small. A second way to handle data is to aggregate it into categories. Humans have a highly developed ability to gather, summarize, and catalog similar bits of data into categories. As we aggregate data in this manner, we can handle more and more data at one time. As our society becomes more information intensive, more and more aggregation will become essential.

How does aggregation occur? We hope according to the laws of mathematics and statistics. Humans are not as good at aggregating numbers and figures as they are at aggregating concepts and philosophical ideas. Further, people may unintentionally group data inappropriately. For example, combining ordinal and ratio data makes the resulting aggregate nonsense. People more skillful at aggregating may attempt to intentionally deceive others with aggregation like the aspirin ad described earlier.

To us, the authors, there seem to be no forces in our society working against the continued development of an information-based society. The speed with which that information travels and the sheer volume of information available will necessarily lead to greater and greater degrees of aggregation. For each of us to understand and deal with such massive amounts of information, we will have to be able to judge the appropriateness of the means used to summarize it. People unaware of the basic tenets of statistical theory will be susceptible to manipulation in their business careers and their personal lives.

We urge you not to become part of the problem. For example, in using the regression commands in Minitab, you can set the amount of printout with a brief command. To generate the maximum printout just to make your analysis look more impressive is an example of the needless creation of information.

## 12-4 ADVANCED CAPABILITIES OF MINITAB

Throughout this course, we have used the Minitab statistical language. There are several good reasons for this selection. Minitab is inexpensive, user-friendly, and accurate. Above all, however, Minitab is designed for the student. This fact brings with it a mixed bag of blessings and curses.

(1) Minitab will do simple tests and procedures with total accuracy, but it will not do many of the more sophisticated tests that are commonly used.

(2) Minitab helps the beginning student to stay out of trouble by not asking questions that you cannot answer. As a result, it will not allow you to modify a particular technique to your own specifications.

If you need more sophistication than Minitab provides, look into other statistical software. Two excellent statistical packages are SPSS (Statistical Package for the Social Sciences) and BMDP (BioMedical Data Package). Why didn't we write this book using SPSS or BMDP? Those statistical packages were made for researchers who are both knowledgeable and sophisticated in statistical analysis, and those programs are far from being user-friendly. However, do not get the impression that Minitab is a simple and thus useless language. In this text, we have only scratched the surface of Minitab's capabilities. Remember, our goal was to use Minitab as a teaching tool for introductory statistics, not to teach you all about Minitab. You might want to know some of the more powerful options within the Minitab system that we have not discussed in the text. These include the following:

(1) *Matrix Operations*. Minitab allows you to add, subtract, multiply, transform, and invert matrices. Through these techniques, you can perform virtually every statistical test used. It will also compute eigenvectors and eigenvalues, which pave the way for principal components analysis and factor analysis (two rather advanced statistical techniques).

(2) *Plotting Options*. Minitab will also produce two-dimensional representations of three-dimensional surfaces. Also, multiple sets of data can be plotted on the same graph. Future releases of Minitab will feature some rather exotic plotting tools.

(3) *Two-Way Analysis of Variance*. In addition to the simple analysis of variance presented in Chapter 9, Minitab will treat two factors simultaneously. This enables you to evaluate more complex solutions to problems.

**(4)** *Stepwise Regression.* With the stepwise regression command, you can evaluate many variables in a regression model and iteratively select those forming the strongest model. Your instructor may feel otherwise, but we feel that this is the most dangerous statistical technique of all. The potential for abuse is high.

**(5)** *Tables.* Minitab supports very extensive cross tabulation techniques. These visual methods help clarify raw data and are typically used in many types of field research.

We have presented Minitab commands as necessary for you to understand the statistical techniques under discussion. In other words, this is a book on statistics, not Minitab. If at this time you are interested in furthering your knowledge about the Minitab system, refer to the *Minitab Handbook*, 2nd ed. (Boston: Duxbury Press, 1985), or *Minitab Handbook for Business and Economics* (Boston: PWS-Kent, 1988).

As you progress through your remaining coursework, you should find this course helpful for two main reasons:

**(1)** You have become familiar with some very powerful ways of handling and analyzing data. We hope that you will apply the tools and techniques you have learned here in your studies of finance, marketing, or psychology.

**(2)** You have learned that the computer can offer you help in many ways. As we said in the first chapter, go play with the system and your data. Look at your data analysis from several angles. You are not confined to using only one technique.

If you see now or in the future that you have benefited from either of these points, we are pleased to have been a part of the process. Good luck!

# APPENDIX A

**Complete Data File on Employees of Go-Fast Running Shoe Company**

| Employee | Age | Years of Schooling | Productivity (units) | Height (inches) | Employee | Age | Years of Schooling | Productivity (units) | Height (inches) |
|---|---|---|---|---|---|---|---|---|---|
| 1 | 35 | 11 | 230 | 69 | 18 | 30 | 11 | 220 | 71 |
| 2 | 33 | 13 | 230 | 69 | 19 | 29 | 14 | 205 | 69 |
| 3 | 33 | 13 | 235 | 69 | 20 | 32 | 12 | 195 | 69 |
| 4 | 36 | 12 | 220 | 67 | 21 | 34 | 11 | 230 | 69 |
| 5 | 31 | 14 | 190 | 70 | 22 | 31 | 12 | 225 | 71 |
| 6 | 29 | 13 | 175 | 72 | 23 | 29 | 13 | 195 | 72 |
| 7 | 30 | 12 | 190 | 70 | 24 | 32 | 13 | 180 | 69 |
| 8 | 33 | 12 | 195 | 69 | 25 | 34 | 10 | 235 | 69 |
| 9 | 35 | 12 | 225 | 66 | 26 | 35 | 11 | 230 | 70 |
| 10 | 34 | 12 | 240 | 70 | 27 | 34 | 12 | 210 | 71 |
| 11 | 33 | 12 | 195 | 68 | 28 | 27 | 13 | 180 | 75 |
| 12 | 38 | 13 | 255 | 66 | 29 | 32 | 11 | 205 | 70 |
| 13 | 31 | 11 | 225 | 73 | 30 | 38 | 12 | 260 | 66 |
| 14 | 30 | 11 | 210 | 72 | 31 | 32 | 11 | 200 | 72 |
| 15 | 34 | 11 | 220 | 69 | 32 | 29 | 12 | 195 | 70 |
| 16 | 38 | 12 | 260 | 66 | 33 | 29 | 11 | 195 | 71 |
| 17 | 34 | 11 | 200 | 69 | 34 | 32 | 12 | 185 | 71 |

*(continued on next page)*

*(Continued)*

| Employee | Age | Years of Schooling | Productivity (units) | Height (inches) | Employee | Age | Years of Schooling | Productivity (units) | Height (inches) |
|---|---|---|---|---|---|---|---|---|---|
| 35 | 35 | 11 | 230 | 69 | 68 | 41 | 14 | 265 | 63 |
| 36 | 31 | 10 | 195 | 73 | 69 | 36 | 13 | 240 | 67 |
| 37 | 31 | 11 | 230 | 71 | 70 | 35 | 12 | 230 | 68 |
| 38 | 35 | 13 | 240 | 68 | 71 | 32 | 12 | 195 | 69 |
| 39 | 34 | 12 | 225 | 68 | 72 | 28 | 13 | 190 | 74 |
| 40 | 29 | 12 | 190 | 71 | 73 | 35 | 10 | 215 | 68 |
| 41 | 22 | 12 | 120 | 76 | 74 | 34 | 12 | 235 | 69 |
| 42 | 33 | 13 | 210 | 68 | 75 | 32 | 14 | 205 | 70 |
| 43 | 26 | 13 | 185 | 74 | 76 | 32 | 13 | 210 | 71 |
| 44 | 33 | 12 | 235 | 69 | 77 | 27 | 13 | 190 | 74 |
| 45 | 28 | 11 | 160 | 74 | 78 | 32 | 11 | 205 | 70 |
| 46 | 32 | 13 | 215 | 71 | 79 | 29 | 10 | 180 | 71 |
| 47 | 19 | 12 | 130 | 79 | 80 | 32 | 12 | 200 | 70 |
| 48 | 30 | 13 | 185 | 71 | 81 | 26 | 14 | 180 | 74 |
| 49 | 29 | 13 | 185 | 71 | 82 | 35 | 12 | 240 | 70 |
| 50 | 29 | 11 | 190 | 71 | 83 | 35 | 13 | 245 | 69 |
| 51 | 36 | 12 | 230 | 69 | 84 | 27 | 12 | 160 | 75 |
| 52 | 29 | 11 | 190 | 72 | 85 | 31 | 15 | 200 | 69 |
| 53 | 34 | 13 | 220 | 67 | 86 | 33 | 11 | 215 | 70 |
| 54 | 36 | 14 | 220 | 68 | 87 | 36 | 14 | 220 | 65 |
| 55 | 24 | 12 | 140 | 76 | 88 | 33 | 12 | 225 | 67 |
| 56 | 26 | 10 | 140 | 74 | 89 | 31 | 13 | 205 | 73 |
| 57 | 25 | 13 | 145 | 73 | 90 | 30 | 12 | 195 | 72 |
| 58 | 25 | 13 | 165 | 75 | 91 | 33 | 11 | 195 | 69 |
| 59 | 38 | 13 | 235 | 66 | 92 | 34 | 11 | 225 | 69 |
| 60 | 36 | 12 | 200 | 67 | 93 | 36 | 13 | 220 | 67 |
| 61 | 28 | 11 | 230 | 73 | 94 | 30 | 11 | 210 | 71 |
| 62 | 36 | 10 | 230 | 69 | 95 | 29 | 12 | 175 | 73 |
| 63 | 38 | 13 | 230 | 65 | 96 | 32 | 11 | 220 | 69 |
| 64 | 31 | 12 | 200 | 72 | 97 | 29 | 11 | 170 | 73 |
| 65 | 32 | 12 | 200 | 70 | 98 | 36 | 12 | 225 | 66 |
| 66 | 31 | 12 | 200 | 72 | 99 | 23 | 13 | 150 | 75 |
| 67 | 33 | 12 | 200 | 69 | 100 | 28 | 12 | 175 | 74 |

# Appendix B:
## Statistical Tables

**TABLE 1**  Normal Curve Areas

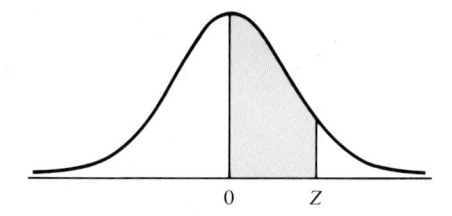

The entries in this table are the probabilities that a random variable having the standard normal distribution assumes a value between 0 and Z; the probability is represented by the area under the curve shaded in the figure. Areas for negative values of Z are obtained by symmetry.

| Z | .00 | .01 | .02 | .03 | .04 | .05 | .06 | .07 | .08 | .09 |
|---|-----|-----|-----|-----|-----|-----|-----|-----|-----|-----|
| 0.0 | .0000 | .0040 | .0080 | .0120 | .0160 | .0199 | .0239 | .0279 | .0319 | .0359 |
| 0.1 | .0398 | .0438 | .0478 | .0517 | .0557 | .0596 | .0636 | .0675 | .0714 | .0753 |
| 0.2 | .0793 | .0832 | .0871 | .0910 | .0948 | .0987 | .1026 | .1064 | .1103 | .1141 |
| 0.3 | .1179 | .1217 | .1255 | .1293 | .1331 | .1368 | .1406 | .1443 | .1480 | .1517 |
| 0.4 | .1554 | .1591 | .1628 | .1664 | .1700 | .1736 | .1772 | .1808 | .1844 | .1879 |
| 0.5 | .1915 | .1950 | .1985 | .2019 | .2054 | .2088 | .2123 | .2157 | .2190 | .2224 |
| 0.6 | .2257 | .2291 | .2324 | .2357 | .2389 | .2422 | .2454 | .2486 | .2517 | .2549 |
| 0.7 | .2580 | .2611 | .2642 | .2673 | .2704 | .2734 | .2764 | .2794 | .2823 | .2852 |
| 0.8 | .2881 | .2910 | .2939 | .2967 | .2995 | .3023 | .3051 | .3078 | .3106 | .3133 |
| 0.9 | .3159 | .3186 | .3212 | .3238 | .3264 | .3289 | .3315 | .3340 | .3365 | .3389 |
| 1.0 | .3413 | .3438 | .3461 | .3485 | .3508 | .3531 | .3554 | .3577 | .3599 | .3621 |
| 1.1 | .3643 | .3665 | .3686 | .3708 | .3729 | .3749 | .3770 | .3790 | .3810 | .3830 |
| 1.2 | .3849 | .3869 | .3888 | .3907 | .3925 | .3944 | .3962 | .3980 | .3997 | .4015 |
| 1.3 | .4032 | .4049 | .4066 | .4082 | .4099 | .4115 | .4131 | .4147 | .4162 | .4177 |
| 1.4 | .4192 | .4207 | .4222 | .4236 | .4251 | .4265 | .4279 | .4292 | .4306 | .4319 |
| 1.5 | .4332 | .4345 | .4357 | .4370 | .4382 | .4394 | .4406 | .4418 | .4429 | .4441 |
| 1.6 | .4452 | .4463 | .4474 | .4484 | .4495 | .4505 | .4515 | .4525 | .4535 | .4545 |
| 1.7 | .4454 | .4564 | .4573 | .4582 | .4591 | .4599 | .4608 | .4616 | .4625 | .4633 |
| 1.8 | .4641 | .4649 | .4656 | .4664 | .4671 | .4678 | .4686 | .4693 | .4699 | .4706 |
| 1.9 | .4713 | .4719 | .4726 | .4732 | .4738 | .4744 | .4750 | .4756 | .4761 | .4767 |
| 2.0 | .4772 | .4778 | .4783 | .4788 | .4793 | .4798 | .4803 | .4808 | .4812 | .4817 |
| 2.1 | .4821 | .4826 | .4830 | .4834 | .4838 | .4842 | .4846 | .4850 | .4854 | .4857 |
| 2.2 | .4861 | .4864 | .4868 | .4871 | .4875 | .4878 | .4881 | .4884 | .4887 | .4890 |
| 2.3 | .4893 | .4896 | .4898 | .4901 | .4904 | .4906 | .4909 | .4911 | .4913 | .4916 |
| 2.4 | .4918 | .4920 | .4922 | .4925 | .4927 | .4929 | .4931 | .4932 | .4934 | .4936 |
| 2.5 | .4938 | .4940 | .4941 | .4943 | .4945 | .4946 | .4948 | .4949 | .4951 | .4952 |
| 2.6 | .4953 | .4955 | .4956 | .4957 | .4959 | .4960 | .4961 | .4962 | .4963 | .4964 |
| 2.7 | .4965 | .4966 | .4967 | .4968 | .4969 | .4970 | .4971 | .4972 | .4973 | .4974 |
| 2.8 | .4974 | .4975 | .4976 | .4977 | .4977 | .4978 | .4979 | .4979 | .4980 | .4981 |
| 2.9 | .4981 | .4982 | .4982 | .4983 | .4984 | .4984 | .4985 | .4985 | .4986 | .4986 |
| 3.0 | .4987 | .4987 | .4987 | .4988 | .4988 | .4989 | .4989 | .4989 | .4990 | .4990 |

Abridged from Table 1 of *Statistical Tables and Formulas*, by A. Hald (New York: John Wiley & Sons, 1952). Reproduced by permission of A. Hald and the publishers, John Wiley & Sons.

## TABLE 2  Critical Values of Student's *t* Distribution

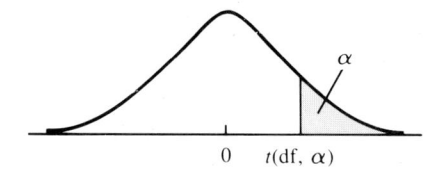

The entries in this table are the critical values for Student's *t* for an area of $\alpha$ in the right-hand tail. Critical values for the left-hand tail are found by symmetry.

| df | \multicolumn{6}{c}{Amount of $\alpha$ in One Tail} |
|---|---|---|---|---|---|---|
|   | 0.25 | 0.10 | 0.05 | 0.025 | 0.01 | 0.005 |
| 1 | 1.000 | 3.08 | 6.31 | 12.7 | 31.8 | 63.7 |
| 2 | 0.816 | 1.89 | 2.92 | 4.30 | 6.97 | 9.92 |
| 3 | 0.765 | 1.64 | 2.35 | 3.18 | 4.54 | 5.84 |
| 4 | 0.741 | 1.53 | 2.13 | 2.78 | 3.75 | 4.60 |
| 5 | 0.727 | 1.48 | 2.02 | 2.57 | 3.37 | 4.03 |
| 6 | 0.718 | 1.44 | 1.94 | 2.45 | 3.14 | 3.71 |
| 7 | 0.711 | 1.42 | 1.89 | 2.36 | 3.00 | 3.50 |
| 8 | 0.706 | 1.40 | 1.86 | 2.31 | 2.90 | 3.36 |
| 9 | 0.703 | 1.38 | 1.83 | 2.26 | 2.82 | 3.25 |
| 10 | 0.700 | 1.37 | 1.81 | 2.23 | 2.76 | 3.17 |
| 11 | 0.697 | 1.36 | 1.80 | 2.20 | 2.72 | 3.11 |
| 12 | 0.695 | 1.36 | 1.78 | 2.18 | 2.68 | 3.05 |
| 13 | 0.694 | 1.35 | 1.77 | 2.16 | 2.65 | 3.01 |
| 14 | 0.692 | 1.35 | 1.76 | 2.14 | 2.62 | 2.98 |
| 15 | 0.691 | 1.34 | 1.75 | 2.13 | 2.60 | 2.95 |
| 16 | 0.690 | 1.34 | 1.75 | 2.12 | 2.58 | 2.92 |
| 17 | 0.689 | 1.33 | 1.74 | 2.11 | 2.57 | 2.90 |
| 18 | 0.688 | 1.33 | 1.73 | 2.10 | 2.55 | 2.88 |
| 19 | 0.688 | 1.33 | 1.73 | 2.09 | 2.54 | 2.86 |
| 20 | 0.687 | 1.33 | 1.72 | 2.09 | 2.53 | 2.85 |
| 21 | 0.686 | 1.32 | 1.72 | 2.08 | 2.52 | 2.83 |
| 22 | 0.686 | 1.32 | 1.72 | 2.07 | 2.51 | 2.82 |
| 23 | 0.685 | 1.32 | 1.71 | 2.07 | 2.50 | 2.81 |
| 24 | 0.685 | 1.32 | 1.71 | 2.06 | 2.49 | 2.80 |
| 25 | 0.684 | 1.32 | 1.71 | 2.06 | 2.49 | 2.79 |
| 26 | 0.684 | 1.32 | 1.71 | 2.06 | 2.48 | 2.78 |
| 27 | 0.684 | 1.31 | 1.70 | 2.05 | 2.47 | 2.77 |
| 28 | 0.683 | 1.31 | 1.70 | 2.05 | 2.47 | 2.76 |
| 29 | 0.683 | 1.31 | 1.70 | 2.05 | 2.46 | 2.76 |
| Z | 0.674 | 1.28 | 1.65 | 1.96 | 2.33 | 2.58 |

*Note*: For df $\geq$ 30, the critical value $t(df,\alpha)$ is approximated by $Z(\alpha)$, given in the bottom row of table.

Adapted from E. S. Pearson and H. O. Hartley, *Biometrika Tables for Statisticians*, vol. I (1966), p. 146. Reprinted by permission of the Biometrika Trustees. The two columns headed "0.10" and "0.01" are taken from Table III (adapted) on p. 46 of Fisher and Yates, *Statistical Tables for Biological, Agricultural and Medical Research*, 6th ed., published by Longman Group Ltd., London, 1974 (previously published by Oliver and Boyd, Edinburgh), and by permission of the authors and publishers.

**TABLE 3a** Critical Values of the $F$ Distribution ($\alpha = .05$)

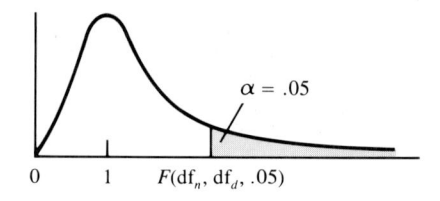

The entries in this table are critical values of $F$ for which the area under the curve to the right is equal to .05.

Degrees of Freedom for Denominator / Degrees of Freedom for Numerator

| df | 1 | 2 | 3 | 4 | 5 | 6 | 7 | 8 | 9 | 10 |
|---|---|---|---|---|---|---|---|---|---|---|
| 1 | 161 | 200 | 216 | 225 | 230 | 234 | 237 | 239 | 241 | 242 |
| 2 | 18.5 | 19.0 | 19.2 | 19.2 | 19.3 | 19.3 | 19.4 | 19.4 | 19.4 | 19.4 |
| 3 | 10.1 | 9.55 | 9.28 | 9.12 | 9.01 | 8.94 | 8.89 | 8.85 | 8.81 | 8.79 |
| 4 | 7.71 | 6.94 | 6.59 | 6.39 | 6.26 | 6.16 | 6.09 | 6.04 | 6.00 | 5.96 |
| 5 | 6.61 | 5.79 | 5.41 | 5.19 | 5.05 | 4.95 | 4.88 | 4.82 | 4.77 | 4.74 |
| 6 | 5.99 | 5.14 | 4.76 | 4.53 | 4.39 | 4.28 | 4.21 | 4.15 | 4.10 | 4.06 |
| 7 | 5.59 | 4.74 | 4.35 | 4.12 | 3.97 | 3.87 | 3.79 | 3.73 | 3.68 | 3.64 |
| 8 | 5.32 | 4.46 | 4.07 | 3.84 | 3.69 | 3.58 | 3.50 | 3.44 | 3.39 | 3.35 |
| 9 | 5.12 | 4.26 | 3.86 | 3.63 | 3.48 | 3.37 | 3.29 | 3.23 | 3.18 | 3.14 |
| 10 | 4.96 | 4.10 | 3.71 | 3.48 | 3.33 | 3.22 | 3.14 | 3.07 | 3.02 | 2.98 |
| 11 | 4.84 | 3.98 | 3.59 | 3.36 | 3.20 | 3.09 | 3.01 | 2.95 | 2.90 | 2.85 |
| 12 | 4.75 | 3.89 | 3.49 | 3.26 | 3.11 | 3.00 | 2.91 | 2.85 | 2.80 | 2.75 |
| 13 | 4.67 | 3.81 | 3.41 | 3.18 | 3.03 | 2.92 | 2.83 | 2.77 | 2.71 | 2.67 |
| 14 | 4.60 | 3.74 | 3.34 | 3.11 | 2.96 | 2.85 | 2.76 | 2.70 | 2.65 | 2.60 |
| 15 | 4.54 | 3.68 | 3.29 | 3.06 | 2.90 | 2.79 | 2.71 | 2.64 | 2.59 | 2.54 |
| 16 | 4.49 | 3.63 | 3.24 | 3.01 | 2.85 | 2.74 | 2.66 | 2.59 | 2.54 | 2.49 |
| 17 | 4.45 | 3.59 | 3.20 | 2.96 | 2.81 | 2.70 | 2.61 | 2.55 | 2.49 | 2.45 |
| 18 | 4.41 | 3.55 | 3.16 | 2.93 | 2.77 | 2.66 | 2.58 | 2.51 | 2.46 | 2.41 |
| 19 | 4.38 | 3.52 | 3.13 | 2.90 | 2.74 | 2.63 | 2.54 | 2.48 | 2.42 | 2.38 |
| 20 | 4.35 | 3.49 | 3.10 | 2.87 | 2.71 | 2.60 | 2.51 | 2.45 | 2.39 | 2.35 |
| 21 | 4.32 | 3.47 | 3.07 | 2.84 | 2.68 | 2.57 | 2.49 | 2.42 | 2.37 | 2.32 |
| 22 | 4.30 | 3.44 | 3.05 | 2.82 | 2.66 | 2.55 | 2.46 | 2.40 | 2.34 | 2.30 |
| 23 | 4.28 | 3.42 | 3.03 | 2.80 | 2.64 | 2.53 | 2.44 | 2.37 | 2.32 | 2.27 |
| 24 | 4.26 | 3.40 | 3.01 | 2.78 | 2.62 | 2.51 | 2.42 | 2.36 | 2.30 | 2.25 |
| 25 | 4.24 | 3.39 | 2.99 | 2.76 | 2.60 | 2.49 | 2.40 | 2.34 | 2.28 | 2.24 |
| 30 | 4.17 | 3.32 | 2.92 | 2.69 | 2.53 | 2.42 | 2.33 | 2.27 | 2.21 | 2.16 |
| 40 | 4.08 | 3.23 | 2.84 | 2.61 | 2.45 | 2.34 | 2.25 | 2.18 | 2.12 | 2.08 |
| 60 | 4.00 | 3.15 | 2.76 | 2.53 | 2.37 | 2.25 | 2.17 | 2.10 | 2.04 | 1.99 |
| 120 | 3.92 | 3.07 | 2.68 | 2.45 | 2.29 | 2.18 | 2.09 | 2.02 | 1.96 | 1.91 |
| $\infty$ | 3.84 | 3.00 | 2.60 | 2.37 | 2.21 | 2.10 | 2.01 | 1.94 | 1.88 | 1.83 |

## TABLE 3a Critical Values of the $F$ Distribution ($\alpha = .05$)

Degrees of Freedom for Numerator

|   | 12 | 15 | 20 | 24 | 30 | 40 | 60 | 120 | $\infty$ |
|---|---|---|---|---|---|---|---|---|---|
| 1 | 244 | 246 | 248 | 249 | 250 | 251 | 252 | 253 | 254 |
| 2 | 19.4 | 19.4 | 19.4 | 19.5 | 19.5 | 19.5 | 19.5 | 19.5 | 19.5 |
| 3 | 8.74 | 8.70 | 8.66 | 8.64 | 8.62 | 8.59 | 8.57 | 8.55 | 8.53 |
| 4 | 5.91 | 5.86 | 5.80 | 5.77 | 5.75 | 5.72 | 5.69 | 5.66 | 5.63 |
| 5 | 4.68 | 4.62 | 4.56 | 4.53 | 4.50 | 4.46 | 4.43 | 4.40 | 4.37 |
| 6 | 4.00 | 3.94 | 3.87 | 3.84 | 3.81 | 3.77 | 3.74 | 3.70 | 3.67 |
| 7 | 3.57 | 3.51 | 3.44 | 3.41 | 3.38 | 3.34 | 3.30 | 3.27 | 3.23 |
| 8 | 3.28 | 3.22 | 3.15 | 3.12 | 3.08 | 3.04 | 3.01 | 2.97 | 2.93 |
| 9 | 3.07 | 3.01 | 2.94 | 2.90 | 2.86 | 2.83 | 2.79 | 2.75 | 2.71 |
| 10 | 2.91 | 2.85 | 2.77 | 2.74 | 2.70 | 2.66 | 2.62 | 2.58 | 2.54 |
| 11 | 2.79 | 2.72 | 2.65 | 2.61 | 2.57 | 2.53 | 2.49 | 2.45 | 2.40 |
| 12 | 2.69 | 2.62 | 2.54 | 2.51 | 2.47 | 2.43 | 2.38 | 2.34 | 2.30 |
| 13 | 2.60 | 2.53 | 2.46 | 2.42 | 2.38 | 2.34 | 2.30 | 2.25 | 2.21 |
| 14 | 2.53 | 2.46 | 2.39 | 2.35 | 2.31 | 2.27 | 2.22 | 2.18 | 2.13 |
| 15 | 2.48 | 2.40 | 2.33 | 2.29 | 2.25 | 2.20 | 2.16 | 2.11 | 2.07 |
| 16 | 2.42 | 2.35 | 2.28 | 2.24 | 2.19 | 2.15 | 2.11 | 2.06 | 2.01 |
| 17 | 2.38 | 2.31 | 2.23 | 2.19 | 2.15 | 2.10 | 2.06 | 2.01 | 1.96 |
| 18 | 2.34 | 2.27 | 2.19 | 2.15 | 2.11 | 2.06 | 2.02 | 1.97 | 1.92 |
| 19 | 2.31 | 2.23 | 2.16 | 2.11 | 2.07 | 2.03 | 1.98 | 1.93 | 1.88 |
| 20 | 2.28 | 2.20 | 2.12 | 2.08 | 2.04 | 1.99 | 1.95 | 1.90 | 1.84 |
| 21 | 2.25 | 2.18 | 2.10 | 2.05 | 2.01 | 1.96 | 1.92 | 1.87 | 1.81 |
| 22 | 2.23 | 2.15 | 2.07 | 2.03 | 1.98 | 1.94 | 1.89 | 1.84 | 1.78 |
| 23 | 2.20 | 2.13 | 2.05 | 2.01 | 1.96 | 1.91 | 1.86 | 1.81 | 1.76 |
| 24 | 2.18 | 2.11 | 2.03 | 1.98 | 1.94 | 1.89 | 1.84 | 1.79 | 1.73 |
| 25 | 2.16 | 2.09 | 2.01 | 1.96 | 1.92 | 1.87 | 1.82 | 1.77 | 1.71 |
| 30 | 2.09 | 2.01 | 1.93 | 1.89 | 1.84 | 1.79 | 1.74 | 1.68 | 1.62 |
| 40 | 2.00 | 1.92 | 1.84 | 1.79 | 1.74 | 1.69 | 1.64 | 1.58 | 1.51 |
| 60 | 1.92 | 1.84 | 1.75 | 1.70 | 1.65 | 1.59 | 1.53 | 1.47 | 1.39 |
| 120 | 1.83 | 1.75 | 1.66 | 1.61 | 1.55 | 1.50 | 1.43 | 1.35 | 1.25 |
| $\infty$ | 1.75 | 1.67 | 1.57 | 1.52 | 1.46 | 1.39 | 1.32 | 1.22 | 1.00 |

Degrees of Freedom for Denominator

From E.S. Pearson and H.O. Hartley, *Biometrika Tables for Statisticians*, vol. I (1958), pp. 159–163. Reprinted by permission of the Biometrika Trustees.

**TABLE 3b** Critical Values of the F Distribution ($\alpha = .025$)

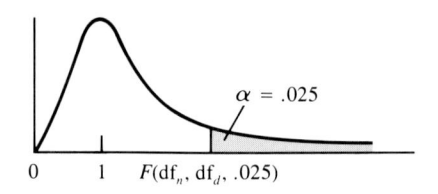

The entries in this table are critical values of $F$ for which the area under the curve to the right is equal to .025.

|   | Degrees of Freedom for Numerator |||||||||| |
|---|---|---|---|---|---|---|---|---|---|---|
| | 1 | 2 | 3 | 4 | 5 | 6 | 7 | 8 | 9 | 10 |
| 1 | 648 | 800 | 864 | 900 | 922 | 937 | 948 | 957 | 963 | 969 |
| 2 | 38.5 | 39.0 | 39.2 | 39.2 | 39.3 | 39.3 | 39.4 | 39.4 | 39.4 | 39.4 |
| 3 | 17.4 | 16.0 | 15.4 | 15.1 | 14.9 | 14.7 | 14.6 | 14.5 | 14.5 | 14.4 |
| 4 | 12.2 | 10.6 | 9.98 | 9.60 | 9.36 | 9.20 | 9.07 | 8.98 | 8.90 | 8.84 |
| 5 | 10.0 | 8.43 | 7.76 | 7.39 | 7.15 | 6.98 | 6.85 | 6.76 | 6.68 | 6.62 |
| 6 | 8.81 | 7.26 | 6.60 | 6.23 | 5.99 | 5.82 | 5.70 | 5.60 | 5.52 | 5.46 |
| 7 | 8.07 | 6.54 | 5.89 | 5.52 | 5.29 | 5.12 | 4.99 | 4.90 | 4.82 | 4.76 |
| 8 | 7.57 | 6.06 | 5.42 | 5.05 | 4.82 | 4.65 | 4.53 | 4.43 | 4.36 | 4.30 |
| 9 | 7.21 | 5.71 | 5.08 | 4.72 | 4.48 | 4.32 | 4.20 | 4.10 | 4.03 | 3.96 |
| 10 | 6.94 | 5.46 | 4.83 | 4.47 | 4.24 | 4.07 | 3.95 | 3.85 | 3.78 | 3.72 |
| 11 | 6.72 | 5.26 | 4.63 | 4.28 | 4.04 | 3.88 | 3.76 | 3.66 | 3.59 | 3.53 |
| 12 | 6.55 | 5.10 | 4.47 | 4.12 | 3.89 | 3.73 | 3.61 | 3.51 | 3.44 | 3.37 |
| 13 | 6.41 | 4.97 | 4.35 | 4.00 | 3.77 | 3.60 | 3.48 | 3.39 | 3.31 | 3.25 |
| 14 | 6.30 | 4.86 | 4.24 | 3.89 | 3.66 | 3.50 | 3.38 | 3.28 | 3.21 | 3.15 |
| 15 | 6.20 | 4.77 | 4.15 | 3.80 | 3.58 | 3.41 | 3.29 | 3.20 | 3.12 | 3.06 |
| 16 | 6.12 | 4.69 | 4.08 | 3.73 | 3.50 | 3.34 | 3.22 | 3.12 | 3.05 | 2.99 |
| 17 | 6.04 | 4.62 | 4.01 | 3.66 | 3.44 | 3.28 | 3.16 | 3.06 | 2.98 | 2.92 |
| 18 | 5.98 | 4.56 | 3.95 | 3.61 | 3.38 | 3.22 | 3.10 | 3.01 | 2.93 | 2.87 |
| 19 | 5.92 | 4.51 | 3.90 | 3.56 | 3.33 | 3.17 | 3.05 | 2.96 | 2.88 | 2.82 |
| 20 | 5.87 | 4.46 | 3.86 | 3.51 | 3.29 | 3.13 | 3.01 | 2.91 | 2.84 | 2.77 |
| 21 | 5.83 | 4.42 | 3.82 | 3.48 | 3.25 | 3.09 | 2.97 | 2.87 | 2.80 | 2.73 |
| 22 | 5.79 | 4.38 | 3.78 | 3.44 | 3.22 | 3.05 | 2.93 | 2.84 | 2.76 | 2.70 |
| 23 | 5.75 | 4.35 | 3.75 | 3.41 | 3.18 | 3.02 | 2.90 | 2.81 | 2.73 | 2.67 |
| 24 | 5.72 | 4.32 | 3.72 | 3.38 | 3.15 | 2.99 | 2.87 | 2.78 | 2.70 | 2.64 |
| 25 | 5.69 | 4.29 | 3.69 | 3.35 | 3.13 | 2.97 | 2.85 | 2.75 | 2.68 | 2.61 |
| 30 | 5.57 | 4.18 | 3.59 | 3.25 | 3.03 | 2.87 | 2.75 | 2.65 | 2.57 | 2.51 |
| 40 | 5.42 | 4.05 | 3.46 | 3.13 | 2.90 | 2.74 | 2.62 | 2.53 | 2.45 | 2.39 |
| 60 | 5.29 | 3.93 | 3.34 | 3.01 | 2.79 | 2.63 | 2.51 | 2.41 | 2.33 | 2.27 |
| 120 | 5.15 | 3.80 | 3.23 | 2.89 | 2.67 | 2.52 | 2.39 | 2.30 | 2.22 | 2.16 |
| ∞ | 5.02 | 3.69 | 3.12 | 2.79 | 2.57 | 2.41 | 2.29 | 2.19 | 2.11 | 2.05 |

Degrees of Freedom for Denominator

**TABLE 3b** Critical Values of the $F$ Distribution ($\alpha = .025$)

|  |  | \multicolumn{9}{c}{Degrees of Freedom for Numerator} |
|---|---|---|---|---|---|---|---|---|---|---|

| | | 12 | 15 | 20 | 24 | 30 | 40 | 60 | 120 | ∞ |
|---|---|---|---|---|---|---|---|---|---|---|
| **Degrees of Freedom for Denominator** | 1 | 977 | 985 | 993 | 997 | 1,001 | 1,006 | 1,010 | 1,014 | 1,018 |
| | 2 | 39.4 | 39.4 | 39.4 | 39.5 | 39.5 | 39.5 | 39.5 | 39.5 | 39.5 |
| | 3 | 14.3 | 14.3 | 14.2 | 14.1 | 14.1 | 14.0 | 14.0 | 13.9 | 13.9 |
| | 4 | 8.75 | 8.66 | 8.56 | 8.51 | 8.46 | 8.41 | 8.36 | 8.31 | 8.26 |
| | 5 | 6.52 | 6.43 | 6.33 | 6.28 | 6.23 | 6.18 | 6.12 | 6.07 | 6.02 |
| | 6 | 5.37 | 5.27 | 5.17 | 5.12 | 5.07 | 5.01 | 4.96 | 4.90 | 4.85 |
| | 7 | 4.67 | 4.57 | 4.47 | 4.42 | 4.36 | 4.31 | 4.25 | 4.20 | 4.14 |
| | 8 | 4.20 | 4.10 | 4.00 | 3.95 | 3.89 | 3.84 | 3.78 | 3.73 | 3.67 |
| | 9 | 3.87 | 3.77 | 3.67 | 3.61 | 3.56 | 3.51 | 3.45 | 3.39 | 3.33 |
| | 10 | 3.62 | 3.52 | 3.42 | 3.37 | 3.31 | 3.26 | 3.20 | 3.14 | 3.08 |
| | 11 | 3.43 | 3.33 | 3.23 | 3.17 | 3.12 | 3.06 | 3.00 | 2.94 | 2.88 |
| | 12 | 3.28 | 3.18 | 3.07 | 3.02 | 2.96 | 2.91 | 2.85 | 2.79 | 2.72 |
| | 13 | 3.15 | 3.05 | 2.95 | 2.89 | 2.84 | 2.78 | 2.72 | 2.66 | 2.60 |
| | 14 | 3.05 | 2.95 | 2.84 | 2.79 | 2.73 | 2.67 | 2.61 | 2.55 | 2.49 |
| | 15 | 2.96 | 2.86 | 2.76 | 2.70 | 2.64 | 2.59 | 2.52 | 2.46 | 2.40 |
| | 16 | 2.89 | 2.79 | 2.68 | 2.63 | 2.57 | 2.51 | 2.45 | 2.38 | 2.32 |
| | 17 | 2.82 | 2.72 | 2.62 | 2.56 | 2.50 | 2.44 | 2.38 | 2.32 | 2.25 |
| | 18 | 2.77 | 2.67 | 2.56 | 2.50 | 2.44 | 2.38 | 2.32 | 2.26 | 2.19 |
| | 19 | 2.72 | 2.62 | 2.51 | 2.45 | 2.39 | 2.33 | 2.27 | 2.20 | 2.13 |
| | 20 | 2.68 | 2.57 | 2.46 | 2.41 | 2.35 | 2.29 | 2.22 | 2.16 | 2.09 |
| | 21 | 2.64 | 2.53 | 2.42 | 2.37 | 2.31 | 2.25 | 2.18 | 2.11 | 2.04 |
| | 22 | 2.60 | 2.50 | 2.39 | 2.33 | 2.27 | 2.21 | 2.14 | 2.08 | 2.00 |
| | 23 | 2.57 | 2.47 | 2.36 | 2.30 | 2.24 | 2.18 | 2.11 | 2.04 | 1.97 |
| | 24 | 2.54 | 2.44 | 2.33 | 2.27 | 2.21 | 2.15 | 2.08 | 2.01 | 1.94 |
| | 25 | 2.51 | 2.41 | 2.30 | 2.24 | 2.18 | 2.12 | 2.05 | 1.98 | 1.91 |
| | 30 | 2.41 | 2.31 | 2.20 | 2.14 | 2.07 | 2.01 | 1.94 | 1.87 | 1.79 |
| | 40 | 2.29 | 2.18 | 2.07 | 2.01 | 1.94 | 1.88 | 1.80 | 1.72 | 1.64 |
| | 60 | 2.17 | 2.06 | 1.94 | 1.88 | 1.82 | 1.74 | 1.67 | 1.58 | 1.48 |
| | 120 | 2.05 | 1.95 | 1.82 | 1.76 | 1.69 | 1.61 | 1.53 | 1.43 | 1.31 |
| | ∞ | 1.94 | 1.83 | 1.71 | 1.64 | 1.57 | 1.48 | 1.39 | 1.27 | 1.00 |

From E. S. Pearson and H. O. Hartley, *Biometrika Tables for Statisticians*, vol. I (1958), pp. 159–163. Reprinted by permission of the Biometrika Trustees.

**TABLE 3c** Critical Values of the $F$ Distribution ($\alpha = .01$)

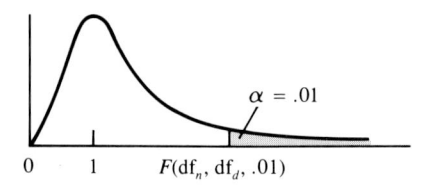

The entries in the table are critical values of $F$ for which the area under the curve to the right is equal to .01.

Degrees of Freedom for Numerator

| df_d | 1 | 2 | 3 | 4 | 5 | 6 | 7 | 8 | 9 | 10 |
|---|---|---|---|---|---|---|---|---|---|---|
| 1 | 4,052 | 5,000 | 5,403 | 5,625 | 5,764 | 5,859 | 5,928 | 5,982 | 6,023 | 6,056 |
| 2 | 98.5 | 99.0 | 99.2 | 99.2 | 99.3 | 99.3 | 99.4 | 99.4 | 99.4 | 99.4 |
| 3 | 34.1 | 30.8 | 29.5 | 28.7 | 28.2 | 27.9 | 27.7 | 27.5 | 27.3 | 27.2 |
| 4 | 21.2 | 18.0 | 16.7 | 16.0 | 15.5 | 15.2 | 15.0 | 14.8 | 14.7 | 14.5 |
| 5 | 16.3 | 13.3 | 12.1 | 11.4 | 11.0 | 10.7 | 10.5 | 10.3 | 10.2 | 10.1 |
| 6 | 13.7 | 10.9 | 9.78 | 9.15 | 8.75 | 8.47 | 8.26 | 8.10 | 7.98 | 7.87 |
| 7 | 12.2 | 9.55 | 8.45 | 7.85 | 7.46 | 7.19 | 6.99 | 6.84 | 6.72 | 6.62 |
| 8 | 11.3 | 8.65 | 7.59 | 7.01 | 6.63 | 6.37 | 6.18 | 6.03 | 5.91 | 5.81 |
| 9 | 10.6 | 8.02 | 6.99 | 6.42 | 6.06 | 5.80 | 5.61 | 5.47 | 5.35 | 5.26 |
| 10 | 10.0 | 7.56 | 6.55 | 5.99 | 5.64 | 5.39 | 5.20 | 5.06 | 4.94 | 4.85 |
| 11 | 9.65 | 7.21 | 6.22 | 5.67 | 5.32 | 5.07 | 4.89 | 4.74 | 4.63 | 4.54 |
| 12 | 9.33 | 6.93 | 5.95 | 5.41 | 5.06 | 4.82 | 4.64 | 4.50 | 4.39 | 4.30 |
| 13 | 9.07 | 6.70 | 5.74 | 5.21 | 4.86 | 4.62 | 4.44 | 4.30 | 4.19 | 4.10 |
| 14 | 8.86 | 6.51 | 5.56 | 5.04 | 4.70 | 4.46 | 4.28 | 4.14 | 4.03 | 3.94 |
| 15 | 8.68 | 6.36 | 5.42 | 4.89 | 4.56 | 4.32 | 4.14 | 4.00 | 3.89 | 3.80 |
| 16 | 8.53 | 6.23 | 5.29 | 4.77 | 4.44 | 4.20 | 4.03 | 3.89 | 3.78 | 3.69 |
| 17 | 8.40 | 6.11 | 5.19 | 4.67 | 4.34 | 4.10 | 3.93 | 3.79 | 3.68 | 3.59 |
| 18 | 8.29 | 6.01 | 5.09 | 4.58 | 4.25 | 4.01 | 3.84 | 3.71 | 3.60 | 3.51 |
| 19 | 8.19 | 5.93 | 5.01 | 4.50 | 4.17 | 3.94 | 3.77 | 3.63 | 3.52 | 3.43 |
| 20 | 8.10 | 5.85 | 4.94 | 4.43 | 4.10 | 3.87 | 3.70 | 3.56 | 3.46 | 3.37 |
| 21 | 8.02 | 5.78 | 4.87 | 4.37 | 4.04 | 3.81 | 3.64 | 3.51 | 3.40 | 3.31 |
| 22 | 7.95 | 5.72 | 4.82 | 4.31 | 3.99 | 3.76 | 3.59 | 3.45 | 3.35 | 3.26 |
| 23 | 7.88 | 5.66 | 4.76 | 4.26 | 3.94 | 3.71 | 3.54 | 3.41 | 3.30 | 3.21 |
| 24 | 7.82 | 5.61 | 4.72 | 4.22 | 3.90 | 3.67 | 3.50 | 3.36 | 3.26 | 3.17 |
| 25 | 7.77 | 5.57 | 4.68 | 4.18 | 3.86 | 3.63 | 3.46 | 3.32 | 3.22 | 3.13 |
| 30 | 7.56 | 5.39 | 4.51 | 4.02 | 3.70 | 3.47 | 3.30 | 3.17 | 3.07 | 2.98 |
| 40 | 7.31 | 5.18 | 4.31 | 3.83 | 3.51 | 3.29 | 3.12 | 2.99 | 2.89 | 2.80 |
| 60 | 7.08 | 4.98 | 4.13 | 3.65 | 3.34 | 3.12 | 2.95 | 2.82 | 2.72 | 2.63 |
| 120 | 6.85 | 4.79 | 3.95 | 3.48 | 3.17 | 2.96 | 2.79 | 2.66 | 2.56 | 2.47 |
| ∞ | 6.63 | 4.61 | 3.78 | 3.32 | 3.02 | 2.80 | 2.64 | 2.51 | 2.41 | 2.32 |

Degrees of Freedom for Denominator

**TABLE 3c** Critical Values of the $F$ Distribution ($\alpha = .01$)

<table>
<tr><th colspan="10" align="center">Degrees of Freedom for Numerator</th></tr>
<tr><th></th><th>12</th><th>15</th><th>20</th><th>24</th><th>30</th><th>40</th><th>60</th><th>120</th><th>$\infty$</th></tr>
<tr><td>1</td><td>6,106</td><td>6,157</td><td>6,209</td><td>6,235</td><td>6,261</td><td>6,287</td><td>6,313</td><td>6,339</td><td>6,366</td></tr>
<tr><td>2</td><td>99.4</td><td>99.4</td><td>99.4</td><td>99.5</td><td>99.5</td><td>99.5</td><td>99.5</td><td>99.5</td><td>99.5</td></tr>
<tr><td>3</td><td>27.1</td><td>26.9</td><td>26.7</td><td>26.6</td><td>26.5</td><td>26.4</td><td>26.3</td><td>26.2</td><td>26.1</td></tr>
<tr><td>4</td><td>14.4</td><td>14.2</td><td>14.0</td><td>13.9</td><td>13.8</td><td>13.7</td><td>13.7</td><td>13.6</td><td>13.5</td></tr>
<tr><td>5</td><td>9.89</td><td>9.72</td><td>9.55</td><td>9.47</td><td>9.38</td><td>9.29</td><td>9.20</td><td>9.11</td><td>9.02</td></tr>
<tr><td>6</td><td>7.72</td><td>7.56</td><td>7.40</td><td>7.31</td><td>7.23</td><td>7.14</td><td>7.06</td><td>6.97</td><td>6.88</td></tr>
<tr><td>7</td><td>6.47</td><td>6.31</td><td>6.16</td><td>6.07</td><td>5.99</td><td>5.91</td><td>5.82</td><td>5.74</td><td>5.65</td></tr>
<tr><td>8</td><td>5.67</td><td>5.52</td><td>5.36</td><td>5.28</td><td>5.20</td><td>5.12</td><td>5.03</td><td>4.95</td><td>4.86</td></tr>
<tr><td>9</td><td>5.11</td><td>4.96</td><td>4.81</td><td>4.73</td><td>4.65</td><td>4.57</td><td>4.48</td><td>4.40</td><td>4.31</td></tr>
<tr><td>10</td><td>4.71</td><td>4.56</td><td>4.41</td><td>4.33</td><td>4.25</td><td>4.17</td><td>4.08</td><td>4.00</td><td>3.91</td></tr>
<tr><td>11</td><td>4.40</td><td>4.25</td><td>4.10</td><td>4.02</td><td>3.94</td><td>3.86</td><td>3.78</td><td>3.69</td><td>3.60</td></tr>
<tr><td>12</td><td>4.16</td><td>4.01</td><td>3.86</td><td>3.78</td><td>3.70</td><td>3.62</td><td>3.54</td><td>3.45</td><td>3.36</td></tr>
<tr><td>13</td><td>3.96</td><td>3.82</td><td>3.66</td><td>3.59</td><td>3.51</td><td>3.43</td><td>3.34</td><td>3.25</td><td>3.17</td></tr>
<tr><td>14</td><td>3.80</td><td>3.66</td><td>3.51</td><td>3.43</td><td>3.35</td><td>3.27</td><td>3.18</td><td>3.09</td><td>3.00</td></tr>
<tr><td>15</td><td>3.67</td><td>3.52</td><td>3.37</td><td>3.29</td><td>3.21</td><td>3.13</td><td>3.05</td><td>2.96</td><td>2.87</td></tr>
<tr><td>16</td><td>3.55</td><td>3.41</td><td>3.26</td><td>3.18</td><td>3.10</td><td>3.02</td><td>2.93</td><td>2.84</td><td>2.75</td></tr>
<tr><td>17</td><td>3.46</td><td>3.31</td><td>3.16</td><td>3.08</td><td>3.00</td><td>2.92</td><td>2.83</td><td>2.75</td><td>2.65</td></tr>
<tr><td>18</td><td>3.37</td><td>3.23</td><td>3.08</td><td>3.00</td><td>2.92</td><td>2.84</td><td>2.75</td><td>2.66</td><td>2.57</td></tr>
<tr><td>19</td><td>3.30</td><td>3.15</td><td>3.00</td><td>2.92</td><td>2.84</td><td>2.76</td><td>2.67</td><td>2.58</td><td>2.49</td></tr>
<tr><td>20</td><td>3.23</td><td>3.09</td><td>2.94</td><td>2.86</td><td>2.78</td><td>2.69</td><td>2.61</td><td>2.52</td><td>2.42</td></tr>
<tr><td>21</td><td>3.17</td><td>3.03</td><td>2.88</td><td>2.80</td><td>2.72</td><td>2.64</td><td>2.55</td><td>2.46</td><td>2.36</td></tr>
<tr><td>22</td><td>3.12</td><td>2.98</td><td>2.83</td><td>2.75</td><td>2.67</td><td>2.58</td><td>2.50</td><td>2.40</td><td>2.31</td></tr>
<tr><td>23</td><td>3.07</td><td>2.93</td><td>2.78</td><td>2.70</td><td>2.62</td><td>2.54</td><td>2.45</td><td>2.35</td><td>2.26</td></tr>
<tr><td>24</td><td>3.03</td><td>2.89</td><td>2.74</td><td>2.66</td><td>2.58</td><td>2.49</td><td>2.40</td><td>2.31</td><td>2.21</td></tr>
<tr><td>25</td><td>2.99</td><td>2.85</td><td>2.70</td><td>2.62</td><td>2.53</td><td>2.45</td><td>2.36</td><td>2.27</td><td>2.17</td></tr>
<tr><td>30</td><td>2.84</td><td>2.70</td><td>2.55</td><td>2.47</td><td>2.39</td><td>2.30</td><td>2.21</td><td>2.11</td><td>2.01</td></tr>
<tr><td>40</td><td>2.66</td><td>2.52</td><td>2.37</td><td>2.29</td><td>2.20</td><td>2.11</td><td>2.02</td><td>1.92</td><td>1.80</td></tr>
<tr><td>60</td><td>2.50</td><td>2.35</td><td>2.20</td><td>2.12</td><td>2.03</td><td>1.94</td><td>1.84</td><td>1.73</td><td>1.60</td></tr>
<tr><td>120</td><td>2.34</td><td>2.19</td><td>2.03</td><td>1.95</td><td>1.86</td><td>1.76</td><td>1.66</td><td>1.53</td><td>1.38</td></tr>
<tr><td>$\infty$</td><td>2.18</td><td>2.04</td><td>1.88</td><td>1.79</td><td>1.70</td><td>1.59</td><td>1.47</td><td>1.32</td><td>1.00</td></tr>
</table>

*Degrees of Freedom for Denominator*

From E.S. Pearson and H.O. Hartley, *Biometrika Tables for Statisticians*, vol. I (1958), pp. 159–163. Reprinted by permission of the Biometrika Trustees.

**TABLE 4** Critical Values of the Chi-Square Distribution

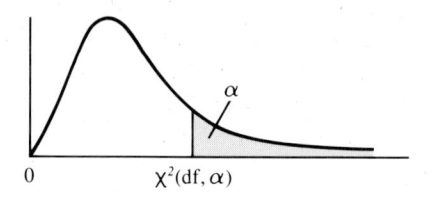

The entries in this table are the critical values for chi-square for which the area to the right under the curve is equal to $\alpha$.

### Amount of $\alpha$ in Right-Hand Tail

| df | 0.995 | 0.990 | 0.975 | 0.950 | 0.900 | 0.100 | 0.050 | 0.025 | 0.010 | 0.005 |
|---|---|---|---|---|---|---|---|---|---|---|
| 1 | 0.0000393 | 0.000157 | 0.000982 | 0.00393 | 0.0158 | 2.71 | 3.84 | 5.02 | 6.64 | 7.88 |
| 2 | 0.0100 | 0.0201 | 0.0506 | 0.103 | 0.211 | 4.61 | 6.00 | 7.38 | 9.21 | 10.6 |
| 3 | 0.0717 | 0.115 | 0.216 | 0.352 | 0.584 | 6.25 | 7.82 | 9.35 | 11.4 | 12.9 |
| 4 | 0.207 | 0.297 | 0.484 | 0.711 | 1.0636 | 7.78 | 9.50 | 11.1 | 13.3 | 14.9 |
| 5 | 0.412 | 0.554 | 0.831 | 1.15 | 1.61 | 9.24 | 11.1 | 12.8 | 15.1 | 16.8 |
| 6 | 0.676 | 0.872 | 1.24 | 1.64 | 2.20 | 10.6 | 12.6 | 14.5 | 16.8 | 18.6 |
| 7 | 0.990 | 1.24 | 1.69 | 2.17 | 2.83 | 12.0 | 14.1 | 16.0 | 18.5 | 20.3 |
| 8 | 1.34 | 1.65 | 2.18 | 2.73 | 3.49 | 13.4 | 15.5 | 17.5 | 20.1 | 22.0 |
| 9 | 1.73 | 2.09 | 2.70 | 3.33 | 4.17 | 14.7 | 17.0 | 19.0 | 21.7 | 23.6 |
| 10 | 2.16 | 2.56 | 3.25 | 3.94 | 4.87 | 16.0 | 18.3 | 20.5 | 23.2 | 25.2 |
| 11 | 2.60 | 3.05 | 3.82 | 4.58 | 5.58 | 17.2 | 19.7 | 21.9 | 24.7 | 26.8 |
| 12 | 3.07 | 3.57 | 4.40 | 5.23 | 6.30 | 18.6 | 21.0 | 23.3 | 26.2 | 28.3 |
| 13 | 3.57 | 4.11 | 5.01 | 5.90 | 7.04 | 19.8 | 22.4 | 24.7 | 27.7 | 29.8 |
| 14 | 4.07 | 4.66 | 5.63 | 6.57 | 7.79 | 21.1 | 23.7 | 26.1 | 29.1 | 31.3 |
| 15 | 4.60 | 5.23 | 6.26 | 7.26 | 8.55 | 22.3 | 25.0 | 27.5 | 30.6 | 32.8 |
| 16 | 5.14 | 5.81 | 6.91 | 7.96 | 9.31 | 23.5 | 26.3 | 28.9 | 32.0 | 34.3 |
| 17 | 5.70 | 6.41 | 7.56 | 8.67 | 10.1 | 24.8 | 27.6 | 30.2 | 33.4 | 35.7 |
| 18 | 6.26 | 7.01 | 8.23 | 9.39 | 10.9 | 26.0 | 28.9 | 31.5 | 34.8 | 37.2 |
| 19 | 6.84 | 7.63 | 8.91 | 10.1 | 11.7 | 27.2 | 30.1 | 32.9 | 36.2 | 38.6 |
| 20 | 7.43 | 8.26 | 9.59 | 10.9 | 12.4 | 28.4 | 31.4 | 34.2 | 37.6 | 40.0 |
| 21 | 8.03 | 8.90 | 10.3 | 11.6 | 13.2 | 29.6 | 32.7 | 35.5 | 39.0 | 41.4 |
| 22 | 8.64 | 9.54 | 11.0 | 12.3 | 14.0 | 30.8 | 33.9 | 36.8 | 40.3 | 42.8 |
| 23 | 9.26 | 10.2 | 11.0 | 13.1 | 14.9 | 32.0 | 35.2 | 38.1 | 41.6 | 44.2 |
| 24 | 9.89 | 10.9 | 12.4 | 13.9 | 15.7 | 33.2 | 36.4 | 39.4 | 43.0 | 45.6 |
| 25 | 10.5 | 11.5 | 13.1 | 14.6 | 16.5 | 34.4 | 37.7 | 40.7 | 44.3 | 46.9 |
| 26 | 11.2 | 12.2 | 13.8 | 15.4 | 17.3 | 35.6 | 38.9 | 41.9 | 45.6 | 48.3 |
| 27 | 11.8 | 12.9 | 14.6 | 16.2 | 18.1 | 36.7 | 40.1 | 43.2 | 47.0 | 49.7 |
| 28 | 12.5 | 13.6 | 15.3 | 16.9 | 18.9 | 37.9 | 41.3 | 44.5 | 48.3 | 51.0 |
| 29 | 13.1 | 14.3 | 16.1 | 17.7 | 19.8 | 39.1 | 42.6 | 45.7 | 49.6 | 52.3 |
| 30 | 13.8 | 15.0 | 16.8 | 18.5 | 20.6 | 40.3 | 43.8 | 47.0 | 50.9 | 53.7 |
| 40 | 20.7 | 22.2 | 24.4 | 26.5 | 29.1 | 51.8 | 55.8 | 59.3 | 63.7 | 66.8 |
| 50 | 28.0 | 29.7 | 32.4 | 34.8 | 37.7 | 63.2 | 67.5 | 71.4 | 76.2 | 79.5 |
| 60 | 35.5 | 37.5 | 40.5 | 43.2 | 46.5 | 74.4 | 79.1 | 83.3 | 88.4 | 92.0 |
| 70 | 43.3 | 45.4 | 48.8 | 51.8 | 55.3 | 85.5 | 90.5 | 95.0 | 100.0 | 104.0 |
| 80 | 51.2 | 53.5 | 57.2 | 60.4 | 64.3 | 96.6 | 102.0 | 107.0 | 112.0 | 116.0 |
| 90 | 59.2 | 61.8 | 65.7 | 69.1 | 73.3 | 108.0 | 113.0 | 118.0 | 124.0 | 128.0 |
| 100 | 67.3 | 70.1 | 74.2 | 77.9 | 82.4 | 114.0 | 124.0 | 130.0 | 136.0 | 140.0 |

Adapted from E. S. Pearson and H. O. Hartley, *Biometrika Tables for Statisticians*, vol. I (1962), pp. 130–131. Reprinted by permission of the Biometrika Trustees.

**TABLE 5** Binomial Probabilities $\left[\binom{n}{x}p^x q^{n-x}\right]$

| | | | | | | | p | | | | | | | |
|---|---|---|---|---|---|---|---|---|---|---|---|---|---|---|
| n | x | 0.01 | 0.05 | 0.10 | 0.20 | 0.30 | 0.40 | 0.50 | 0.60 | 0.70 | 0.80 | 0.90 | 0.95 | 0.99 | x |
| 2 | 0 | 980 | 902 | 810 | 640 | 490 | 360 | 250 | 160 | 090 | 040 | 010 | 002 | 0+ | 0 |
| | 1 | 020 | 095 | 180 | 320 | 420 | 480 | 500 | 480 | 420 | 320 | 180 | 095 | 020 | 1 |
| | 2 | 0+ | 002 | 010 | 040 | 090 | 160 | 250 | 360 | 490 | 640 | 810 | 902 | 980 | 2 |
| 3 | 0 | 970 | 857 | 729 | 512 | 343 | 216 | 125 | 064 | 027 | 008 | 001 | 0+ | 0+ | 0 |
| | 1 | 029 | 135 | 243 | 384 | 441 | 432 | 375 | 288 | 189 | 096 | 027 | 007 | 0+ | 1 |
| | 2 | 0+ | 007 | 027 | 096 | 189 | 288 | 375 | 432 | 441 | 384 | 243 | 135 | 029 | 2 |
| | 3 | 0+ | 0+ | 001 | 008 | 027 | 064 | 125 | 216 | 343 | 512 | 729 | 857 | 970 | 3 |
| 4 | 0 | 961 | 815 | 656 | 410 | 240 | 130 | 062 | 026 | 008 | 002 | 0+ | 0+ | 0+ | 0 |
| | 1 | 039 | 171 | 292 | 410 | 412 | 346 | 250 | 154 | 076 | 026 | 004 | 0+ | 0+ | 1 |
| | 2 | 001 | 014 | 049 | 154 | 265 | 346 | 375 | 346 | 265 | 154 | 049 | 014 | 001 | 2 |
| | 3 | 0+ | 0+ | 004 | 026 | 076 | 154 | 250 | 346 | 412 | 410 | 292 | 171 | 039 | 3 |
| | 4 | 0+ | 0+ | 0+ | 002 | 008 | 026 | 062 | 130 | 240 | 410 | 656 | 815 | 961 | 4 |
| 5 | 0 | 951 | 774 | 590 | 328 | 168 | 078 | 031 | 010 | 002 | 0+ | 0+ | 0+ | 0+ | 0 |
| | 1 | 048 | 204 | 328 | 410 | 360 | 259 | 156 | 077 | 028 | 006 | 0+ | 0+ | 0+ | 1 |
| | 2 | 001 | 021 | 073 | 205 | 309 | 346 | 312 | 230 | 132 | 051 | 008 | 001 | 0+ | 2 |
| | 3 | 0+ | 001 | 008 | 051 | 132 | 230 | 312 | 346 | 309 | 205 | 073 | 021 | 001 | 3 |
| | 4 | 0+ | 0+ | 0+ | 006 | 028 | 077 | 156 | 259 | 360 | 410 | 328 | 204 | 048 | 4 |
| | 5 | 0+ | 0+ | 0+ | 0+ | 002 | 010 | 031 | 078 | 168 | 328 | 590 | 774 | 951 | 5 |
| 6 | 0 | 941 | 735 | 531 | 262 | 118 | 047 | 016 | 004 | 001 | 0+ | 0+ | 0+ | 0+ | 0 |
| | 1 | 057 | 232 | 354 | 393 | 303 | 187 | 094 | 037 | 010 | 002 | 0+ | 0+ | 0+ | 1 |
| | 2 | 001 | 031 | 098 | 246 | 324 | 311 | 234 | 138 | 060 | 015 | 001 | 0+ | 0+ | 2 |
| | 3 | 0+ | 002 | 015 | 082 | 185 | 276 | 312 | 276 | 185 | 082 | 015 | 002 | 0+ | 3 |
| | 4 | 0+ | 0+ | 001 | 015 | 060 | 138 | 234 | 311 | 324 | 246 | 098 | 031 | 001 | 4 |
| | 5 | 0+ | 0+ | 0+ | 002 | 010 | 037 | 094 | 187 | 303 | 393 | 354 | 232 | 057 | 5 |
| | 6 | 0+ | 0+ | 0+ | 0+ | 001 | 004 | 016 | 047 | 118 | 262 | 531 | 735 | 941 | 6 |
| 7 | 0 | 932 | 698 | 478 | 210 | 082 | 028 | 008 | 002 | 0+ | 0+ | 0+ | 0+ | 0+ | 0 |
| | 1 | 066 | 257 | 372 | 367 | 247 | 131 | 055 | 017 | 004 | 0+ | 0+ | 0+ | 0+ | 1 |
| | 2 | 002 | 041 | 124 | 275 | 318 | 261 | 164 | 077 | 025 | 004 | 0+ | 0+ | 0+ | 2 |
| | 3 | 0+ | 004 | 023 | 115 | 227 | 290 | 273 | 194 | 097 | 029 | 003 | 0+ | 0+ | 3 |
| | 4 | 0+ | 0+ | 003 | 029 | 097 | 194 | 273 | 290 | 227 | 115 | 023 | 004 | 0+ | 4 |
| | 5 | 0+ | 0+ | 0+ | 004 | 025 | 077 | 164 | 261 | 318 | 275 | 124 | 041 | 002 | 5 |
| | 6 | 0+ | 0+ | 0+ | 0+ | 004 | 017 | 055 | 131 | 247 | 367 | 372 | 257 | 066 | 6 |
| | 7 | 0+ | 0+ | 0+ | 0+ | 0+ | 002 | 008 | 028 | 082 | 210 | 478 | 698 | 932 | 7 |
| 8 | 0 | 923 | 663 | 430 | 168 | 058 | 017 | 004 | 001 | 0+ | 0+ | 0+ | 0+ | 0+ | 0 |
| | 1 | 075 | 279 | 383 | 336 | 198 | 090 | 031 | 008 | 001 | 0+ | 0+ | 0+ | 0+ | 1 |
| | 2 | 003 | 051 | 149 | 294 | 296 | 209 | 109 | 041 | 010 | 001 | 0+ | 0+ | 0+ | 2 |
| | 3 | 0+ | 005 | 033 | 147 | 254 | 279 | 219 | 124 | 047 | 009 | 0+ | 0+ | 0+ | 3 |
| | 4 | 0+ | 0+ | 005 | 046 | 136 | 232 | 273 | 232 | 136 | 046 | 005 | 0+ | 0+ | 4 |
| | 5 | 0+ | 0+ | 0+ | 009 | 047 | 124 | 219 | 279 | 254 | 147 | 033 | 005 | 0+ | 5 |
| | 6 | 0+ | 0+ | 0+ | 001 | 010 | 041 | 109 | 209 | 296 | 294 | 149 | 051 | 003 | 6 |
| | 7 | 0+ | 0+ | 0+ | 0+ | 001 | 008 | 031 | 090 | 198 | 336 | 383 | 279 | 075 | 7 |
| | 8 | 0+ | 0+ | 0+ | 0+ | 0+ | 001 | 004 | 017 | 058 | 168 | 430 | 663 | 923 | 8 |

*(continued on next page)*

**TABLE 5** *(Continued)*

| n | x | 0.01 | 0.05 | 0.10 | 0.20 | 0.30 | 0.40 | 0.50 | 0.60 | 0.70 | 0.80 | 0.90 | 0.95 | 0.99 | x |
|---|---|------|------|------|------|------|------|------|------|------|------|------|------|------|---|
| 9 | 0 | 914 | 630 | 387 | 134 | 040 | 010 | 002 | 0+ | 0+ | 0+ | 0+ | 0+ | 0+ | 0 |
|   | 1 | 083 | 299 | 387 | 302 | 156 | 060 | 018 | 004 | 0+ | 0+ | 0+ | 0+ | 0+ | 1 |
|   | 2 | 003 | 063 | 172 | 302 | 267 | 161 | 070 | 021 | 004 | 0+ | 0+ | 0+ | 0+ | 2 |
|   | 3 | 0+ | 008 | 045 | 176 | 267 | 251 | 164 | 074 | 021 | 003 | 0+ | 0+ | 0+ | 3 |
|   | 4 | 0+ | 001 | 007 | 066 | 172 | 251 | 246 | 167 | 074 | 017 | 001 | 0+ | 0+ | 4 |
|   | 5 | 0+ | 0+ | 001 | 017 | 074 | 167 | 246 | 251 | 172 | 066 | 007 | 001 | 0+ | 5 |
|   | 6 | 0+ | 0+ | 0+ | 003 | 021 | 074 | 164 | 251 | 267 | 176 | 045 | 008 | 0+ | 6 |
|   | 7 | 0+ | 0+ | 0+ | 0+ | 004 | 021 | 070 | 161 | 267 | 302 | 172 | 063 | 003 | 7 |
|   | 8 | 0+ | 0+ | 0+ | 0+ | 0+ | 004 | 018 | 060 | 156 | 302 | 387 | 299 | 083 | 8 |
|   | 9 | 0+ | 0+ | 0+ | 0+ | 0+ | 0+ | 002 | 010 | 040 | 134 | 387 | 630 | 914 | 9 |
| 10 | 0 | 904 | 599 | 349 | 107 | 028 | 006 | 001 | 0+ | 0+ | 0+ | 0+ | 0+ | 0+ | 0 |
|    | 1 | 091 | 315 | 387 | 268 | 121 | 040 | 010 | 002 | 0+ | 0+ | 0+ | 0+ | 0+ | 1 |
|    | 2 | 004 | 075 | 194 | 302 | 233 | 121 | 044 | 011 | 001 | 0+ | 0+ | 0+ | 0+ | 2 |
|    | 3 | 0+ | 010 | 057 | 201 | 267 | 215 | 117 | 042 | 009 | 001 | 0+ | 0+ | 0+ | 3 |
|    | 4 | 0+ | 001 | 011 | 088 | 200 | 251 | 205 | 111 | 037 | 006 | 0+ | 0+ | 0+ | 4 |
|    | 5 | 0+ | 0+ | 001 | 026 | 103 | 201 | 246 | 201 | 103 | 026 | 001 | 0+ | 0+ | 5 |
|    | 6 | 0+ | 0+ | 0+ | 006 | 037 | 111 | 205 | 251 | 200 | 088 | 011 | 001 | 0+ | 6 |
|    | 7 | 0+ | 0+ | 0+ | 001 | 009 | 042 | 117 | 215 | 267 | 201 | 057 | 010 | 0+ | 7 |
|    | 8 | 0+ | 0+ | 0+ | 0+ | 001 | 011 | 044 | 121 | 233 | 302 | 194 | 075 | 004 | 8 |
|    | 9 | 0+ | 0+ | 0+ | 0+ | 0+ | 002 | 010 | 040 | 121 | 268 | 387 | 315 | 091 | 9 |
|    | 10 | 0+ | 0+ | 0+ | 0+ | 0+ | 0+ | 001 | 006 | 028 | 107 | 349 | 599 | 904 | 10 |
| 11 | 0 | 895 | 569 | 314 | 086 | 020 | 004 | 0+ | 0+ | 0+ | 0+ | 0+ | 0+ | 0+ | 0 |
|    | 1 | 099 | 329 | 384 | 236 | 093 | 027 | 005 | 001 | 0+ | 0+ | 0+ | 0+ | 0+ | 1 |
|    | 2 | 005 | 087 | 213 | 295 | 200 | 089 | 027 | 005 | 001 | 0+ | 0+ | 0+ | 0+ | 2 |
|    | 3 | 0+ | 014 | 071 | 221 | 257 | 177 | 081 | 023 | 004 | 0+ | 0+ | 0+ | 0+ | 3 |
|    | 4 | 0+ | 001 | 016 | 111 | 220 | 236 | 161 | 070 | 017 | 002 | 0+ | 0+ | 0+ | 4 |
|    | 5 | 0+ | 0+ | 002 | 039 | 132 | 221 | 226 | 147 | 057 | 010 | 0+ | 0+ | 0+ | 5 |
|    | 6 | 0+ | 0+ | 0+ | 010 | 057 | 147 | 226 | 221 | 132 | 039 | 002 | 0+ | 0+ | 6 |
|    | 7 | 0+ | 0+ | 0+ | 002 | 017 | 070 | 161 | 236 | 220 | 111 | 016 | 001 | 0+ | 7 |
|    | 8 | 0+ | 0+ | 0+ | 0+ | 004 | 023 | 081 | 177 | 257 | 221 | 071 | 014 | 0+ | 8 |
|    | 9 | 0+ | 0+ | 0+ | 0+ | 001 | 005 | 027 | 089 | 200 | 295 | 213 | 087 | 005 | 9 |
|    | 10 | 0+ | 0+ | 0+ | 0+ | 0+ | 001 | 005 | 027 | 093 | 236 | 384 | 329 | 099 | 10 |
|    | 11 | 0+ | 0+ | 0+ | 0+ | 0+ | 0+ | 0+ | 004 | 020 | 086 | 314 | 569 | 895 | 11 |
| 12 | 0 | 886 | 540 | 282 | 069 | 014 | 002 | 0+ | 0+ | 0+ | 0+ | 0+ | 0+ | 0+ | 0 |
|    | 1 | 107 | 341 | 377 | 206 | 071 | 017 | 003 | 0+ | 0+ | 0+ | 0+ | 0+ | 0+ | 1 |
|    | 2 | 006 | 099 | 230 | 283 | 168 | 064 | 016 | 002 | 0+ | 0+ | 0+ | 0+ | 0+ | 2 |
|    | 3 | 0+ | 017 | 085 | 236 | 240 | 142 | 054 | 012 | 001 | 0+ | 0+ | 0+ | 0+ | 3 |
|    | 4 | 0+ | 002 | 021 | 133 | 231 | 213 | 121 | 042 | 008 | 001 | 0+ | 0+ | 0+ | 4 |
|    | 5 | 0+ | 0+ | 004 | 053 | 158 | 227 | 193 | 101 | 029 | 003 | 0+ | 0+ | 0+ | 5 |
|    | 6 | 0+ | 0+ | 0+ | 016 | 079 | 177 | 226 | 177 | 079 | 016 | 0+ | 0+ | 0+ | 6 |
|    | 7 | 0+ | 0+ | 0+ | 003 | 029 | 101 | 193 | 227 | 158 | 053 | 004 | 0+ | 0+ | 7 |
|    | 8 | 0+ | 0+ | 0+ | 001 | 008 | 042 | 121 | 213 | 231 | 133 | 021 | 002 | 0+ | 8 |
|    | 9 | 0+ | 0+ | 0+ | 0+ | 001 | 012 | 054 | 142 | 240 | 236 | 085 | 017 | 0+ | 9 |

*(continued on next page)*

**TABLE 5** Binomial Probabilities

**TABLE 5** (*Continued*)

| n | x | 0.01 | 0.05 | 0.10 | 0.20 | 0.30 | 0.40 | 0.50 | 0.60 | 0.70 | 0.80 | 0.90 | 0.95 | 0.99 | x |
|---|---|------|------|------|------|------|------|------|------|------|------|------|------|------|---|
|   | 10 | 0+ | 0+ | 0+ | 0+ | 0+ | 002 | 016 | 064 | 168 | 283 | 230 | 099 | 006 | 10 |
|   | 11 | 0+ | 0+ | 0+ | 0+ | 0+ | 0+ | 003 | 017 | 071 | 206 | 377 | 341 | 107 | 11 |
|   | 12 | 0+ | 0+ | 0+ | 0+ | 0+ | 0+ | 0+ | 002 | 014 | 069 | 282 | 540 | 886 | 12 |
| 13 | 0 | 878 | 513 | 254 | 055 | 010 | 001 | 0+ | 0+ | 0+ | 0+ | 0+ | 0+ | 0+ | 0 |
|    | 1 | 115 | 351 | 367 | 179 | 054 | 011 | 002 | 0+ | 0+ | 0+ | 0+ | 0+ | 0+ | 1 |
|    | 2 | 007 | 111 | 245 | 268 | 139 | 045 | 010 | 001 | 0+ | 0+ | 0+ | 0+ | 0+ | 2 |
|    | 3 | 0+ | 021 | 100 | 246 | 218 | 111 | 035 | 006 | 001 | 0+ | 0+ | 0+ | 0+ | 3 |
|    | 4 | 0+ | 003 | 028 | 154 | 234 | 184 | 087 | 024 | 003 | 0+ | 0+ | 0+ | 0+ | 4 |
|    | 5 | 0+ | 0+ | 006 | 069 | 180 | 221 | 157 | 066 | 014 | 001 | 0+ | 0+ | 0+ | 5 |
|    | 6 | 0+ | 0+ | 001 | 023 | 103 | 197 | 209 | 131 | 044 | 006 | 0+ | 0+ | 0+ | 6 |
|    | 7 | 0+ | 0+ | 0+ | 006 | 044 | 131 | 209 | 197 | 103 | 023 | 001 | 0+ | 0+ | 7 |
|    | 8 | 0+ | 0+ | 0+ | 001 | 014 | 066 | 157 | 221 | 180 | 069 | 006 | 0+ | 0+ | 8 |
|    | 9 | 0+ | 0+ | 0+ | 0+ | 003 | 024 | 087 | 184 | 234 | 154 | 028 | 003 | 0+ | 9 |
|    | 10 | 0+ | 0+ | 0+ | 0+ | 001 | 006 | 035 | 111 | 218 | 246 | 100 | 021 | 0+ | 10 |
|    | 11 | 0+ | 0+ | 0+ | 0+ | 0+ | 001 | 010 | 045 | 139 | 268 | 245 | 111 | 007 | 11 |
|    | 12 | 0+ | 0+ | 0+ | 0+ | 0+ | 0+ | 002 | 011 | 054 | 179 | 367 | 351 | 115 | 12 |
|    | 13 | 0+ | 0+ | 0+ | 0+ | 0+ | 0+ | 0+ | 001 | 010 | 055 | 254 | 513 | 878 | 13 |
| 14 | 0 | 869 | 488 | 229 | 044 | 007 | 001 | 0+ | 0+ | 0+ | 0+ | 0+ | 0+ | 0+ | 0 |
|    | 1 | 123 | 359 | 356 | 154 | 041 | 007 | 001 | 0+ | 0+ | 0+ | 0+ | 0+ | 0+ | 1 |
|    | 2 | 008 | 123 | 257 | 250 | 113 | 032 | 006 | 001 | 0+ | 0+ | 0+ | 0+ | 0+ | 2 |
|    | 3 | 0+ | 026 | 114 | 250 | 194 | 085 | 022 | 003 | 0+ | 0+ | 0+ | 0+ | 0+ | 3 |
|    | 4 | 0+ | 004 | 035 | 172 | 229 | 155 | 061 | 014 | 001 | 0+ | 0+ | 0+ | 0+ | 4 |
|    | 5 | 0+ | 0+ | 008 | 086 | 196 | 207 | 122 | 041 | 007 | 0+ | 0+ | 0+ | 0+ | 5 |
|    | 6 | 0+ | 0+ | 001 | 032 | 126 | 207 | 183 | 092 | 023 | 002 | 0+ | 0+ | 0+ | 6 |
|    | 7 | 0+ | 0+ | 0+ | 009 | 062 | 157 | 209 | 157 | 062 | 009 | 0+ | 0+ | 0+ | 7 |
|    | 8 | 0+ | 0+ | 0+ | 002 | 023 | 092 | 183 | 207 | 126 | 032 | 001 | 0+ | 0+ | 8 |
|    | 9 | 0+ | 0+ | 0+ | 0+ | 007 | 041 | 122 | 207 | 196 | 086 | 008 | 0+ | 0+ | 9 |
|    | 10 | 0+ | 0+ | 0+ | 0+ | 001 | 014 | 061 | 155 | 229 | 172 | 035 | 004 | 0+ | 10 |
|    | 11 | 0+ | 0+ | 0+ | 0+ | 0+ | 003 | 022 | 085 | 194 | 250 | 114 | 026 | 0+ | 11 |
|    | 12 | 0+ | 0+ | 0+ | 0+ | 0+ | 001 | 006 | 032 | 113 | 250 | 257 | 123 | 008 | 12 |
|    | 13 | 0+ | 0+ | 0+ | 0+ | 0+ | 0+ | 001 | 007 | 041 | 154 | 356 | 359 | 123 | 13 |
|    | 14 | 0+ | 0+ | 0+ | 0+ | 0+ | 0+ | 0+ | 001 | 007 | 044 | 229 | 488 | 869 | 14 |
| 15 | 0 | 860 | 463 | 206 | 035 | 005 | 0+ | 0+ | 0+ | 0+ | 0+ | 0+ | 0+ | 0+ | 0 |
|    | 1 | 130 | 366 | 343 | 132 | 031 | 005 | 0+ | 0+ | 0+ | 0+ | 0+ | 0+ | 0+ | 1 |
|    | 2 | 009 | 135 | 267 | 231 | 092 | 022 | 003 | 0+ | 0+ | 0+ | 0+ | 0+ | 0+ | 2 |
|    | 3 | 0+ | 031 | 129 | 250 | 170 | 063 | 014 | 002 | 0+ | 0+ | 0+ | 0+ | 0+ | 3 |
|    | 4 | 0+ | 005 | 043 | 188 | 219 | 127 | 042 | 007 | 001 | 0+ | 0+ | 0+ | 0+ | 4 |
|    | 5 | 0+ | 001 | 010 | 103 | 206 | 186 | 092 | 024 | 003 | 0+ | 0+ | 0+ | 0+ | 5 |
|    | 6 | 0+ | 0+ | 002 | 043 | 147 | 207 | 153 | 061 | 012 | 001 | 0+ | 0+ | 0+ | 6 |
|    | 7 | 0+ | 0+ | 0+ | 014 | 081 | 177 | 196 | 118 | 035 | 003 | 0+ | 0+ | 0+ | 7 |
|    | 8 | 0+ | 0+ | 0+ | 003 | 035 | 118 | 196 | 177 | 081 | 014 | 0+ | 0+ | 0+ | 8 |
|    | 9 | 0+ | 0+ | 0+ | 001 | 012 | 061 | 153 | 207 | 147 | 043 | 002 | 0+ | 0+ | 9 |

From Frederick Mosteller, Robert E. K. Rourke, and George B. Thomas, Jr., *Probability with Statistical Applications,* 2nd ed., © 1970, Addison-Wesley Publishing Company, Reading, Mass., pp. 475–477. Reprinted with permission.

# Appendix C:
# Alphabetical List of Minitab Commands Used in This Book

```
ACF for series in C
ADD E to E, put in E
AOVOneway on data in C,...,C
ARIMa p=K, d=K, q=K [seasonal values sp=K, sd=K, sq=K, s=K];
[[FOREcast start after period K] forecast length K periods].
BOXPlot of data in C,...,C
CDF for values in C [put results in C];
   BINOmial with n = K and p = K.
   POISson with mean = K.
   BERNoulli with p = K.
   INTEgers with uniform on K to K.
   DISCrete with values in C, probabilities in C.
   NORMal with mu = K, sigma = K.
   UNIForm with continuous uniform on K to K.
CHISquare on table in C,...,C
CODE [K,...,K] to K for C,...,C, put in C,...,C
```

# APPENDIX C

```
COPY C,...,C into C,...,C;
   USE rows K,...,K;
   USE C = values K,...,K;
   OMIT rows K,...,K;
   OMIT C = values K,..,K.
CORRelation coefficient of C,...,C
COUNt data in C, [put in K]
DELEte
DESCribe data in C,...,C
DIFFerence [of lag K] for data in C, put in C
DIVIde E by E, put in E
DOTPlot of data in C,...,C
END
EXECute
HISTogram of data C,...,C
INSErt
LAG by K, data in C, put in C
LET
MAXimum data in C, [put in K]
MEAN data in C, [put in K]
MEDIan data in C, [put in K]
MINimum data in C, [put in K]
MULTiply E by E, put in E
N data in C, [put in K]
NAME C 'YOURNAME'
NMISs data in C, [put in K]
PACF for series in C
PARSum of data in C, put in C
PDF for values in C [put results in C];
   BINOmial with n = K and p = K.
   POISson with mean = K.
   BERNoulli with p = K.
   INTEgers with uniform on K to K.
   DISCrete with values in C, probabilities in C.
   NORMal with mu = K, sigma = K.
   UNIForm with continuous uniform on K to K.
PLOT values in C vs C
PRINt E,...,E
RANDom K observations, put into C,...,C;
   BINOmial with n = K and p = K.
   POISson with mean = K.
   BERNoulli with p = K.
   INTEgers with uniform on K to K.
   DISCrete with values in C, probabilities in C.
   NORMal with mu = K, sigma = K.
   UNIForm with continuous uniform on K to K.
RCOUnt C,...,C, put in C
READ data into C,...,C
```

```
REGRess y-value in C on K predictors in C,...,C;
   DW;
   PREDict y-value for x-value set E,...,E.
RMAX C,...,C, put in C
RMEAn C,...,C, put in C
RMEDian C,...,C, put in C
RMIN C,...,C, put in C
RN C,...,C, put in C
RNMIs C,...,C, put in C
ROUNd C, put in C
RSSQ C,...,C, put in C
RSTDev C,...,C, put in C
RSUM C,...,C, put in C
SET data into C
SORT data in C, put in C
SQRT of E, put in E
SSQ data in C, [put in K]
STACk E on top of E, put in C
STDEv data in C, [put in K]
STEM-and-leaf of data in C,...,C
STORe
SUBTract E from E, put in E
SUM data in C, [put in K]
TABLe data in C,...,C;
   CHISquare [K].
TINTerval with conf level = K, data in C
TSPLot [with K periods], data in C
TTESt mu = K, data in C;
   ALTernative hypothesis = K.
TWOSample t, K percent confidence, data in C and C;
   ALTernative hypothesis = K;
   POOLed.
UNSTack C into C,...,C;
   SUBScripts in C.
ZINTerval, confidence level = K, sigma = K, data in C
ZTESt mu = K, sigma = K, data in C;
   ALTernative hypothesis = K.
```

# INDEX

**ACF**, 271, 290–293
**ADD**, 3, 12–13
Alternative hypothesis codes, 175
Analysis of variance
 description of, 210–213
 hand calculation of, 213–215
 two-way, 308
**AOVOneway**, 209
Arguments to commands, 56
**ARIMa**
 command, 271, 293–295
 description of, 293–295
 **FOREcast** subcommand, 271, 293–295
Autocorrelation, 245
Autocorrelation function, 290
Average. *See* Mean

Bayes' theorem, 85
Bernoulli distribution, 109–111
Bernoulli population, 106
**BERNoulli** subcommand, 105, 107–110
Beta one, 244
Binomial distribution
 description of, 113–119
 mean of, 116
Binomial population, 125–132
Binomial probability
 density function, 116
 description of, 116–119
 distribution function, 116
 table, 117, 323
**BINOmial** subcommand, 105, 107–109
BMDP (BioMedical Data Package), 308
Box–Jenkins forecasting, 290–295. *See also* Time series analysis
**BOXPlot**, 47, 60–61
Business cycle, 277

**CDF**
 **BERNoulli**, 105, 109
 **BINOmial**, 105, 109, 115, 119
 description of, 105, 109, 119
 **DISCrete**, 105, 109
 **INTEgers**, 105, 109
 **NORMal**, 105, 109
 **POISson**, 105, 109, 122
 **UNIForm**, 105, 109
Central limit theorem
 binomial example, 125–131
 description of, 122, 124–138
 Poisson example, 131–136
 uniform example, 136–138
**CHISquare**
 description of, 209, 215–220

# INDEX

**CHISquare** (*cont.*)
   hand calculation, 220–222
   table, 322
Classical decomposition, 273–289
**CODE**, 209, 216
Coefficient of determination ($r^2$),
   244. *See also under*
   Regression analysis
Columns, 6, 9, 25–27
Commas, 8
Communications satellites, 307
Confidence interval
   description of, 151–157
   estimates, 151
   hand calculations, 158
**COPY**, 183, 190–191
Correlation analysis
   description of, 237–239
   spurious, 246–248
**CORRelation** coefficient
   description of, 236–237, 251
   hand calculation, 255–256
   range of values, 238
**COUNt**, 47, 53
Cumulative probability density
   function. *See* **CDF**

Data correction, 14–16
Data entry, 7–9
Decision theory, 92
Decision trees
   discussion of, 88–92
   expected monetary value, 89–91
   risk aversion, 91–92
Degrees of freedom, 185
**DELEte**, 4, 15–16
**DESCribe**, 46, 51–53
**DIFFerence**, 271, 290
**DISCrete** subcommand, 105,
   107–109
Distribution-free tests, 215
Distributions
   bimodal, 49
   Poisson, 49
   symmetric, 49
**DIVIde**, 4, 13
**DOTPLot**
   arguments for, 61
   command, 47, 60
   discussion of, 60
Durbin–Watson statistic, 245

EMV. *See* Expected monetary
   value

**END**, 4, 7
Errors, Type I and Type II, 169–170
Ethics
   information ethics, 302–304
   issues in statistics, 5, 57, 176, 301
   misuse of statistics, 305–307
**EXECute**, 105, 138
Expected monetary value, 89–91

$F$ table, 316
$F$ test, 210–213
**FOREcast** subcommand, 271, 293–295
Forecasting
   Box–Jenkins, 290–295
   classical decomposition, 273–289
   time series analysis
     cyclical component, 272, 276–281, 287
     description of, 272–295
     making projections, 287–289, 293–295
     random fluctuation, 272
     seasonal component, 272, 281–287
     trend component, 272, 275–276

Go-Fast Running Shoe Company
   case study of, 1–2
   raw data on employees, 311–312
Gossett, W. S., 185

**HISTogram**
   arguments, 56
   command, 3, 8, 55–58
   description of, 55–58
Hypothesis
   development of, 167–169
   discussion of, 167–173
   errors in testing, 169–170
   tests, 171–172

Independence, 87–88
Index numbers
   aggregate, 36–38
     simple aggregate, 37–38
     simple average, 37
   Laspeyres-type, 38
   Paasche-type, 39
   simple, 34–36

Information float, 307
**INSErt**, 4, 15
Insufficient reason, 88
**INTEgers** subcommand, 105, 107–109
Interval data, 48
Interval estimates, 148

**LAG**, 271
Law of large numbers, 111–113
**LET**, 4, 14–15, 23, 24–25

Matrix operations, 308
**MAX**, 47, 52
Mean
   definition of, 8, 49
   hand calculation, 54
**MEAN**, 3, 8, 12, 14, 46, 49, 52
**MEDIan**, 46, 49, 52
Median
   definition of, 49
   hand calculation, 54
Metric data, 48
**MIN**, 47, 52
Missing observations, count of, 46
Misuse of statistics, 305–307
Mode
   definition of, 49
   hand calculation, 53
Moving average, 277–281
**MTB >**, 7
Multicollinearity, 250
**MULTiply**, 3, 13

**N**, 46, 52, 53
**NAME**, 23, 28–29
**NMISs**, 46, 52, 53
Nominal data, 48
Nonmissing observations, count of, 46
Nonparametric tests, 215, 222–223
Normal curve, 145–149
Normal distribution, 122–124
**NORMal** subcommand, 105, 107–109

One-tail tests, 150
One-way analysis of variance, 210–213
Operations, 9
Ordinal data, 48

**PACF**, 271, 290–293
Paired $t$ test, 188–190

# INDEX

**PARSum**, 105, 111–112
Partial autocorrelation function, 290–293
**PDF**
  **BERNoulli**, 105, 108
  **BINOmial**, 105, 108, 118
  description of, 105, 108, 118
  **DISCrete**, 105, 108
  **INTEgers**, 105, 108
  **NORMal**, 105, 108
  **POISson**, 105, 108, 120
  **UNIForm**, 105, 108
Pearson product-moment coefficient of correlation
  definition of, 237
  hand calculation, 255
**PLOT**, 23, 31
Plotting, 31–34, 309
Point estimates, 144–145
Poisson distribution, 120
**POISson** subcommand, 105, 107–109
Poisson trials, 120
Pooled *t* test, 191
Power of a test, 170–171
**PREDict** subcommand, 253
**PRINt**, 4, 14
Printout, 5–6
Probability density function. *See* **PDF**
Probability theory
  axioms
    addition theorem, 86
    Bayes' theorem, 85
    multiplication theorem, 85
    simple probability, 80
  insufficient reason, 88
  statistical independence, 87
  types
    logical, 75
    relative, 75
    subjective, 75–76
Probability tree, 113–114
Prompts, 7

Quartiles
  concepts of, 52
  interquartile, 52
  Q1 (first), 52
  Q3 (third), 52

$r^2$. *See* Coefficient of determination

**RANDom**
  **BERNoulli**, 105, 107
  **BINOmial**, 105, 107, 115
  description of, 105–107
  **DISCrete**, 105, 107
  **INTEgers**, 105, 107
  **NORMal**, 105–107
  **POISson**, 105, 107, 120
  **UNIForm**, 105, 107
Random number generator, 106
Range, 65
Ratio data, 48
**READ**, 3, 10–11
Regression analysis
  beta one, 244
  coefficient of determination ($r^2$)
    description of, 244
    hand calculation, 256
  multiple linear, 248–255
  residuals, 243
  simple linear
    description of, 239–248
    hand calculation, 256–257
  stepwise, 309
**REGRess**
  description of, 236, 241
  **DW** subcommand, 236, 241
  **PREDict** subcommand, 236, 253
Repetitive data
  entry of, 27
  set command for, 27
Risk aversion, 91–92
**ROUNd**, 23
Row commands
  **RCOUnt**, 143, 161
  **RMAX**, 143, 161
  **RMEAn**, 143, 161
  **RMEDian**, 143, 161
  **RMIN**, 143, 161
  **RN**, 143, 160
  **RNMIss**, 143, 160
  **RSTDev**, 143, 161
  **RSSQ**, 144, 161
  **RSUM**, 144, 161
Rows, 9

Sampling
  discussion of, 92–97
  nonrepresentativeness, 94
  simple random, 93–94
  survey sampling
    cluster, 95
    stratified, 96
Scatter diagram, 240

**SET**, 3, 6, 10
Sets (in probability)
  description of, 76–80
  intersection, 80
  union, 81
Slope of a line, 240
**SORT**, 23, 28
SPSS (Statistical Package for the Social Sciences), 308
Spurious correlation, 246–248
**SQRT**, 4, 13
**SSQ**, 47, 53
**STACk**, 23, 27
Standard deviation
  description of, 47, 63–64
  hand calculation, 65–66, 195–196
Standard error of the mean
  description of, 158
  hand calculation, 196–197
Standard normal curve, 149–151. *See also* Normal curve
**STDEv**, 47, 52, 53, 64
**STEM-and-leaf**
  arguments, 61
  description of, 47, 58–60
Stepwise multiple regression, 309
**STOP**, 9
**STORe**, 105, 138
Student's *t* test
  description of, 185–187
  hand calculations, 197–198
  table, 35
**SUBTract**, 3, 13
**SUM**, 3, 8, 47
Sum of squares, 47

**TABLe**
  description of, 209, 217–220
  **CHISquare** subcommand, 209, 217–220
Time series analysis
  Box–Jenkins, 290–295
  classical decomposition, 272–289
  cyclical component, 272, 276–281, 287
  making projections, 287–290, 293–295
  random fluctuation, 272
  seasonal component, 272, 281–287
  trend component, 272, 275–276
Time series plot, 23, 32–33

**TINTerval**, 183, 187
Trimmed mean, 52, 64
**TSPLot**, 23, 32, 271, 274
*t* table, 315
*t* test
   **TTESt**, 183, 185
   assumptions, 184–185
   interval estimates, 187–188
   paired *t* test, 188–190
   pooled *t* test
      description of, 191–193
      hand calculations, 198–199
   Student's *t*
      description of, 185–187

      hand calculations, 197–198
   two-sample *t* tests, 190–194
   unpooled *t* test,
      description of, 191, 193–194
      hand calculations, 199–201
Two-tail tests, 150
**TWOSample**, 183
Two-way analysis of variance, 308

Unpooled *t* test, 191, 193–194
**UNIForm** subcommand, 105, 107–109
**UNSTack**, 271, 284–285

Variance
   description of, 62–63
   hand calculation, 65
Venn diagrams, 76–80. *See also*
   Sets

**ZINTerval**, 143, 154
*Z* score, 149
*Z* statistic, 149–151
*Z* table, 153
**ZTEST**
   description of, 167, 173–176
   hand calculations, 176–177

**TABLE 1** Normal Curve Areas

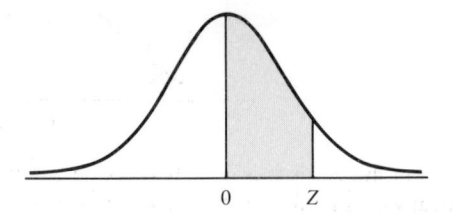

The entries in this table are the probabilities that a random variable having the standard normal distribution assumes a value between 0 and Z; the probability is represented by the area under the curve shaded in the figure. Areas for negative values of Z are obtained by symmetry.

| Z | .00 | .01 | .02 | .03 | .04 | .05 | .06 | .07 | .08 | .09 |
|---|---|---|---|---|---|---|---|---|---|---|
| 0.0 | .0000 | .0040 | .0080 | .0120 | .0160 | .0199 | .0239 | .0279 | .0319 | .0359 |
| 0.1 | .0398 | .0438 | .0478 | .0517 | .0557 | .0596 | .0636 | .0675 | .0714 | .0753 |
| 0.2 | .0793 | .0832 | .0871 | .0910 | .0948 | .0987 | .1026 | .1064 | .1103 | .1141 |
| 0.3 | .1179 | .1217 | .1255 | .1293 | .1331 | .1368 | .1406 | .1443 | .1480 | .1517 |
| 0.4 | .1554 | .1591 | .1628 | .1664 | .1700 | .1736 | .1772 | .1808 | .1844 | .1879 |
| 0.5 | .1915 | .1950 | .1985 | .2019 | .2054 | .2088 | .2123 | .2157 | .2190 | .2224 |
| 0.6 | .2257 | .2291 | .2324 | .2357 | .2389 | .2422 | .2454 | .2486 | .2517 | .2549 |
| 0.7 | .2580 | .2611 | .2642 | .2673 | .2704 | .2734 | .2764 | .2794 | .2823 | .2852 |
| 0.8 | .2881 | .2910 | .2939 | .2967 | .2995 | .3023 | .3051 | .3078 | .3106 | .3133 |
| 0.9 | .3159 | .3186 | .3212 | .3238 | .3264 | .3289 | .3315 | .3340 | .3365 | .3389 |
| 1.0 | .3413 | .3438 | .3461 | .3485 | .3508 | .3531 | .3554 | .3577 | .3599 | .3621 |
| 1.1 | .3643 | .3665 | .3686 | .3708 | .3729 | .3749 | .3770 | .3790 | .3810 | .3830 |
| 1.2 | .3849 | .3869 | .3888 | .3907 | .3925 | .3944 | .3962 | .3980 | .3997 | .4015 |
| 1.3 | .4032 | .4049 | .4066 | .4082 | .4099 | .4115 | .4131 | .4147 | .4162 | .4177 |
| 1.4 | .4192 | .4207 | .4222 | .4236 | .4251 | .4265 | .4279 | .4292 | .4306 | .4319 |
| 1.5 | .4332 | .4345 | .4357 | .4370 | .4382 | .4394 | .4406 | .4418 | .4429 | .4441 |
| 1.6 | .4452 | .4463 | .4474 | .4484 | .4495 | .4505 | .4515 | .4525 | .4535 | .4545 |
| 1.7 | .4454 | .4564 | .4573 | .4582 | .4591 | .4599 | .4608 | .4616 | .4625 | .4633 |
| 1.8 | .4641 | .4649 | .4656 | .4664 | .4671 | .4678 | .4686 | .4693 | .4699 | .4706 |
| 1.9 | .4713 | .4719 | .4726 | .4732 | .4738 | .4744 | .4750 | .4756 | .4761 | .4767 |
| 2.0 | .4772 | .4778 | .4783 | .4788 | .4793 | .4798 | .4803 | .4808 | .4812 | .4817 |
| 2.1 | .4821 | .4826 | .4830 | .4834 | .4838 | .4842 | .4846 | .4850 | .4854 | .4857 |
| 2.2 | .4861 | .4864 | .4868 | .4871 | .4875 | .4878 | .4881 | .4884 | .4887 | .4890 |
| 2.3 | .4893 | .4896 | .4898 | .4901 | .4904 | .4906 | .4909 | .4911 | .4913 | .4916 |
| 2.4 | .4918 | .4920 | .4922 | .4925 | .4927 | .4929 | .4931 | .4932 | .4934 | .4936 |
| 2.5 | .4938 | .4940 | .4941 | .4943 | .4945 | .4946 | .4948 | .4949 | .4951 | .4952 |
| 2.6 | .4953 | .4955 | .4956 | .4957 | .4959 | .4960 | .4961 | .4962 | .4963 | .4964 |
| 2.7 | .4965 | .4966 | .4967 | .4968 | .4969 | .4970 | .4971 | .4972 | .4973 | .4974 |
| 2.8 | .4974 | .4975 | .4976 | .4977 | .4977 | .4978 | .4979 | .4979 | .4980 | .4981 |
| 2.9 | .4981 | .4982 | .4982 | .4983 | .4984 | .4984 | .4985 | .4985 | .4986 | .4986 |
| 3.0 | .4987 | .4987 | .4987 | .4988 | .4988 | .4989 | .4989 | .4989 | .4990 | .4990 |

Abridged from Table 1 of *Statistical Tables and Formulas*, by A. Hald (New York: John Wiley & Sons, 1952). Reproduced by permission of A. Hald and the publishers, John Wiley & Sons.